The Lost Notebook of ENRICO FERMI

Francesco Guerra · Nadia Robotti

The Lost Notebook of ENRICO FERMI

The True Story of the Discovery of Neutron-Induced Radioactivity

Società Italiana di Fisica

Springer

Francesco Guerra
Dipartimento di Fisica
Università di Roma La Sapienza
INFN Sezione di Roma, Rome, Italy
Museo Storico della Fisica e Centro
Studi e, Ricerche Enrico Fermi
Rome
Italy

Nadia Robotti
Dipartimento di Fisica
Università di Genova
INFN Sezione di Genova, Genoa, Italy
Museo Storico della Fisica e Centro
Studi e, Ricerche Enrico Fermi
Rome
Italy

ISBN 978-3-319-69253-1 ISBN 978-3-319-69254-8 (eBook)
https://doi.org/10.1007/978-3-319-69254-8

Translated from Italian by Christine V. Pennison

Library of Congress Control Number: 2017958727

Translation from the Italian language edition: *Enrico Fermi e il quaderno ritrovato* by F. Guerra and N. Robotti,

The photograph of Enrico Fermi in the cover is courtesy of Fondazione Domus Galileiana, Pisa, Italy

Printed on acid-free paper

This Springer imprint is published by Springer Nature
The registered company is Springer International Publishing AG
The registered company address is: Gewerbestrasse 11, 6330 Cham, Switzerland

The Original Version of this book was revised.
An erratum to this book can be found at
https://doi.org/10.1007/978-3-319-69254-8_13

Foreword and acknowledgements

Thanks to Enrico Fermi some extraordinary developments took place in the period December 1933 - October 1934 at the Regio Istituto Fisico (Royal Physics Institute) of the Royal University of Rome in via Panisperna. These developments in nuclear physics led first to the creation of an extremely advanced theoretical model to explain beta nuclear decay and then to the discovery of radioactivity induced by neutron bombardment, together with the extraordinary effects of slow neutrons on the activation of some important nuclear reactions. These developments soon made the Istituto Fisico in Rome the foremost centre for nuclear physics research at an international level. Full recognition of this came when Enrico Fermi was awarded the Nobel Prize for Physics in 1938.

It is fascinating to understand what the conditions were that made this kind of miracle possible and exactly how these important discoveries were made. In particular, by using Fermi's first laboratory notebook, which we have identified at the Fondazione Oscar D'Agostino in Avellino, it is possible to follow in real time the whole process that led to the discovery of neutron-induced radioactivity.

In this book we also study the beginning of nuclear physics research in Rome in the national and international contexts. We see that in national research planning for this sector, organised by the Consiglio Nazionale delle Ricerche (National Research Council) in 1933, Rome was initially assigned a rather marginal task, directed towards nuclear gamma spectroscopy, which fitted well with Rome's research tradition. But very soon, through an incredible sequence of events, Enrico Fermi completely disrupted this line of planning and let Rome acquire a key role of strategic relevance in neutron physics which would then become the basis for all future developments in this sector.

The success of the nuclear structure model developed by Ettore Majorana during his stay in Leipzig in 1933, which considerably improved the proton and neutron model previously introduced by Werner Heisenberg, also played an important role in Fermi's drastic change to the direction of lines of research in Rome.

In 1938 Fermi, after being awarded the Nobel Prize in the ceremony in Stockholm in December, carried out his decision to emigrate permanently to the United States. A few months earlier, at the end of March, Ettore Majorana suddenly disappeared in circumstances which are still not entirely clear. Within a few months Italy lost its two greatest experts in nuclear physics.

In the course of our research we have consulted archive material belonging to several institutions, including the "Archivio Centrale dello Stato" in Rome, the "Domus Galilaeana" in Pisa, the Special Collections Research Center of the University of Chicago, the "Institut Curie" in Paris, the "Accademia Nazionale dei Lincei" in Rome, the

"Accademia Nazionale delle Scienze detta dei XL" in Rome, the "Fondazione Oscar D'Agostino" in Avellino, the Department of Physics of the University of Rome "La Sapienza", the "Archivio Occhialini-Dilworth" at the University of Milan, the Churchill Archives Centre in Cambridge, the California Institute of Technology in Pasadena, the Duke University in Durham, the University of São Paulo in Brazil, the Nobel Foundation in Stockholm, the "Massimiliano Massimo" Institute in Rome, the University of Rome "La Sapienza", the University of Palermo. To the directors and staff of all these institutions we express our grateful thanks for their courteous welcome, collaboration and assistance.

Special thanks go to:
– Carla Onesti of the University of Rome "La Sapienza", Francesco Marchetti and Maura Beghé of the Domus Galilaeana, Angelo Poggio and Mario Pomodoro of the Library of the Department of Physics of the University of Genoa, Natalie Pigeard and Anaïs Massiot of the Musée Curie in Paris, for their enthusiasm and friendly help in tracing material,
– Giovanni Battimelli, Luisa Bonolis, Mauro Giannini, Alessandro Paoletti, for interesting and perceptive discussions,
– Matteo Leone for ongoing useful discussions and for the joy of collaboration,
– Ettore Majorana, Jr. for fruitful exchanges of opinion on Ettore Majorana's scientific, cultural and human personality, and for his generous willingness to allow us access to important documents,
– the Presidents of the Fondazione Oscar D'Agostino, who have succeeded each other over time, Michele Cardellicchio, Paolino Marotta, Pietro Caterini, who allowed us to access the documents and to reproduce them,
– Luisa Cifarelli for her friendly and affectionate encouragement.

This work was made possible by support from MIUR (Ministry of Education, Universities and Research), from INFN (National Institute for Nuclear Physics), from the University of Rome "La Sapienza", from the University of Genoa, from the Società Italiana di Fisica (Italian Physical Society), from the Museo Storico della Fisica and Centro di Studi e Ricerche "Enrico Fermi" in Rome.

FRANCESCO GUERRA and NADIA ROBOTTI

Buonabitacolo - Genova - Roma
10 July, 2015

In the present English version we have included some additional material coming from the ongoing research.

It is a pleasure to thank Angela Oleandri and Christine V. Pennison for the marvellous job they have done, for the deep cultural sensibility, and for the infinite patience.

FG & NR
10 September, 2017

Contents

1 Introduction

As is well known, in Rome in March 1934 Enrico Fermi (Rome, 29 September 1901 - Chicago, 28 November 1954) discovered neutron-induced radioactivity. For this discovery, and for the connected discovery in October that same year of the effect of the slowing down of neutrons, Fermi was awarded the Nobel Prize for Physics in 1938. The motivation read as follows: "for his demonstrations of the existence of new radioactive elements produced by neutron irradiation, and for his related discovery of nuclear reactions brought about by slow neutrons".

The discovery of neutron induced radioactivity and the consequent study of neutron physics opened up new frontiers for humanity, not only in the field of pure scientific knowledge but also in the field of applications to exploit nuclear energy, with significant consequences, for good and evil.

Much has been written about this discovery (made public by Fermi on 25 March 1934) by his closest collaborators (Edoardo Amaldi (1908-1989), Oscar D'Agostino (1901-1975), Bruno Pontecorvo (1913-1993), Franco Rasetti (1901-2001), Emilio Segrè (1905-1989)), by professional historians, and also by his wife Laura Fermi (née Capon, 1907-1977). With the exception of the book by Laura Fermi *Atoms in the Family: My Life with Enrico Fermi* [69], which was written while Fermi was still alive, all the other reconstructions were presented after Fermi's untimely death in November 1954.

We can now explain the reason for this book of ours.

All the historical reconstructions produced so far, which often contradict each other, are based essentially on direct or indirect testimony, often gathered years later, with all the typical drawbacks of personal recollections, and with a natural tendency, consciously or unconsciously, to stress aspects linked to the person's own contribution, variously perceived, and to consider events in relation to the person's own role. In July 2002, as we shall describe in detail later, through a series of happy circumstances, we identified one of Enrico Fermi's noteooks in the Library of the Istituto Tecnico per Geometri(*) "Oscar D'Agostino" in Avellino. This notebook, which covers all the work leading to the discovery of neutron-induced radioactivity, we can define as the "Nobel Prize first notebook". Together with the notebook we also identified a bundle of 16 sheets of paper which date to about the same period [3,4].

The laboratory procedures relating to the discovery of neutron-induced radioactivity

(*) A technical high school [Translator's Note].

F. Guerra and N. Robotti, *The Lost Notebook of ENRICO FERMI*,
https://doi.org/10.1007/978-3-319-69254-8_1

Fig. 1. – The Royal Institute of Physics in via Panisperna - DFUR.

are very simple and direct, and completely documented in the notebook. In this book we will carry out a detailed analysis of the contents of the notebook, especially of the most significant passages, to arrive at a complete reconstruction of the discovery. This reconstruction allows us to understand in depth the conceptual and operational procedure followed by Fermi during this first experimental undertaking in advanced nuclear physics research. Moreover it also helps us to understand how a laboratory notebook was kept in those days, and what we can glean from it. The methodology followed by Fermi, in its efficient simplicity, allowed him to arrive at results that were outstanding at an international level, even though he had very limited means available to him, such as those that existed at the time in via Panisperna (Fig. 1)(*). From this point of view, Fermi's progress provides a universal lesson, and one which is completely relevant today. The achievement of momentous results with very limited means constitutes the triumph of intelligence and creativity. A lesson to be pondered.

Naturally, for a complete appreciation of Fermi's investigations and the scope of his discoveries it is first necessary to offer a comprehensive description of the state of nuclear physics in those years. So the first seven chapters are devoted to a reconstruction of the developments in nuclear physics from the 1920s, when the neutron had not yet been discovered, to the spring of 1934 when Fermi began his experimental adventure with the neutron. Particular attention has been paid to the situation in Italy, above all in Rome where Fermi worked until 1938 before moving to the United States of America after

(*) The acronyms in the figure captions are all specified in the list of credits on page 256.

receiving the Nobel Prize in Stockholm in December 1938. His was an entirely Italian Nobel Prize!

In tackling this part, and also while analysing Fermi's notebook, we have tried to provide all the essential information necessary so that the contents of our book can also be used for teaching purposes, both in specific university courses, for example in a course on the History of Physics or of History of Science, but also in introductory modules of a historical nature on Nuclear Physics for other courses. The simplicity and intuitive characteristics of Fermi's experimental process should allow our description to be used also in the final years of secondary school education.

For a reconstruction of the history of the Physics Institute in Rome before Enrico Fermi's arrival we refer the reader to the book by Giovanni Battimelli and Maria Grazia Ianniello [15].

2 Fermi in Italy: his manuscripts

2.1 The "Fermi Archive" at the Domus Galilaeana in Pisa

As is well known, from his time in high school (1918) to his departure for Stockholm and then for the United States of America (1938), Enrico Fermi carried out intense and productive scientific research in Italy. The results obtained by Fermi in this "Italian" period are sensational. In a special Section, at the end of the volume, we provide a list of all Fermi's publications(*) in his Italian period, including minor works such as reports to conferences and contributions to non-scientific journals, expanding the list given in "Note e Memorie" (Collected Papers) [6], correcting some mistakes in the titles, and following chronological order. In some cases these minor publications are very useful in order to understand why Fermi made his strategic physics choices and why he adopted particular formulations. Fermi's activity in disseminating scientific culture is really stunning, as is borne witness to by the numerous articles in the "Atti delle Riunioni della Società Italiana per il Progresso delle Scienze" (Proceedings of the meetings of the Italian Society for the Progress of Science), by lectures, and by contributions to journals aimed at the general public, such as Sapere [F145] and Gerarchia [F95].

Let us briefly run through some points in Fermi's scientific output in Italy.

Let us first recall his work in 1926 "Sulla quantizzazione del gas perfetto monoatomico" (On the Quantization of the Monoatomic Ideal Gas) [F42], regarding the foundations of the quantum statistics of an ideal monoatomic electron gas, and more generally of a gas composed of particles that obey Pauli's exclusion principle. Fermi's statistics are historically called Fermi-Dirac statistics, due to the publication of similar results by Paul Adrien Maurice Dirac (1902-1984) a few months later, the result of independent research [63].

In 1927 the formulation [F56] of the statistical model of the atom appeared, followed by further developments and applications. This provides a semiquantitative method of calculating various atomic properties in the field of quantum mechanics. This model, which is still of great use, including in the field of astrophysics, is called the "Thomas-Fermi" model. Indeed Llewellyn Hilleth Thomas (1903-1992) had published analogous results a few months earlier [161] as far as the formulation of the model was concerned but limited only to the fundamental level.

(*) In this volume we refer to Fermi's publications by their number in the list, preceded by F, *i.e.* [F1], [F2] etc., to distinguish them from the ordinary bibliography.

F. Guerra and N. Robotti, *The Lost Notebook of ENRICO FERMI*,
https://doi.org/10.1007/978-3-319-69254-8_2

Fermi also arrived at the formulation of a quantitative theory of the hyperfine structure of spectrum lines (1930) from which the magnetic moments of many nuclei could be deduced [F73].

Furthermore Fermi presented an extremely simple and efficient formulation of quantum electrodynamics [F85] which formed the basis of the cultural and scientific training of an entire generation of Physicists in this area.

At the end of 1933 a theory of nuclear beta decay was proposed [F115], based on the hypothesis of the existence of the neutrino and on the remarkable idea that electrons did not pre-exist in the nucleus but were created, together with neutrinos, at the moment of their emission. This would then reveal itself to be the basis of all weak interactions between elementary particles. In March 1934 came the discovery of neutron-induced radioactivity [F120, 121, 122] and later the discovery of the effect of slowing them down [F140], followed by an in depth analysis of the properties of diffusion, absorption and slowing of neutrons in various material substances. For the discovery of neutron-induced radioactivity and the effects of slowing, seen also as the crowning achievement of his previous scientific activity, Fermi was awarded the Nobel Prize for Physics in 1938.

We are lucky enough to have here in Italy, at the Domus Galilaeana in Pisa, almost all the original manuscript documents relating to this intense research activity [106].

Altogether there are 27 notebooks, 9 laboratory notebooks, about 600 cards recording data, 38 manuscripts and typescripts, 39 letters, as well as a miscellaneous collection of personal documents. Most of this material, as Edoardo Amaldi tells us, was left behind by Fermi at the Physics Institute of the University of Rome before he left Italy permanently on 6 December 1938. After Fermi's death in Chicago on 28 November 1954 Amaldi, the only one of Fermi's associates still in Rome, after consulting the President of the Accademia dei Lincei Francesco Giordani (1896-1961) and various Colleagues, including Enrico Persico (1900-1969), Franco Rasetti, Emilio Segrè, decided that this material, together with other material recovered later in the house of Fermi's sister, should be entrusted to the Domus Galilaeana in Pisa, designated in the early 1940s as the key Italian institution for the study of the History of Science. This material is still housed today at the Domus Galilaeana, in safe keeping in two sturdy safes on the first floor of the splendid *palazzo* at 26 via Santa Maria, and it constitutes the so-called "Fermi Archive".

Amaldi observed in 1959, at the end of the article in which he presented "The Fermi Manuscripts at the Domus Galilaeana" [8] to the scientific community: "I am glad that these manuscripts are now kept at the Domus Galilaeana for future generations, beside other documents of men who have also honoured mankind with their thoughts and experiments".

To complete the picture we also make a few brief references to Fermi's academic career during his Italian period. He graduated in Physics at the University of Pisa on 4 July 1922 and was awarded the Diploma of the Scuola Normale Superiore on the 7 July immediately following. In 1925 he obtained the *Libera Docenza*(*) in Mathematical Physics. He was one of the three winners in the competition for the chair in Mathematical Physics at the University of Cagliari which concluded on 21 January 1926. He came second behind Giovanni Giorgi (1871-1950) and ahead of Rocco Serini (1886-1864). There was a head to head contest for first place between Giorgi, who received three votes from the board members Giovanni Guglielmo (1853-1935), Carlo Somigliana (1860-1955) and Roberto Marcolongo (1862-1943), and Fermi who received only two votes from the board members Vito Volterra (1860-1940) and Tullio Levi-Civita (1863-1941). According to a belief widely held in the via Panisperna group "Fermi was rather disappointed and upset by the outcome of the competition which he thought was unfair", as claimed by Emilio Segrè in the introduction to "Note e Memorie" [6]. In fact the only advantage of coming first of the three would have been that of being appointed immediately as Professor at Cagliari, as happened to Giorgi. Instead Fermi actually renounced the privileges of his second place, very chivalrously, in order that Rocco Serini, who had come third, could be appointed at Pavia.

In any case Fermi was not left without a chair. Indeed Orso Mario Corbino (1876-1937), a powerful Senator and Director of the Physics Institute of the University of Rome, immediately managed to launch a competition for the post of Professor of Theoretical Physics at the University of Rome. It was the first competition in Italy for this discipline. The examining board, made up of Gian Antonio Maggi (1856-1937), Michele Cantone (1857-1932), Antonio Garbasso (1871-1933), Quirino Majorana (1871-1957) and Orso Mario Corbino, unanimously put Enrico Fermi in first place of the three, followed by Enrico Persico in second place, with a majority of three votes out of five, and Aldo Pontremoli (1896-1928) in third place with a unanimous vote. Fermi immediately took up his post as Professor in Rome on 1 January 1927.

By Royal Decree on 18 March 1929, proposed by the Head of Government Benito Mussolini (1883-1945), Fermi was nominated as a member of the newly created Accademia d'Italia. Great satisfaction for this event is expressed in a page of notebook No. 2 at the Domus Galilaeana where Fermi noted "A VII - 18 - 3 - 29 - Incipit vita nova – Gaudeamus igitur!" (A New Life begins - So let us Rejoice) (Fig. 2). It is one of the very few personal displays in a research notebook of Fermi's where he echoes the words of a well-known student song: "Gaudeamus igitur, Iuvenes dum sumus; ...". Finally Fermi received the 1938 Nobel Prize for Physics from the hands of King Gustav V in the ceremony in Stockholm on 10 December 1938.

On a more personal note, which is relevant for our understanding of later events, we recall that in a civil ceremony on 17 July 1928 Fermi married Laura Capon, a member of Rome's Jewish community, the daughter of Admiral Augusto Capon (1872-1943), an extraordinary character with firm personal convictions who was also a nationalist writer using the pseudonym Adriacus. His life was interwoven with the history of the Italian navy. Augusto Capon's life came to a tragic end following the roundup of Jews on 16

(*) Qualification entitling Fermi to teach at university level [Translator's Note].

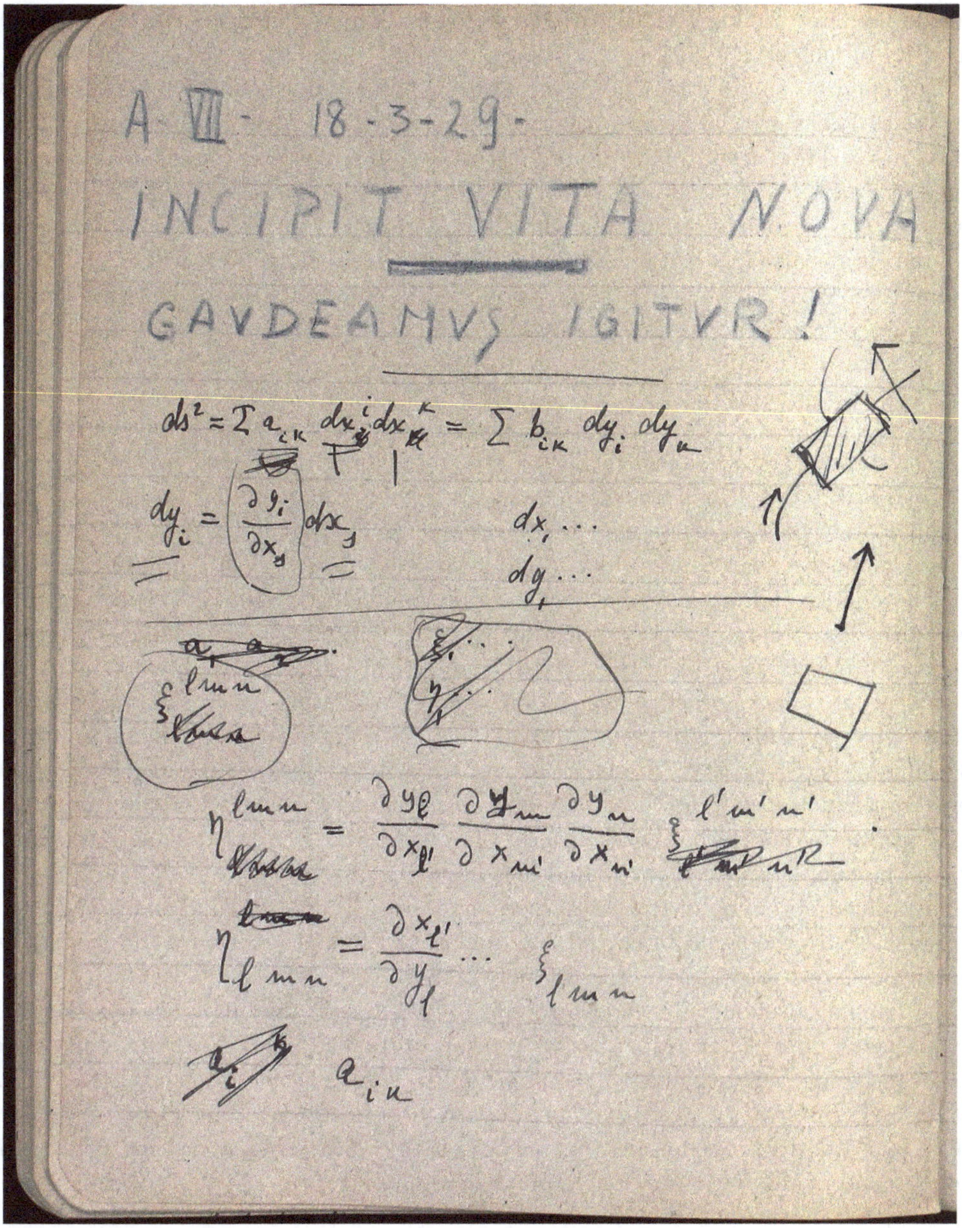

Fig. 2. – Fermi comments with satisfaction on his nomination as a member of the Accademia d'Italia (18 March 1929) - FDG.

October 1943 in Rome by the occupying German forces. Fermi's sister Maria told him about the dramatic day of Capon's arrest in a highly emotional letter, dated 4 July 1944 immediately after the liberation of Rome, which is preserved in the Fermi Archive at the University of Chicago.

Their children Nella (1931-1995), born on 31 January 1931, and Giulio (1936-1997), born on 16 February 1936, were baptized on 23 February 1936 in the church of Saint Bonosa in via Tirso according to a note in the Zanghi Archive at Sapienza University of Rome.

On the eve of their departure for Stockholm and the United States, on 5 December 1938, the Fermis also married according to Catholic rites, officiated by Monsignor Ernesto Ruffini (1888-1967), at the parish church of St. Roberto Bellarmino in Piazza Ungheria after Laura received the required baptism.

2.2 A gap in the Archive

A few years ago, when we began to study the birth and development of nuclear physics in Italy, consulting and reorganising the Fermi Archive in the Domus Galilaeana, we immediately noticed a strange gap.

The documents concerning neutron-induced radioactivity consist in all of nine laboratory notebooks and more than 600 cards of recorded data. They provide complete coverage of the experimental work carried out by Fermi and his team (Amaldi, D'Agostino, Pontecorvo, Rasetti, Segrè) in the period from 20 April 1934 until May 1935, when the research group was in practice disbanded, while Fermi's discovery of neutron-induced radioactivity was announced on 25 March 1934. So it seemed that more than a month of Fermi's research activity, in the decisive phase of the discovery, had left no written documentation. Recently however we found in Irpinia, at Avellino, two manuscript documents written by Fermi that cover exactly the period March-April 1934 and which allow us to fill in completely the gap we had met before.

It is a lined notebook of the same type as those entrusted to the Domus Galilaeana, characterized by the same brick red coloured cover and produced by the same Dutch company, together with a packet of 16 loose sheets.

The Irpinia notebook, we choose to call it that way to highlight its provenance, is lined, is made up of 78 pages and is written on both sides. The front of the notebook, which can be identified thanks to the paper mill's stamp at the bottom of the cover, is written by Fermi over 15 pages, numbered with circled numbers but without any dates, and is dedicated to problems relating to beta decay. The cover and some pages of the notebook and of the loose pages are shown in chapter 11, pp. 179-227. We indicate by Qf# the pages in the front of the notebook, by Q# the pages numbered from the back, and by S# (S#a) the front (back) pages of the loose sheets.

After the first fifteen pages, where these theoretical calculations are reported, the notebook was turned upside down so it could be used as a laboratory record for Fermi's subsequent experiments on neutron-induced radioactivity. This side of the notebook is composed of 141 pages, numbered by Fermi, with the first date, 27 March 1934, written at the top of page 44 (!), and the last, 24 April, written on page 140. This notebook contains both the preparatory procedures for the experiments and the entries of the measurements taken during the months of March and April 1934.

So the dates, but above all the contents, allow us to identify this notebook as Fermi's first notebook on neutron-induced radioactivity and therefore the notebook of the discovery.

The loose sheets however start with the date 7 April 1934 and end on 20 April 1934 and therefore they concern the research carried out immediately after the discovery. They are written on both sides and, as we have been able to establish, were ripped out of another laboratory notebook, also on neutron-induced radioactivity, stored at the Domus Galilaeana. This notebook is identical to the Irpinia notebook, produced by the same firm, and was mainly compiled by Segrè so that it is conventionally known as the "Segrè Notebook", indicating "20-4-34" as the first date written at the beginning of the notebook. The notebook clearly shows that the first part had been forcibly ripped out. Moreover on the first page is the deep impression of some numbers, heavily traced in pencil on the last page but one of the 16 Irpinia pages.

So Fermi, after using and filling up the first Dutch notebook, presumably bought during his visit to Leiden a few years before, moved on to recording data in the second notebook. Around 20 April he tore out the 16 pages, which are the consistent continuation of the first notebook, and he passed the second notebook to Segrè so that he could record the measurements that had been assigned to him. In an initial phase these concerned the evaluation of the absorption of radiation in matter and are not directly connected to neutron-induced radioactivity.

Moreover, at the end of the last page of the bundle of 16 pages, "end of measurements / Segrè Notebook" is written. The handwriting of this note is Fermi's. Very probably Fermi intended to indicate with these words that the first phase of the research had been concluded. The results were reported in the first three publications under his name alone: the two letters to "La Ricerca Scientifica" on 25 March [F120] and in early April [F121] and the letter to Nature on 10 April [F122]. A new phase was opening up which directly involved his collaborators.

The story of the Irpinia notebook and of these 16 pages is very remarkable, as is that of their discovery.

Ever since way back in 1978, the Library of the Istituto Tecnico per Geometri "Oscar D'Agostino" in Avellino, later merged into the "De Sanctis - D'Agostino" Istituto Superiore di Istruzione Secondaria, has housed all the archive material that had belonged to Oscar D'Agostino (1901-1975), a chemist born in Avellino who had cooperated with Fermi in his research into neutron-induced radioactivity in the period 1934-1935.

This material was donated by Mrs Sofia Melograni, D'Agostino's widow, when the school was named after her husband in October 1978. On this occasion the "Oscar D'Agostino Foundation" was also established with the task of preserving this material and of offering scholarships in Oscar D'Agostino's name to particularly deserving students at the school.

Over the following years the "Oscar D'Agostino Archive" has been the subject of exhibitions and of some publications, in which some parts of it have also been reproduced. A preliminary inventory was compiled by Giovanni Acocella, a scholar of the History of Physics from Avellino, connected with the "Federico II" University in Naples, and presented by him at the XXII National Conference on the History of Physics and Astronomy which was held in Genova-Chiavari on 6-8 June 2002 [2]. Following this presentation, in July the same year, since at that time we were investigating Fermi's discovery of neutron-induced radioactivity, we joined Giovanni Acocella at Avellino in order to study in depth the contents of the D'Agostino Archive, hoping to find information about the chemical aspects of the research carried out in Rome into neutron-induced radioactivity since D'Agostino was acknowledged as the "group's Chemist". It was then that, to our great surprise, we discovered Fermi's notebook and the bundle of 16 pages amongst D'Agostino's papers. These documents had been ascribed to D'Agostino and had been classified as "Notebook No. 3" and "Notebook No. 4", respectively. Moreover some pages of "Notebook No. 3" had already been published as "D'Agostino's Notebook", even if the handwriting and the contents are clearly not D'Agostino's but Fermi's.

It is difficult to establish why these two documents, which represent the initial missing gap in the Fermi Archive at the Domus Galilaeana, ended up in D'Agostino's personal archive. Perhaps a possible answer is suggested by the Irpinia notebook itself which shows us, on the basis of the handwriting, that Fermi's first collaborator after the discovery was actually D'Agostino. So one might think that D'Agostino's initial, and as we shall see later, constant presence, above all in the early stages of the research, is a reason why these documents are now in Avellino. It is possible that they were given to D'Agostino personally by Fermi, as a testament to the contribution made by this Chemist, who is often relegated to a subordinate and marginal position in historiographical reconstructions, but whom Fermi however always recalled in his writings and thanked meaningfully. Another possibility is that after Fermi's death the notebook and the 16 pages were handed to D'Agostino by Amaldi so that he could have a memento of the period of his collaboration with Fermi.

2.3 How to proceed

The Irpinia notebook is very important, apart from its historical significance, because it represents the only direct and structured testimony of Fermi's discovery of neutron-induced radioactivity and it allows us to reconstruct the initial phase completely on the basis of objective documentation.

With regard to this first phase, the only document available to us was Fermi's Letter dated 25 March 1934 to "La Ricerca Scientifica", which at that time was the Official Journal of the recently founded Consiglio Nazionale delle Ricerche-C.N.d.R. (National Research Council), in which he announced the discovery [F120]. However this Letter is very concise, as we shall see. It only contains the essential information about the results without reference to the procedures followed. The same concision is to be found both in the second Letter to "La Ricerca Scientifica" [F121], and in the Letter to "Nature" [F122], written by Fermi immediately after the discovery, and also in the first extensive article written on the subject by Fermi and sent to "Il Nuovo Cimento" in May 1934

[F131].

Instead the Irpinia notebook affords a detailed reconstruction of all the experimental steps taken by Fermi, of his working rhythm, of his choices in methodology as well as in strategy, in the culminating phase of the discovery.

Moreover, together with the 16 loose sheets and the last 22 pages of a notebook of Amaldi's which we will discuss in chapter 8, it allows us to reconstruct the birth around Fermi of his first working group and the way it operated in this initial moment.

Finally, by analysing the Irpinia notebook, in the light of another of Fermi's laboratory notebooks stored in the Domus Galilaeana which Fermi himself named "Thesaurus Elementorum Radioactivorum" (Fig. 3), it is possible to establish the exact day and hour when Fermi discovered neutron-induced radioactivity: it was Tuesday 20 March 1934, around two o'clock in the afternoon, after a feverish night spent fine tuning the counter and the amplifier and repeated background measurements.

In any case, before proceeding with an analysis of Fermi's discovery which is the main focus of our book, we thought it was appropriate, as we have already anticipated in the introduction, to set this discovery both in the framework of the Physics of the time, giving ample space to the development of Nuclear Physics, starting with the first International Conference on Nuclear Physics organised in Rome in 1931, to the discovery by Frédéric Joliot (1900-1958) and Irène Curie (1897-1956) of alpha particle-induced radiation announced on 15 January 1934 [59], and also in the more specific context of the research carried out in Rome with regard to the atomic nucleus, from the first beginnings with the *Laurea*(*) Thesis of Ettore Majorana (1906-1939) in July 1929, to the creation of a bismuth crystal spectrograph to study gamma radiation (November 1933) and the production of a source of radium D from which polonium could be periodically extracted (December 1933), which in practice were never fully used.

(*) The *Laurea* is the degree issued by the Italian Universities that includes a research thesis and leads to the title "Dottore". It is therefore of a higher level than a Bachelor degree in American or British Universities.

Fig. 3. – The laboratory notebook that Fermi named "Thesaurus Elementorum Radioactivorum" (1934) - FDG.

3 The beginning of Nuclear Physics in Rome

3.1 Fermi and Nuclear Physics: the first steps

From 1928, immediately after the formulation of the statistical model of the atom (later called the "Thomas-Fermi" model), until nearly the whole of 1932, Fermi was mainly interested in quantum electrodynamics, a cutting edge topic in those years, and studied in particular by Werner Heisenberg (1901-1976), Pascual Jordan (1902-1980), Wolfgang Pauli (1900-1958), P. A. M. Dirac. Between 1928 and 1932, Fermi published no less than seven papers on the subject, one of them in collaboration with Hans Albrecht Bethe (1906-2005) [F97].

The reason for Fermi's new interest is to complete atomic physics, which at this point was firmly based on the quantum mechanics of Heisenberg, Schrödinger (1887-1961) and Dirac, also with regard to the problems of radiation. In a talk given in Rome on 31 March 1931 at the XXIV Adunanza Generale della Società Italiana di Fisica (XXIV General Assembly of the Italian Physical Society), of which an abstract was published in "Il Nuovo Cimento" [F94], Fermi "expounds the problem of constructing quantum electrodynamics and illustrates its crucial conceptual interest bringing together in a single edifice the quantum theory of radiation on the one hand and the mechanical laws to which charges are subject and Maxwell's laws on the other".

In these papers Fermi succeeded in writing, both in the non-relativistic and the relativistic case, the equations of a system made up of an electromagnetic field and by any number of charges, introducing a new, in his view "simpler", formalism compared to that used by Heisenberg and by Pauli. The method followed by Fermi "consists basically of considering the coefficients of the expansion into harmonic components of the scalar and vector potentials at a given time as dynamic variables that characterise the electromagnetic field" [F85]. Fermi justified his own paper as follows: "Since the methods followed by these Authors are essentially different from mine, I believe that it would not be pointless to publish my results too".

This attitude was actually typical of Fermi, consisting of elaborating original and direct methods to treat problems at the frontiers of research, openly confronting the structures set by other authors. One of the recurring characteristics of Fermi's methods was their fruitful simplicity. This style would also be fully displayed in his later experimental activity.

In his second paper devoted to quantum electrodynamics Fermi wrote as follows: "The

F. Guerra and N. Robotti, *The Lost Notebook of ENRICO FERMI*,
https://doi.org/10.1007/978-3-319-69254-8_3

final form in which the results of this paper will be expressed is particularly simple. Indeed we will find that the Hamiltonian which, according to the correspondence principle, represents the natural quantum translation of classical electrodynamics is obtained by simply adding to the Hamiltonian of Dirac's theory of radiation a term which represents the electrostatic energy of the system of electric charges; so that, in the present form, quantum electrodynamics is not in any way more complicated than Dirac's theory".

These papers had, amongst other things, the undeniable importance of making accessible topics that were formally very complicated, so much so that one of the articles in Italian [F85] was chosen to be included in the volume edited by Julian Schwinger (1918-1994) [159] which contains a collection of classic articles on quantum electrodynamics. In any case Fermi, in approaching quantum electrodynamics, in operational terms seized on the methods, of second quantisation and quantum field theory, such as had been developed at the time. As we shall see in chapter 6, it would be these methods that, once applied to the problem of beta decay, would allow Fermi in 1933 to arrive at his famous theory and to head towards the discovery of neutron-induced radioactivity.

Fermi however, while he was concerned with quantum electrodynamics, also began to express interest in Nuclear Physics. The first time that his name appeared linked to Nuclear Physics themes was in 1929 when he acted as supervisor for Ettore Majorana's thesis for his laurea in Physics. Majorana's thesis indeed concerned alpha nuclear decay (see [83]).

In 1929, when Majorana was developing his thesis, Nuclear Physics was a completely new topic compared to the lines of research that had been followed in Rome until then. These in fact had mainly been themes of classical physics or of quantum mechanics applied to the atom, in relation to spectroscopy, statistics and, more recently, as we have just said, in relation to electromagnetic radiation. The only exception is a note by Giovanni Gentile Jr. (1906-1942) (at that time an assistant in Rome for a six month period) which appeared in the "Rendiconti dell'Accademia dei Lincei" in February 1928 [79], concerning a model of the nucleus, still based on classical mechanics, which Ernest Rutherford (1871-1937) had posited the year before to explain alpha decay [149], and to which Giovannino Gentile put forward a series of very profound criticisms.

Majorana's thesis, with the title "Sulla meccanica dei nuclei radioattivi" (On the mechanics of radioactive nuclei), presents the result of research into alpha decay of nuclei, presumably begun in 1929 judging by the bibliography indicated, which amongst others quotes a very recent article by George Gamow (1904-1968) and Friedrich G. Houtermans (1903-1966) "Zur Quantenmechanik des Radioaktiven Kerns" [74], accepted on 29/10/1928 and published in issues 7-8, closed on 17/12/1928, of "Zeitschrift für Physik". The thesis was presented on 6 July 1929 to obtain a *Laurea* in Physics. In the report of his scientific activity, presented by Majorana in 1932 attached to his application for the *Libera Docenza* in Theoretical Physics, the title of the thesis was changed significantly with the addition of the word "Quantum" to become "On the Quantum Mechanics of Radioactive Nuclei" (see [83]). There are two original typescript copies of the thesis, with formulas added by hand by Majorana: a bound copy kept in the Family archive, the other in loose pages in the "Giovanni Gentile Jr." archive in the Department of

Physics at Sapienza University of Rome which also contains a note written by Gentile.

The subject was at that time extremely topical. Indeed a few months earlier George Gamow (1904-1968) [73], and independently Ronald Wilfred Gurney (1898-1953) and Edward Uhler Condon (1902-1974) [87], had succeeded in giving an explanation for alpha decay compatible with the experimental data in the setting of quantum mechanics, on the basis of the recently discovered tunnel effect, thus showing for the first time that this new mechanics could also be applied to nuclear phenomena.

Majorana's Thesis is the first work on the application of quantum mechanics to Nuclear Physics carried out in Rome, and more generally in Italy. In it a rigorous mathematical justification is given of Gamow's treatment of alpha decay which had received some serious criticisms in the scientific press. The thesis had a certain success. For example it was asked for (through Giovannino Gentile) by the nuclear physicist János (Johann) Kudar in Berlin who was very interested in the subject. However, even if original and very interesting results were arrived at, fully competitive at an international level, they did not lead to any outlet in official publications. Even the subject dealt with by Majorana would not be tackled at Rome again. Until 1933 Majorana himself would not concern himself again, at least officially, with Nuclear Physics. Nevertheless his results in 1933, which we will discuss later, are connected conceptually to the problems dealt with in his thesis, in reference to the possibility of applying quantum mechanics to nuclear structure. In the case of alpha decay this happened because the alpha particle, which is a heavy particle, was considered phenomenologically as having its own individuality in the nucleus even before decay. As for nuclear structure generally, it would be necessary to wait for the definitive elimination of the supposed "nuclear" electrons from the nucleus. But this could only happen after the discovery of the neutron in 1932.

In Fig. 4 we see Ettore Majorana's *libretto*(*).

Anyway in the years immediately following Majorana's thesis Fermi began to devote himself in person to subjects connected in some way to Nuclear Physics, turning his attention to the study of the hyperfine structure of atomic spectra. This choice is clearly linked to his previous activity in atomic physics, and in particular to his experience acquired in spectroscopy. Hyperfine structure, as Fermi observed in [F83], could be counted among the "phenomena that describe, so to speak, properties outside the nuclei", because due to the interaction of the nuclear magnetic moment with the orbital magnetic field of the electrons. It did not therefore directly involve "events that occur inside the nucleus", and could certainly be described in the setting of quantum mechanics.

As well as the famous fine structure, shown by the splitting of the energy levels due to the spin of the electrons and the relativistic corrections of their motion, the spectral lines, above all those emitted by the heaviest atoms, if observed with a high-resolution spectrometer, also show in some cases a hyperfine structure. These spectral lines appear split into several very close components separated from each other. The effects of the

(*) The booklet issued by Italian universities to their students for identification purposes [Translator's Note].

Fig. 4. – Ettore Majorana in his university *libretto* - FDG.

hyperfine structure are about three orders of magnitude smaller than those of the fine structure.

Already in 1924 hyperfine structure, excluding that which varied with the variation of the isotopic composition of the elements, had been explained by Pauli in the setting of old quantum mechanics, assuming that the nucleus had an intrinsic angular momentum (spin) associated with a magnetic moment.

Because of the coupling of this nuclear magnetic moment to the magnetic field produced by the external electrons, each energy level splits into various sublevels, giving rise to the various components that constitute the hyperfine structure of a spectral line, as had been seen in the experimental context. On the basis of this explanation, it was possible in principle to deduce, by means of a comparison between the hyperfine structures of the experimental lines and of the theoretical ones, the value of the angular momentum and of the magnetic moment of the nucleus, and thus obtain precious information about its inner composition.

Rita Brunetti (1890-1942) for example, in a review article which appeared in "Il Nuovo Cimento" in 1930 [30] wrote about the possibilities and the advantages of hyperfine structure studies: "Faced with the nuclear problem the physicist of today was perplexed for a while; but he is now gladdened that he can lift, at least partially, the veil that surrounds the material part of the atom without new technical inventions, without resorting to dissolving forces which for now seem inaccessible".

Fermi worked on hyperfine structure from the end of 1929 and throughout 1930, publishing three papers, including a Letter to Nature [F72, F73, F74], and delivering a talk to the XIX meeting of the Società Italiana per il Progresso delle Scienze (SIPS) in 1930, with the title "Sul momento magnetico del nucleo" (On the magnetic moment of the atomic nucleus) [F86]. This was never published however.

In these papers Fermi developed a quantum theory of hyperfine structure which could be used to evaluate the magnetic moments of many atomic nuclei. Examples of the application of this theory were then given by Fermi himself for the nuclei of sodium, rubidium and caesium. Fermi would return to the subject of hyperfine structure in 1933, publishing in collaboration with Segrè, a definitive work "sulla teoria delle strutture iperfini" (On the theory of hyperfine structures), in which, amongst other things, certain anomalies, seen experimentally in some hyperfine structures, were interpreted as a perturbative effect. In the version written in Italian for the Memorie dell'Accademia d'Italia [F107] the authors thanked Majorana for his theoretical contribution. The acknowledgement was omitted in the contemporary slightly modified version written in German for the international journal "Zeitschrift für Physik" [F108].

In the end the study of the hyperfine structure proved not to be a very accurate method to obtain information about the nucleus. Indeed Fermi himself observed in 1932 [F100] that "calculations of this type can so far only give hints of the order of magnitude of the magnetic moment of the atomic nucleus". Even so the results obtained by Fermi in this sector were such as to earn him two important invitations: the first, to the Solvay Conference in 1930 devoted to "Le Magnétism", where Fermi presented a report "Sur les Moments Magnétiques des Noyaux" [F87], the second to the 186th regular Meeting of the American Physical Society, which took place in Chicago in June 1933 and was defined in the Proceedings of the Conference [36] as "Perhaps until then the most important scientific session in its history", where Fermi gave a public lecture on the "Theory of hyperfine structures" [F113].

In any case, apart from these results in the area of hyperfine structure, Fermi's name in the nuclear field would soon become famous for another reason: as the organiser of an International Conference in Rome which made a mark in the history of Nuclear Physics.

3.2 The first International Conference on Nuclear Physics: Rome, October 1931

From 11 to 18 October 1931 an International Conference on Nuclear Physics was organised in Rome at the Physics Institute of the Royal University. This was, without exception, the first International Conference in this field. The opening session was held instead in Villa Farnesina, headquarters of the Accademia d'Italia. The history of this Conference, reconstructed through the documents preserved in the archive of the Accademia d'Italia, now kept at the Accademia dei Lincei, is particularly significant.

The year before this Conference (1930), on the initiative of the Società Generale Italiana di Elettricità - Edison in Milan and with financing from the same company, the

"Fondazione Alessandro Volta" was established and affiliated to the Reale Accademia d'Italia. The "Academic" council of this Foundation was chaired by the Nobel prize winner for Physics and Senator of the Kingdom Guglielmo Marconi (1874-1937). As a representative of the Società Edison, Orso Mario Corbino, then Director of the Physics Institute of the University of Rome and also a Senator of the Kingdom, was a member of the council. The Foundation's "most important aim" was to "convene annual meetings of scientists and scholars to discuss a topic that every two years must reflect subjects that fall within the competence of the Class of Physical Sciences and, in the intervening years, subjects within the domain of the other classes: Moral and Historical Sciences, Literature, Art". the privileged position reserved for the Class of Physical Sciences compared to the other classes must be stressed.

For the organisation of the first Conference — the inaugural Conference to be held in 1931 — Enrico Fermi, then Professor of Theoretical Physics at the University of Rome and a member of the Accademia d'Italia, was chosen as General Secretary with Guglielmo Marconi as Honorary President, Orso Mario Corbino as Effective President, and Bruno Rossi (1905-1993) (Florence), Antonio Carrelli (1900-1980) (Catania) and Gleb Wataghin (1899-1986) (Turin) as Secretaries.

For this Conference — as can be seen in particular in the letters sent by Fermi to the various participants and their replies (Figs. 5-12) — the theme initially chosen was "Nuclei and Electrons".

Even though we read the word "Electrons" in the title, and even if we find famous names in this field amongst the various invited scientists, such as John S. E. Townsend (1868-1957), John A. Fleming (1849-1945), Robert A. Millikan (1868-1953), Arthur H. Compton (1892-1962), and also Joseph J. Thomson (1856-1940) and Paul A. M. Dirac (who however were unable to participate), the central point of the Conference in the intentions of the organisers was reserved for the nucleus. Indeed the Proceedings of the Meeting [1], published in 1932, appeared with the title "Convegno di Fisica Nucleare" (Nuclear Physics Conference) (Fig. 13), and the reference made to "Electrons" throughout the Conference mainly concerns the electrons which, at that time as will be amply explained later, were believed to exist inside the nucleus.

According to the organisers in that precise moment in history the importance of delving into the study of nuclear phenomena is essentially linked to two questions. On the one hand, the instruments with which this field could be addressed from the theoretical point of view seemed to be available, at least in part. This was thanks to the complete development of non relativistic quantum mechanics and to the beginning of the relativistic extension of the wave equation and of the debate on the interaction between radiation and particles (that is to say with the beginning of quantum electrodynamics), A recent example of this was the success achieved by G. Gamow in 1928 [73], and independently by R. W. Gurney and E. U. Condon [87], in developing a theory for alpha nuclear decay based on the tunnel effect in quantum mechanics. On the other hand, after Rutherford arrived at the first nuclear disintegration in 1919, obtained by bombarding nitrogen with alpha particles, and with the consequent discovery of the proton [147], research into the artificial transmutations provoked by alpha particles had already been firmly launched.

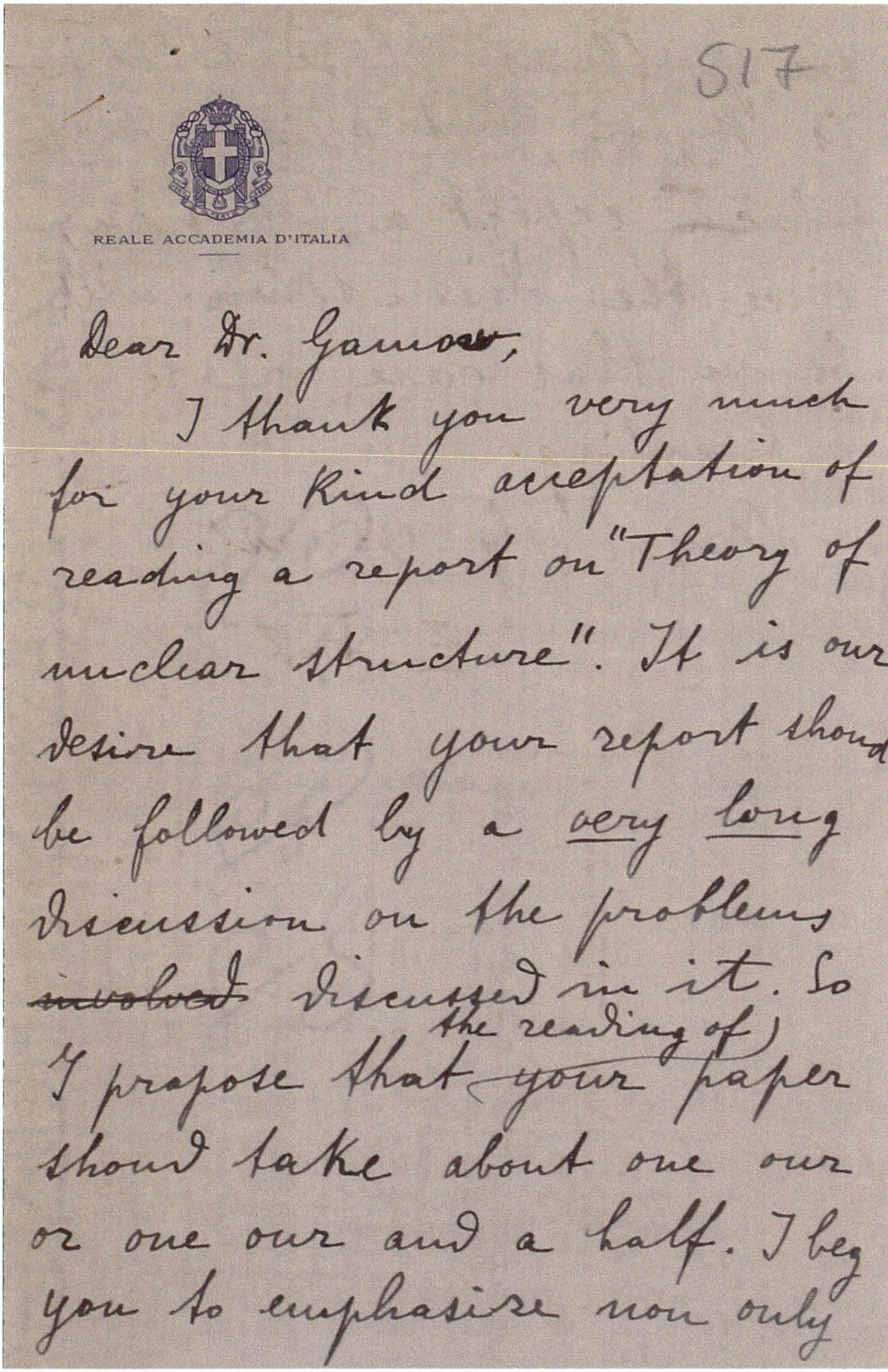

517

REALE ACCADEMIA D'ITALIA

Dear Dr. Gamow,

I thank you very much for your kind acceptation of reading a report on "Theory of nuclear structure". It is our desire that your report shoud be followed by a very long discussion on the problems ~~involved~~ discussed in it. So I propose that ~~your~~ the reading of paper shoud take about one our or one our and a half. I beg you to emphasize non only

Fig. 5. – Handwritten copy of Fermi's letter to George Gamow (undated, first page) - BANLC.

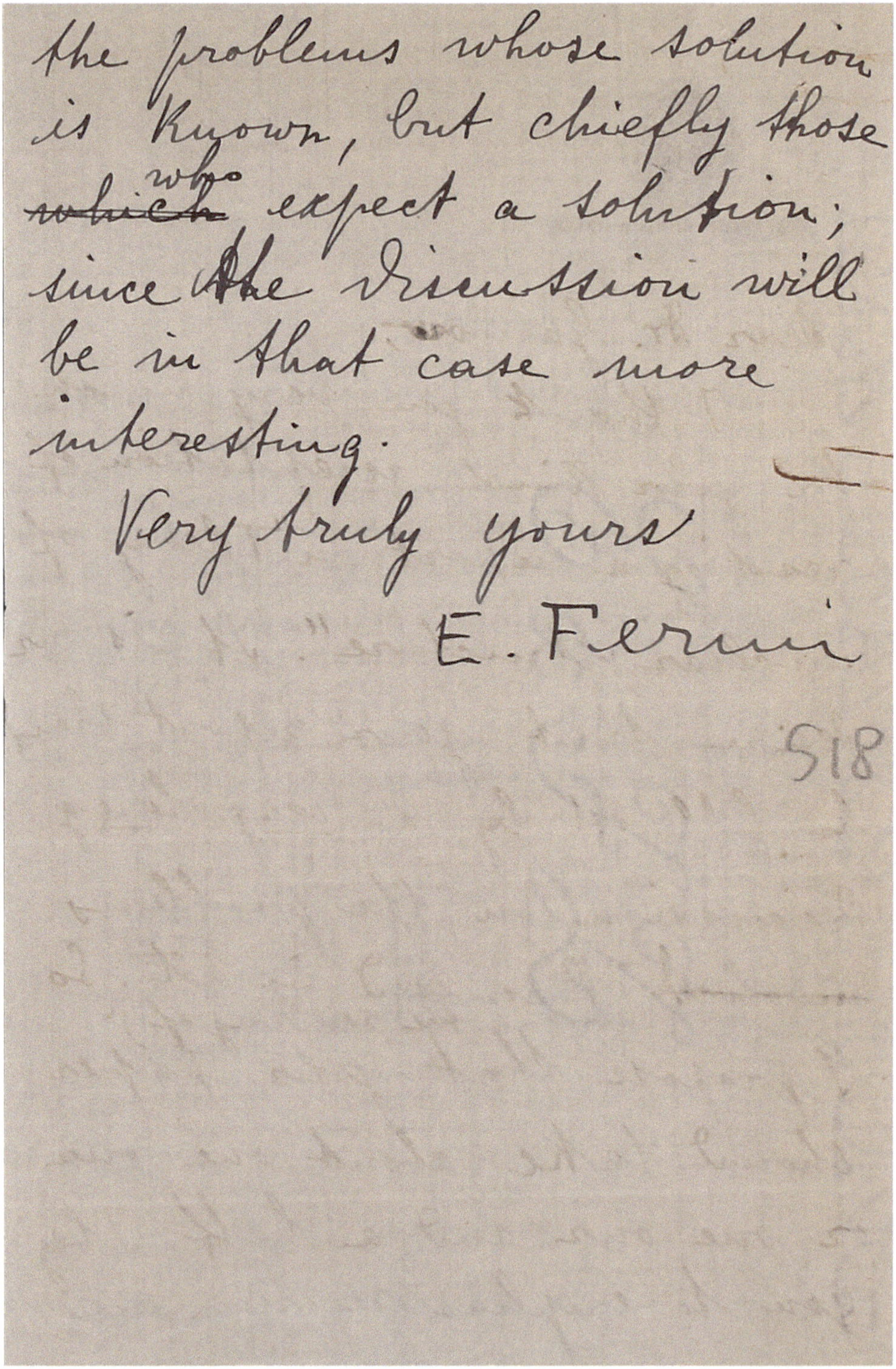
the problems whose solution
is known, but chiefly those
~~which~~ who expect a solution;
since the discussion will
be in that case more
interesting.

Very truly yours

E. Fermi

518

Fig. 6. – Handwritten copy of Fermi's letter to George Gamow (undated, second page).

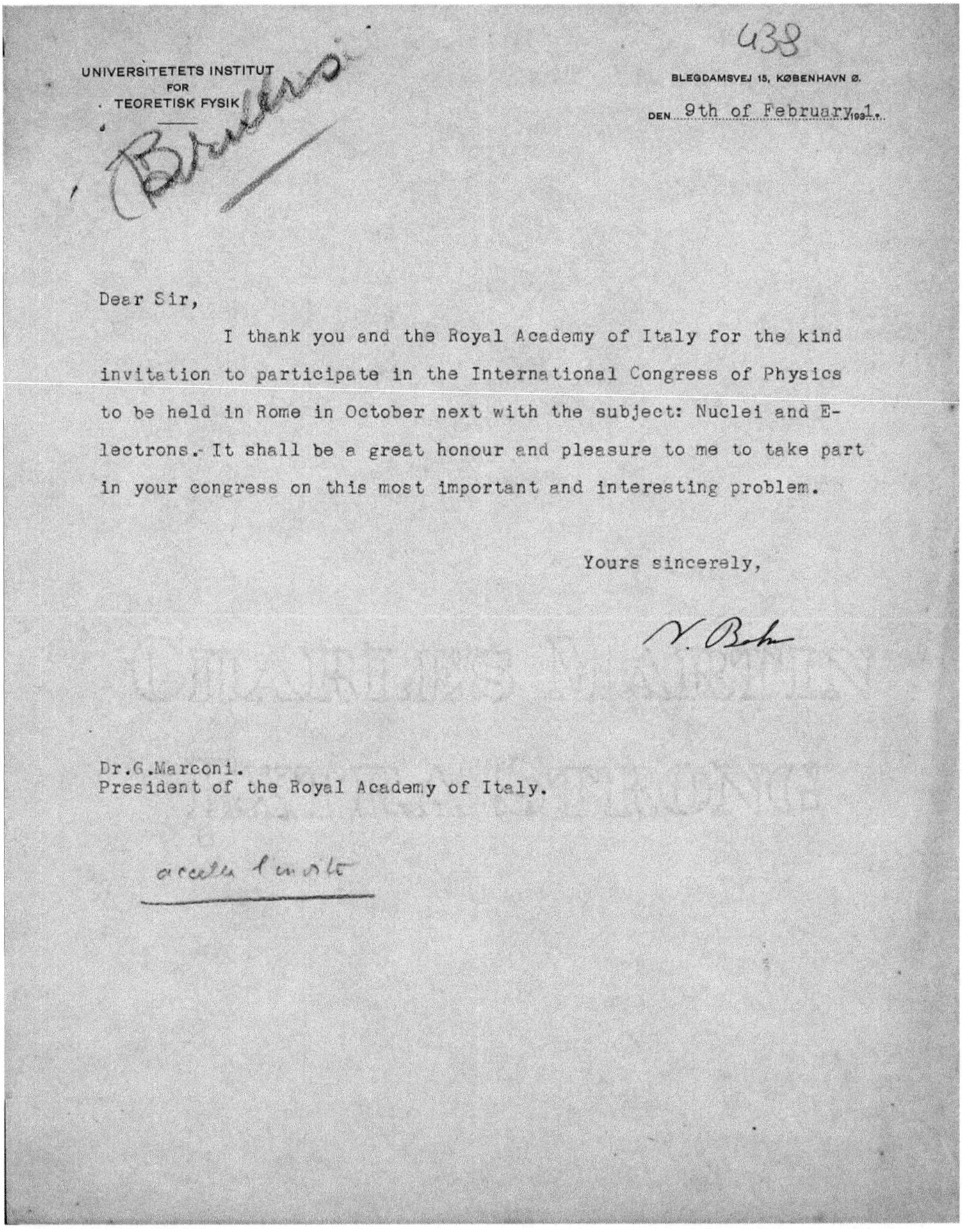

UNIVERSITETETS INSTITUT
FOR
TEORETISK FYSIK

438

BLEGDAMSVEJ 15, KØBENHAVN Ø.

DEN 9th of February 1931.

Dear Sir,

I thank you and the Royal Academy of Italy for the kind invitation to participate in the International Congress of Physics to be held in Rome in October next with the subject: Nuclei and Electrons. It shall be a great honour and pleasure to me to take part in your congress on this most important and interesting problem.

Yours sincerely,

N. Bohr

Dr.G.Marconi.
President of the Royal Academy of Italy.

accetta l'invito

Fig. 7. – Niels Bohr accepts the invitation (9 February 1931) - BANLC.

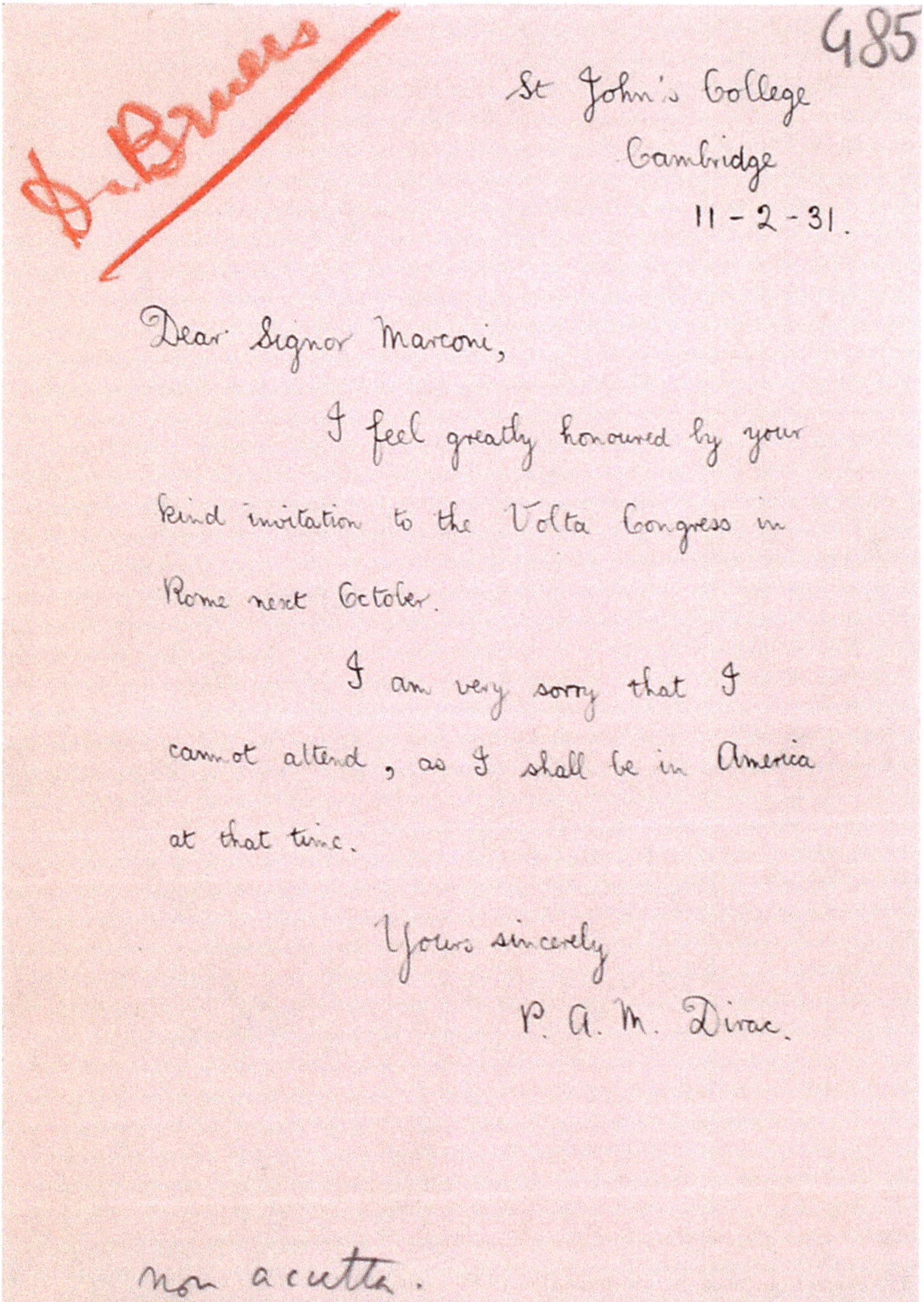

485

St John's College
Cambridge
11-2-31.

Dear Signor Marconi,

I feel greatly honoured by your kind invitation to the Volta Congress in Rome next October.

I am very sorry that I cannot attend, as I shall be in America at that time.

Yours sincerely
P. A. M. Dirac.

non accetta.

Fig. 8. – Paul Dirac's reply (11-2-31) - BANLC.

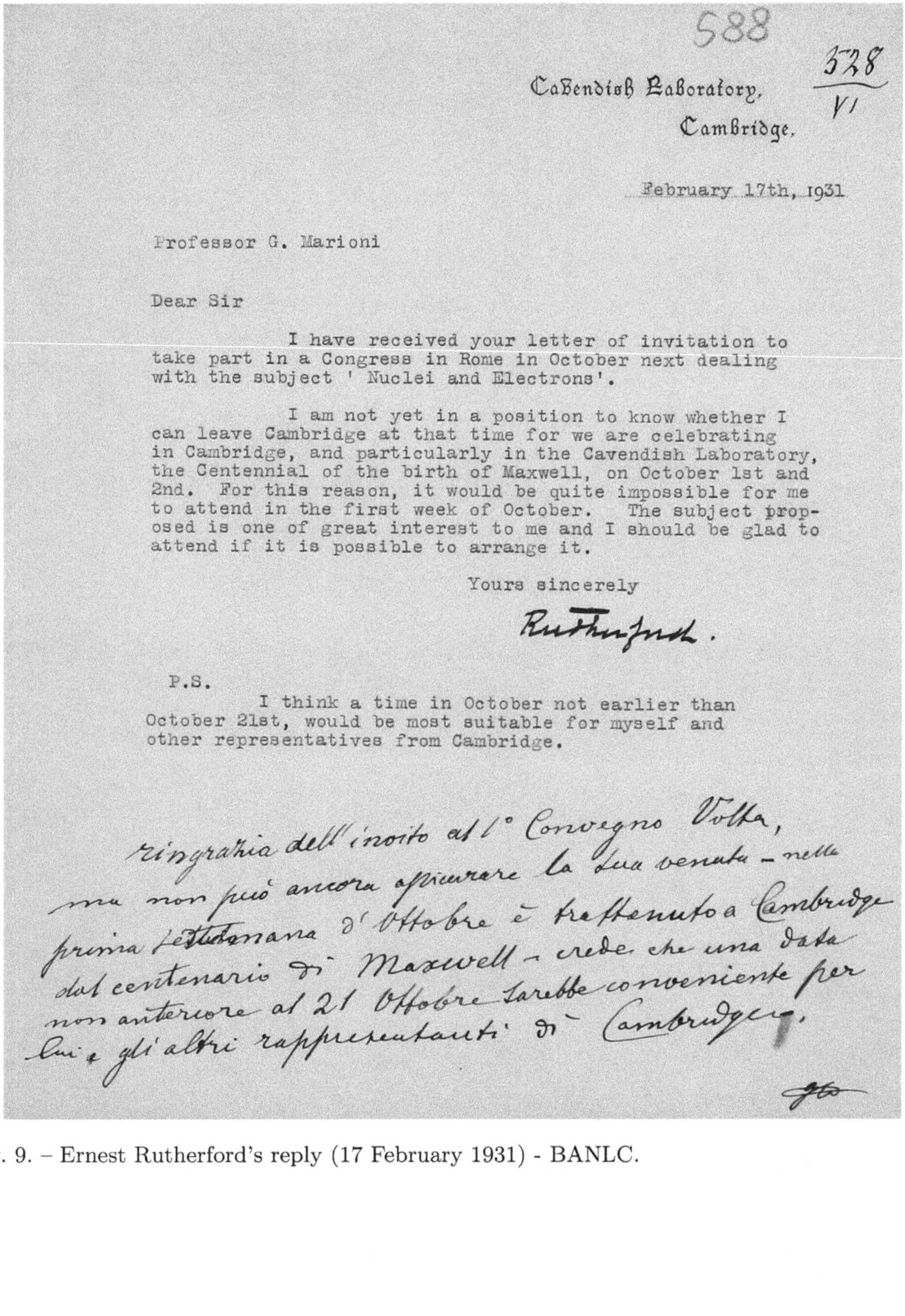
588

528
VI

Cavendish Laboratory,
Cambridge.

February 17th, 1931

Professor G. Marioni

Dear Sir

I have received your letter of invitation to take part in a Congress in Rome in October next dealing with the subject ' Nuclei and Electrons'.

I am not yet in a position to know whether I can leave Cambridge at that time for we are celebrating in Cambridge, and particularly in the Cavendish Laboratory, the Centennial of the birth of Maxwell, on October 1st and 2nd. For this reason, it would be quite impossible for me to attend in the first week of October. The subject proposed is one of great interest to me and I should be glad to attend if it is possible to arrange it.

Yours sincerely

Rutherford.

P.S.

I think a time in October not earlier than October 21st, would be most suitable for myself and other representatives from Cambridge.

ringrazia dell'invito al 1° Convegno Volta,
ma non può ancora assicurare la Sua venuta - nella
prima settimana d'Ottobre è trattenuto a Cambridge
dal centenario di Maxwell - crede che una data
non anteriore al 21 Ottobre sarebbe conveniente per
lui e gli altri rappresentanti di Cambridge.

Fig. 9. – Ernest Rutherford's reply (17 February 1931) - BANLC.

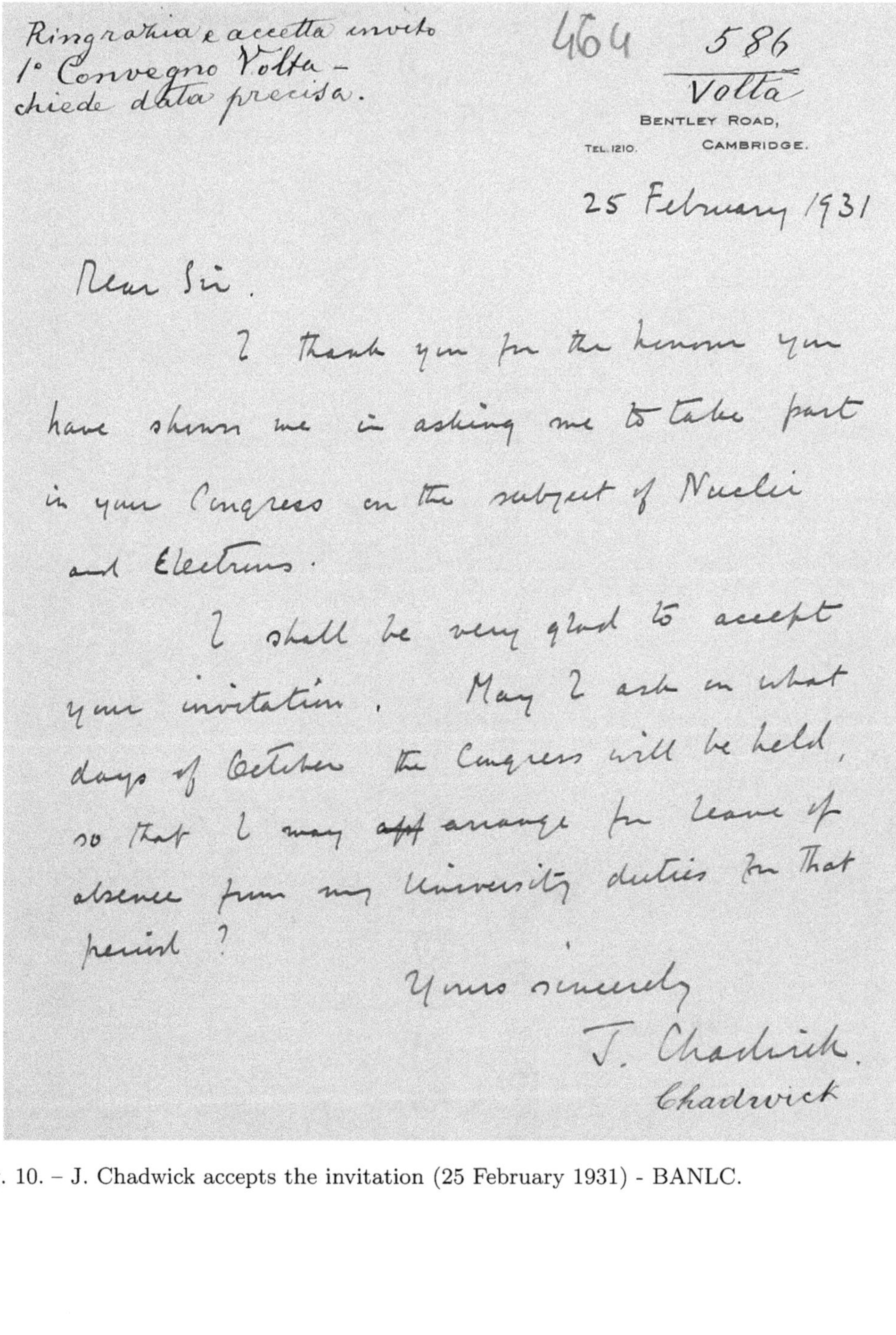

Ringrazia e accetta invito 1° Convegno Volta – chiede data precisa.

464 586

Volta

BENTLEY ROAD,
CAMBRIDGE.
TEL. 1210.

25 February 1931

Dear Sir.

I thank you for the honour you have shown me in asking me to take part in your Congress on the subject of Nuclei and Electrons.

I shall be very glad to accept your invitation. May I ask on what days of October the Congress will be held, so that I may ~~app~~ arrange for leave of absence from my University duties for that period?

Yours sincerely

J. Chadwick.

Chadwick

Fig. 10. – J. Chadwick accepts the invitation (25 February 1931) - BANLC.

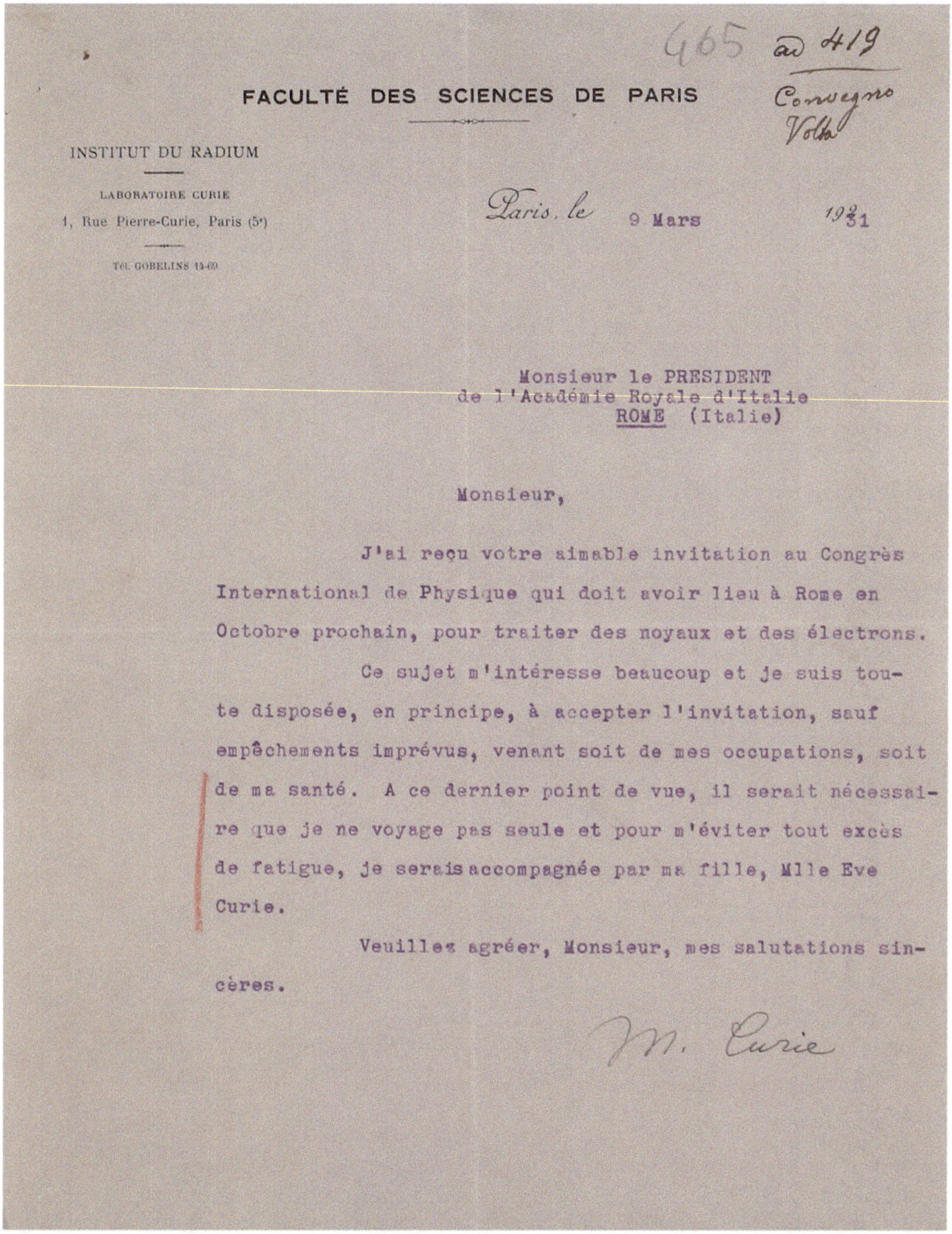

465 ad 419
Convegno Volta

FACULTÉ DES SCIENCES DE PARIS

INSTITUT DU RADIUM

LABORATOIRE CURIE
1, Rue Pierre-Curie, Paris (5e)

TÉL. GOBELINS 14-69

Paris, le 9 Mars 1931

Monsieur le PRESIDENT
de l'Académie Royale d'Italie
ROME (Italie)

Monsieur,

J'ai reçu votre aimable invitation au Congrès International de Physique qui doit avoir lieu à Rome en Octobre prochain, pour traiter des noyaux et des électrons.

Ce sujet m'intéresse beaucoup et je suis toute disposée, en principe, à accepter l'invitation, sauf empêchements imprévus, venant soit de mes occupations, soit de ma santé. A ce dernier point de vue, il serait nécessaire que je ne voyage pas seule et pour m'éviter tout excès de fatigue, je serais accompagnée par ma fille, Mlle Eve Curie.

Veuillez agréer, Monsieur, mes salutations sincères.

M. Curie

Fig. 11. – Marie Curie accepts the invitation (9 March 1931) - BANLC.

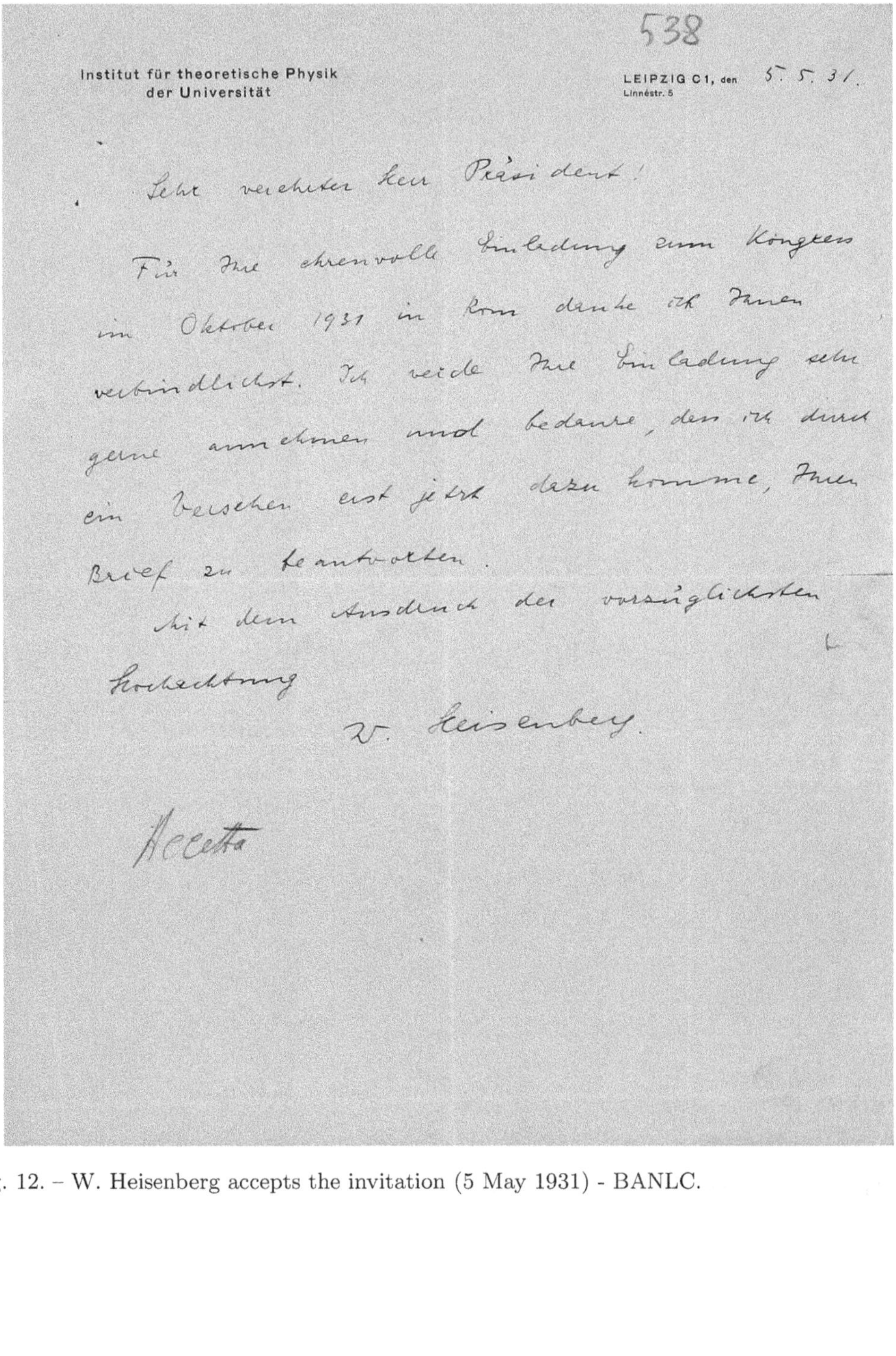

538

Institut für theoretische Physik
der Universität

LEIPZIG C1, den 5. 5. 31.
Linnéstr. 5

Sehr verehrter Herr Präsident!

Für Ihre ehrenvolle Einladung zum Kongress im Oktober 1931 in Rom danke ich Ihnen verbindlichst. Ich werde Ihre Einladung sehr gerne annehmen und bedaure, dass ich durch ein Versehen erst jetzt dazu komme, Ihren Brief zu beantworten.

Mit dem Ausdruck der vorzüglichsten Hochachtung

W. Heisenberg.

Accetta

Fig. 12. – W. Heisenberg accepts the invitation (5 May 1931) - BANLC.

REALE ACCADEMIA D'ITALIA

FONDAZIONE ALESSANDRO VOLTA

ATTI DEI CONVEGNI

1

CONVEGNO

DI

FISICA NUCLEARE

Ottobre 1931-IX

ROMA

REALE ACCADEMIA D'ITALIA

1932-X

Fig. 13. – Proceedings of the Nuclear Physics Conference, Rome, October 1931 - in [1] (BANLC).

These transmutations, moreover, had been confirmed by Patrick Maynard Stuart Blackett (1897-1974) in 1924, with an apparatus developed by Charles Thomson Rees Wilson (1869-1959) in 1912, also at the Cavendish Laboratory, and defined by Rutherford as "the most original and wonderful instrument in scientific history". It was the so-called "Wilson chamber", which allowed the visualisation of the trace of a particle or ionising radiation passing through the gas contained in the chamber (Fig. 14). In particular, on the basis of the traces left by alpha particles while passing through various gases, it was possible to establish that the alpha particles could be absorbed by the nucleus when it was hit, with the consequent emission of a proton, that is to say that they provoked an (α,p) type nuclear transmutation, in which the initial nucleus (with atomic number Z) was transformed into a new nucleus with atomic number $(Z+1)$. Such a transformation can be seen in Fig. 15, obtained by Blackett in a Wilson chamber filled with nitrogen [21], where a trace of the alpha particle splits into two traces, one of which, the thinner one, is due to the expelled proton while the other thicker and shorter one is due to the transmuted nucleus, which is oxygen ${}^{17}_{8}\mathrm{O}$, according to the reaction ${}^{2}_{4}\alpha + {}^{14}_{7}\mathrm{N} \rightarrow {}^{17}_{8}\mathrm{O} + {}^{1}_{1}p$.

In a document of the Accademia d'Italia, in which the Conference was officially presented (Fondo Reale Accademia D'Italia, VIII-16-43-2), we find written: "All the properties and the structure of the different chemical atoms, as far as the outer parts are concerned, can today be considered to be perfectly known. So the problem that we are faced with today is that of studying the structure of the nucleus itself. [...] Until a few years ago radioactive phenomena remained almost totally incomprehensible: only now, with the mechanics of the microscopic world developed over the past four years, do we have the beginnings of an explanation. Recently it has been possible also to produce artificially, albeit on a very small scale, the transformations of one element into another, for example nitrogen into oxygen, accomplishing the dream of the ancient alchemists, even if only for an infinitesimal quantity of matter. Now we have a theoretical explanation for these phenomena too.

For these reasons the research field of Nuclear Physics, opened up to scientists' enquiries only in the past few years, is today one of the most important and fascinating." On the topic and the aims of the Conference Guglielmo Marconi expressed himself thus in his opening speech:

"Man's assault on the Atom, a minute, ultramicroscopic but powerful fortress which has remained unassailable for centuries, has already given encouraging and surprising results.

The outer defences have begun to yield and we now believe that we know how they are constructed; but the central part, the atomic nucleus, still remains a mystery, at least in part. Many are fully convinced that immense treasures are hidden there and that they may, in the near or distant future, be used to the benefit of humanity. The conquest of the nucleus is certainly a shining goal on which it would be worthwhile for the efforts, studies and research of so many amongst the greatest physicists and scientists in the world to converge.

Those attending are called upon to weave their work on this topic".

Clearly the organisers' choice of Nuclear Physics as the subject for the first Interna-

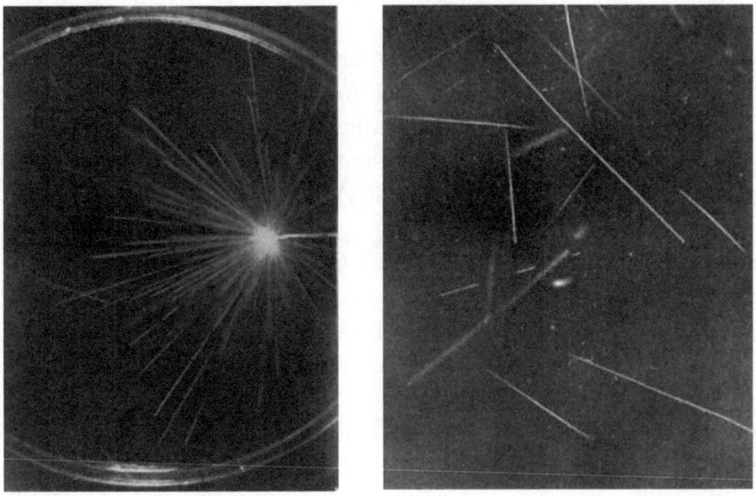

Fig. 14. – Photographs taken in the Wilson chamber of alpha particles emitted by a Radium source (on the left) and by radium emanation (radon) (on the right) - in [167] (PRSL).

tional Conference of the "Fondazione Alessandro Volta" is also indicative of an inclination to engage in research in this field. In reality, a similar idea had been put forward by Corbino in his lecture "I compiti nuovi della fisica sperimentale" (The new tasks of experimental physics), given in Florence on 21 September 1929 during the XVIII Congresso della Società Italiana per il Progresso delle Scienze (SIPS) [42]. Corbino observed on that occasion:

"The only possibility for great new discoveries in Physics lies (therefore) in the chance that the inner nucleus of the atom may be successfully modified. And that will be the truly worthy task of future physics. [...] we can therefore conclude that, while great progress in experimental physics in its ordinary domain appears improbable, many possibilities have opened up on the path to attacking the atomic nucleus; the most

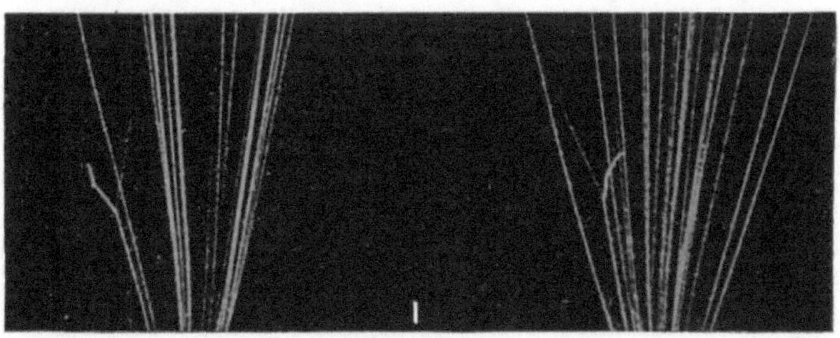

Fig. 15. – Disintegrations of nitrogen by alpha particles (Blackett 1925) - in [21] (PRSL).

enticing field in tomorrow's Physics".

At the time of the Rome Conference, the activity concerning Nuclear Physics previously carried out at the Regio Istituto Fisico in Rome, as we indicated before, includes, in chronological order, Gentile's work criticising Rutherford's satellite model, Majorana's *laurea* thesis, Fermi's papers on hyperfine structure and, finally, a paper on hyperfine structure in 1931 by Giulio Racah (1909-1965) (then a young assistant at the Istituto Fisico in Rome) [131].

It is clear that the main intent of the Rome Conference was to sum up the results and problems current at that time, prevalently from the point of view of scientific knowledge. But the organisers must also have been well aware of the great prospects for practical applications, especially in the energy sector. Indeed in the document presenting the Conference, mentioned above, we can read:

"This interest (in Nuclear Physics) is now purely speculative but it is also possible that one day nuclear phenomena may acquire crucial importance for practical applications, given the enormous energies that come into play in nuclear processes, millions of times greater than in ordinary chemical reactions".

Some points in Corbino's introductory speech, which stress the great energy possibilities of nuclear mass defects, are particularly significant in this regard. Corbino observed: "So about two hundred thousand kilowatt hours are required to break down a gramme of helium into hydrogen; but just as many would be developed if a gramme of hydrogen is successfully transformed into helium. For this synthesis of an element as precious as helium the necessary raw material, hydrogen, is widely available in nature; with only fifty transformed kilogrammes, the ten billion kilowatt hours of electricity used in Italy in a year would be obtained". (Note that at that time the prevailing opinion was that the nucleus was composed of protons and electrons.) Remember that Corbino was on the Board of the Società Generale Italiana Edison di Elettricità and that, even more significantly, the conference was organised under the auspices of the "Fondazione Alessandro Volta", affiliated to the Accademia d'Italia but set up and financed by the Società Edison di Elettricità.

In the conclusion of Corbino's speech we can read the following passage, expressed in accordance with the rhetoric of the age but not lacking in prophetic significance and certainly indicative of the expectations associated with Nuclear Physics research:

"Primitive Man in practice distinguished himself from the animals on the day he mastered fire. Today Man is preparing to master and take control of a far more powerful fire, that can be estimated at billions of degrees: In the new and mysterious crucible the artificial transmutation of the elements and the liberation of masses of energy will be achieved, with grades and quantities incomparably higher than those we have handled thus far".

With these aims, the Conference managed to bring together in Rome the most important physicists in the world working on these topics or potentially interested in them. Overall there were more than fifty invited participants present including, as the organisers pointed out, 7 Nobel Prize winners: Francis William Aston (1877-1945) (Chemistry 1922), Niels Bohr (1885-1962) (Physics 1922), Marie Curie (1867-1934) (Physics 1903,

Chemistry 1911), Arthur H. Compton (Physics 1927), Guglielmo Marconi (Physics 1909), Robert Millikan (Physics 1923), Jean Baptiste Perrin (1870-1942) (Physics 1926).

Amongst the foreigners attending (many of them future Nobel Prize winners) we can mention Nevill Francis Mott (1905-1996, Nobel Prize for Physics in 1977) from Manchester, John S. E. Townsend from Oxford, Patrick M. S. Blackett (Nobel Prize for Physics in 1948), Charles Drummond Ellis and Ralph Howard Fowler (1889-1944) from Cambridge, Werner Heisenberg (Nobel Prize for Physics in 1932), Peter Debye (1884-1966) (Nobel Prize for Chemistry in 1936) and Guido Beck (1903-1988) from Leipzig, Walther Bothe (1891-1956) (Nobel Prize for Physics in 1954) from Giessen, Lise Meitner (1878-1968) from Berlin, Otto Stern (1888-1969) from Hamburg (Nobel Prize for Physics in 1943), Léon Rosenfeld (1904-1974) from Copenaghen, Wolfgang Pauli (Nobel Prize for Physics in 1945) from Zurich, Hans Wilhelm Geiger (1882-1945) from Tübingen. Emil Rupp (1898-1979) from A. E. G. in Berlin, whom we will talk about later, was also invited.

Paul A. M. Dirac (future Nobel Prize for Physics in 1933), Max Planck (1858-1947, Nobel Prize for Physics in 1918), Ernest Rutherford (Nobel Prize for Chemistry in 1908) and Erwin Schrödinger (future Nobel Prize for Physics in 1933) had been invited but had not been able to attend.

Many Italian scientists were also present. As well as Fermi, Corbino and Marconi there were Antonio Carrelli, Bruno Rossi and Gleb Wataghin who were given official roles. Moreover attending the conference there were: Giovanni Battista Bonino (1899-1985) and Quirino Majorana (from Bologna), Ugo Bordoni (1884-1952), Tullio Levi-Civita, Antonino Lo Surdo (1880-1949), Nicola Parravano (1883-1938), Franco Rasetti, Giulio Cesare Trabacchi (1884-1959) (from Rome), Michele Cantone, Francesco Giordani (from Naples), Antonio Garbasso (from Florence), Enrico Persico, Giancarlo Vallauri (1882-1957) (from Turin), Giuseppe Gianfranceschi (1875-1934) (from the Vatican City).

There is no documentation showing that Ettore Majorana was present at the Conference. He was believed to have been recognised in some photographs of the participants but that was certainly P. M. S. Blackett, who did display a certain ressemblance to Majorana which led to a mistaken identification.

The great absent figure at the Conference was Ernest Rutherford (Fig. 16), recently raised to the hereditary peerage as Baron Rutherford of Nelson by royal decree on 22 January 1931, as reported in the The London Gazette: "Whitehall, 23 January 1931. The KING has been pleased, by Letters Patent under the Great Seal of the Realm, bearing date the 22nd instant, to confer the dignity of a Baron of the United Kingdom upon Sir Ernest Rutherford, Knight, Order of Merit, and the heirs male of his body lawfully begotten, by the name, style and title of Baron Rutherford of Nelson, of Cambridge in the County of Cambridge". Unfortunately the title of baron, which brought membership of the House of Lords, ceased to exist on Rutherford's death in 1937 because of a lack of heirs.

As Rutherford himself explained in a letter written on 17 February 1931 (Fig. 9), addressed to "Professor G. Marioni" (by which he certainly meant "Marconi"), he was detained at the Cavendish Laboratory in Cambridge by institutional commitments con-

Fig. 16. – Ernest Rutherford holding the apparatus with which he discovered the proton in 1919 - CUL.

nected to the celebrations for the centenary of Maxwell's birth. He was extremely interested in the subject of the conference and he even proposed that it be postponed to a date after 21 October. He was sent a telegram as a sign of esteem which significantly reads: "The members of the first Conference of the Volta Foundation, meeting in Rome to discuss Nuclear Physics, express their regret for the absence of the founder of this science and express their most devout respects".

A well-known group photograph of the partecipants is included at the beginning of the Proceedings [1]. Instead here we reproduce a photograph (Fig. 17), from the "Fondo Enrico Persico" at the Department of Physics of the Sapienza University of Rome, which on the back shows the signatures of some of the partecipants (Fig. 18).

Corbino in his concluding speech (Archive of the Accademia d'Italia) explained the

Fig. 17. – Photograph of the participants at the Nuclear Physics Conference, Rome 1931 - DFUR.

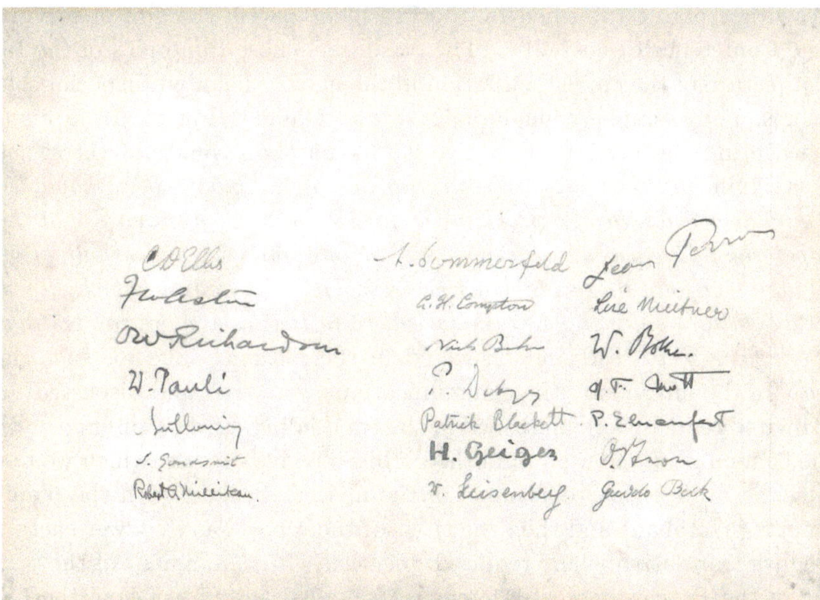

Fig. 18. – Signatures of the participants on the back of the preceding photograph - DFUR.

choice of organising a Conference of such great size, as well as of such high calibre, with the following words:

"We can say that the strength of the major centres studying Physics today is due not only to the greater means available but above all to the presence together in the same setting of many researchers who exchange ideas, reflections, the results of their reading, critical observations. If the validity of research grows exponentially with the number of individuals living in the same laboratory, you can imagine what will be the effect of this Conference where the number of partecipants is great and their worth is greater, as never seen in previous meetings."

Naturally setting up the Rome Conference required an enormous organisational and also economical effort. According to the archive documents the cost was approximately 200,000 Lire. To have an idea of what that figure meant, remember that 1,000 Lire in those days represented an excellent salary, as the song famous in the Thirties recalled: "se potessi avere mille lire al mese ..." ("if I could have a thousand Lire a month ..."). Those invited to attend, and those accompanying them, were offered luxury treatment, housed in the Hotel del Quirinale, with an extensive social programme and banquets in Rome and Tivoli. Generous railway discounts were also arranged in order to allow tourism and cultural trips even before and after the Meeting.

3.3 Accurate scientific planning

Above all the accurate scientific planning should be noted.

The Conference was organised in six Reports, "written by invited authors and in accordance with a plan established by the President", with the aim of serving as "the basis for the Conference discussions". The organisers chose the topics of the talks so as to provide a picture of the current situation in the nuclear field, with special attention to open problems. For example, what Fermi wrote to Gamow (Fig. 5, 6), in his thankyou letter for accepting the invitation to give a talk on the "Quantum theory of nuclear structure" is significant in this regard: *It is our desire that your report should be followed by a very long discussion on the problems discussed in it. So I propose that the reading of your paper should take about one our* [sic] *or one our* [sic] *and a half. I beg you to emphasize non* [sic] *only those problems whose solution is known but chiefly those who* [sic] *expect a solution, since the discussion will be in that case more interesting.*

The six reports were read and discussed in the morning (one for each day of the Conference). In the afternoon "the communications were read and discussed".

All the invited reports were published in the Proceedings of the Conference, while only some of the communications were published. Here we give the list, which we have drawn up following the conference programme, of the invited speakers with the planned titles of their reports (in Italian) and then the titles of the reports as they were actually given and the English translation where required, together with a summary of the programme. The layout in the Proceedings is different. A detailed account of how the Conference developed was prepared by Giulio Racah [132]. Another account was drawn up by Bruno Rossi in the journal "L'Elettricista" [144]. Fermi himself would then publish a summary of the scientific contents of the Conference in the political journal "Gerarchia" [F95] founded by Benito Mussolini.

Monday 12 October: speaker N. F. Mott, *Stato attuale della teoria dell'elettrone*, "On the Present Status of the Theory of the Electron". N. Bohr and L. Rosenfeld participated. In the afternoon Emil Rupp and Otto Stern gave their communications on new applications of Stern and Gerlach's classical experiment.

Tuesday 13 October: S. Goudsmit, *Spettroscopia e momenti nucleari*, "Present Difficulties in the Theory of Hyperfine Structure". N. Bohr, W. Pauli, and Walter Heitler (1904-1981) participated. Franco Rasetti "reported on an experiment he had carried out on the Raman effect in nitrogen which, for the first time, showed the disagreement between the observed nuclear moment and the expected one". O. Stern reported on experiments under way for the direct measurement of nuclear moments. In the afternoon R. A. Millikan reported on some characteristics of the Compton effect while R. H. Fowler described a new quantum model of the nucleus.

Wednesday 14 October: B. Rossi, *La radiazione penetrante*, "Il Problema della Radiazione Penetrante" (The Problem of Penetrating Radiation). There were contributions from H. Geiger on experiments aimed at demonstrating soft corpuscular radiation and from M. Curie on experiments by Dmitri V. Skobelzyn (1892-1990).

Thursday 15 October: W. Bothe, *Recenti ricerche sulla radioattività*. Parte I: *Raggi α, trasformazioni e eccitazioni artificiali dei nuclei, Isotopi*, "α-Strahlen, Künstliche Kernumwandlung und -Anregung, Isotope" (α-Rays, Artificial transformations and excitations of nuclei, Isotopes). In the afternoon L. Meitner spoke on the diffusion of very hard γ rays, and C. D. Ellis spoke on analogous experiments carried out at the Cavendish Laboratory.

Friday 16 October: C. D. Ellis, *Recenti ricerche sperimentali sulla radioattività*. Parte II: *Raggi β e γ*, ("β-Rays and γ-Rays").

Saturday 17 October: speaker G. Gamow, *Teoria quantistica della struttura del nucleo*, "Quantum Theory of Nuclear Structure" (the paper was read by Dr. Delbrück). In the Proceedings the following contributions were also published:
N. Bohr: "Atomic stability and conservation laws",
L. Rosenfeld: "Über die quantentheoretische Behandlung der Strahlungs-probleme",
A. Sommerfeld: "Vereinfachte Ableitung des Thomas-Faktors",
E. Rupp: "Ueber einen experimentellen Nachweis polarisierter Elektronen",
W. Bothe: "Bemerkungen über die Ultra-Korpuskularstrahlung",
A. Sommerfeld: "Ueber den Packungseffekt",
R. H. Fowler: "Γ-rays and a possible model nucleus",
G. Beck: "Über das Niveauschema des Ra C$'$".

Amongst the contributions that are not given in the Proceedings we recall those by O. Stern, F. Rasetti, R. Millikan, H. Geiger, M. Curie, L. Meitner. The summing up "masterful speech" by Debye at the end of the Conference is also missing.

The planning of the Conference events clearly shows us that the organisers, above all Fermi, were extremely up to date on the state of their discipline. The choice of topics, like the choice of the speakers, is indeed extremely shrewd. The topics cover the various branches in which Nuclear Physics had developed until then, both with success and with great difficulty. Bruno Rossi's report brought into the field of Nuclear Physics cosmic rays, then called "penetrating radiation", and should be noted. The speakers were chosen from amongst those who had achieved the most recent and most important results on the subjects they were called to discuss. The opening talk was devoted to the "Stato attuale della teoria dell'elettrone" (Present status of the theory of the electron), as if to stress that at that moment one of the crucial points of the Physics of the nucleus was the electron, and it was entrusted to Neville Francis Mott.

It had clearly been planned for Dirac, who had been one of the first to be invited to the Conference, as shown by the Archive documents but, because of previous study commitments in the United States, he had been obliged to decline the invitation. Indeed it is to Dirac that the relativistic equation of the electron is due, first proposed in 1927 [64] and completed later [65] with regard to the significance to be attributed to the negative energy solutions that this equation (together with the positive energy ones) predicted for the electron.

In any case Neville F. Mott, also from Cambridge and chosen to replace Dirac, proved to be a great expert in this field. He had been trying since 1929 to use Dirac's equation to interpret both the scattering of fast electrons with atomic nuclei and the collision between two electrons. He had also applied himself to the interpretation of Dirac's equation in the two-electron case, publishing a series of papers on the subject in the Proceedings of the Royal Society (as Dirac had done) of which we recall [116].

Walther Bothe was one of the greatest experts on the technology of point counters, also known as Geiger counters. To him and Geiger we owe, amongst other things, the first use of the coincidence method to study the Compton effect [27]. As we shall see, the Geiger counter in its new, more efficient version (the "tube counter" or "electron counter"), proposed by Geiger and Walther Müller (1905-1979) in 1928 [77, 78], and which we will discuss in chapter 9, the so-called "Geiger-Müller counter", would become the fundamental instrument both for the discovery of alpha-particle–induced radiation and for neutron-induced radiation.

The year before the Rome Conference Walther Bothe, together with his young assistant Herbert Becker, had made a very significant contribution to research into the bombardment of light elements with alpha particles [28]. Using a very sensitive Geiger-Müller counter, he had discovered that these particles, in the case of some light elements (Li, Be, B, F, Na, Mg, Al), could not only, as was known at that time, provoke a transmutation of the nucleus with the expulsion of a proton and the formation of a new stable nucleus, that is to say an $(\alpha$,p) reaction, but could also provoke a new type of nuclear reaction in which a "very penetrating radiation" was emitted, especially in the case of beryllium. This emission was explained by Bothe by the fact that the alpha bombardment could provoke "the artificial excitation of the nucleus", which then decayed emitting a gamma radiation.

Starting from these results, presented for the first time at an international level by Bothe at the Rome Conference, and in particular from those concerning the "penetrating radiation of beryllium", within just over a year the neutron would be discovered, a new

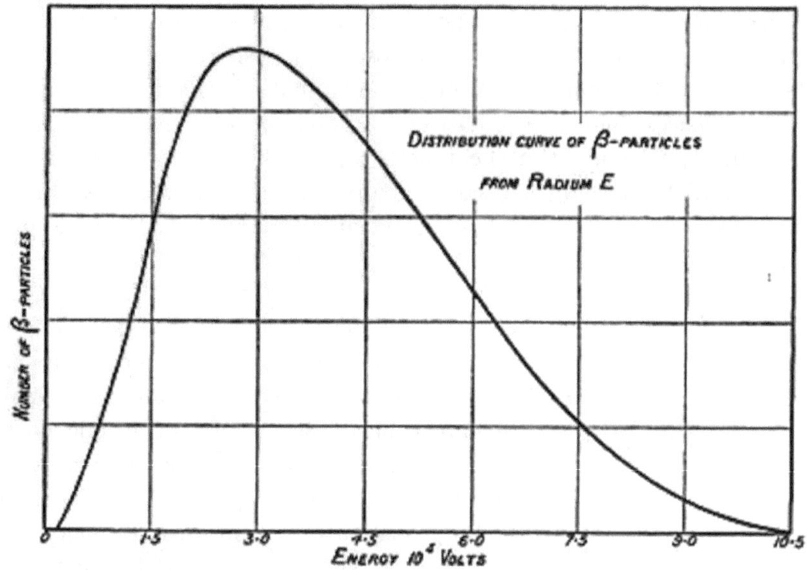

Fig. 19. – Spectrum of the beta particles emitted by Radium E (Bismuth210) - in [66] (PRSL).

particle immediately recognised as a constituent of the nucleus, with the consequent possible elimination of the nuclear electrons.

The year before Charles D. Ellis had worked together with Ernest Rutherford and James Chadwick on the writing of the book "Radiations from Radioactive Substances" [151], which had then become a reference text for experimental Nuclear Physics. A copy of this book would be bought by the Istituto Fisico in Rome on 31 October 1933, as can be seen by the registration entry, kept in the Istituto's Archive. Towards the end of the 1920s, together with W. A. Wooster (1903-1984), then a student at Cambridge, Ellis had obtained very important results [66] on the energy spectrum of beta rays, in particular with regard to the fact that this spectrum, unlike the alpha or gamma spectra, was continuous, so that the energy of the electrons emitted varied in a continuous way within broad limits. Moreover the distribution had the same tendency for the various elements, with an ill defined maximum and a rather clear upper limit (Fig. 19). Ellis and Woosterr [66], with calorimetric measurements, confirmed by Lise Meitner and Wilhelm Orthmann in 1930 [112], had shown unequivocally that the continuity of the beta spectrum was a property of the disintegrating nucleus and not a secondary effect, that is to say that the beta particles exited directly from the nucleus all with different energies.

Samuel A. Goudsmit (1902-1978), who in 1925 had discovered the spin of the electron [162] together with George E. Uhlenbeck (1900-1988), was instead an accomplished spectroscopist, particularly active in the field of hyperfine structure, above all its interpretative aspects. Goudsmit, faced with some difficulties connected to nuclear moments, would put forward Pauli's idea of the neutrino at the Rome Conference, thus presenting this hypothetical new particle for the first time in an appropriate international setting such as the first International Conference on Nuclear Physics.

George Gamow at that time was an important point of reference for the theoretical side of Nuclear Physics. Indeed in 1928, as we indicated earlier, at the same time as Condon and Gurney, he had been the first to apply the new quantum mechanics to Nuclear Physics and to develop a theory of alpha decay. Moreover he had recently turned to the constitution of the nucleus, proposing a nuclear model made up of alpha particles, protons and electrons.

On the first of May of the same year as the Rome Conference (1931) Gamow had published a book [76] in a prestigious series of the University of Oxford, one could say the first of this type, devoted to the "Constitution of Atomic Nuclei and Radioactivity", with the aim, as the author wrote "to give as complete an account as possibile of our present experimental and theoretical knowledge of the nature of atomic nuclei", with special consideration however for the theoretical aspects. In fact in this book we find a snapshot of the situation of Nuclear Physics, as it was in 1931, with all the solved problems, such as for example the isotopic constitution of the various nuclei, as emerged from the most recent measurements by Francis W. Aston (1877-1945), or the alpha and gamma type radioactive decay, and all the more disturbing problems (including, above all, the question of the presence of electrons inside the nucleus). In the book, the points that dealt with this problem were actually flagged by a special warning symbol. This book was only bought by the Istituto Fisico in Rome on 30 May 1932, as can be seen by the registration entry kept in the Archive of the Istituto.

The only Italian speaker invited to make a report to the Conference was Bruno Rossi. To Rossi goes the credit not only of having opened up a new research field on cosmic rays in Italy in 1929, in particular in Florence, gathering around him young researchers such as Gilberto Bernardini (1906-1995), Daria Bocciarelli (1910-2006) and Giuseppe P. S. Occhialini (1907-1993), but also of having made a great contribution from the scientific point of view with the invention of a multiple coincidence electronic circuit, which allowed the simultaneous passage of a particle through three or more Geiger-Müller counters to be detected, the so called "Rossi circuit" [143].

It is surprising that Emil Rupp also appeared amongst the participants at the Conference, with even a contribution published in the Proceedings of his claimed experiments with beams of polarised electrons. Indeed this researcher, active between 1925 and 1935, was the protagonist in a notorious case of ongoing scientific fraud. In any case, well before 1931, he had already been the subject of intense controversy for his alleged experimental results on the observation of the interference properties of light emitted from canal rays. Interviewed by Thomas Kuhn on 18 February 1963 Walther Gerlach (1889-1979) expressed himself very frankly when he stated:

"He was a man who was considered to be the <u>most important</u> and able experimental physicist in the Twenties, late Twenties, early Thirties. He had done the <u>most incredible</u> things. For example: [...] And so the fraud was discovered, and later it was found that everything he had published, <u>everything</u> had been falsified. Including also his interferences on electrons, because he had taken images of Röntgen rays from the literature, he had repeatedly photographed them and copied them until in the end they were how he wanted them to be, of the right sizes. A real - madman, yes I believe that he really was. But all this went on for ten years, ten years." For an investigation into the "strange case of Emil Rupp" we refer you to [70].

3.4 "Nuclear" electrons and the "neutrino" at the Rome Conference

At the time of the Rome Conference, and more generally after the discovery of the proton in 1919 up to 1932, the year the neutron was discovered, the nucleus was considered to be made up of protons, with an angular momentum (spin) equal to $I = 1/2$ (in units $h/2\pi$), as required by the study of the spectral band of the hydrogen molecule, and by electrons, also with an angular momentum equal to 1/2. Protons and electrons could join together to form alpha particles in the nucleus. According to this model, for example, the alpha particle was made up of 4 protons and 2 electrons, so as to guarantee a charge equal to two positive units.

If on the one hand, with this model of the nucleus, the nuclear origin of the electrons present in the beta decay was immediately explained, on the other hand as soon as the problem of the presence of electrons in the nucleus was tackled from the theoretical point of view, and also the mechanism of their emission in radioactive processes, serious problems arose. It was these very problems that formed one of the main themes of the Rome Conference.

As can be seen from the various contributions, the difficulties raised by the presence of the electrons in the nucleus concerned the possible confinement mechanism of an electron in the nucleus and the problems linked to spin, to the magnetic moment, and to statistics(*).

On the basis of Heisenberg's uncertainty principle, electrons, unlike protons which have a mass about 2000 times greater, if compelled by forces as yet unknown to move along an orbit with dimensions of the order of nuclear ones (10^{13} cm), would have acquired a velocity practically equal to that of light, for example $v = 0,\ 9998$ c [76], and as a result "an energy enormously greater than that which they possess in reality", for example greater than those with which electrons were emitted in beta decay. Moreover, because of these enormous kinetic energies, in the end the mass of the electron could no longer have been negligible and should have become perceptible in the atomic weight of the element, which would be in contrast with the experimental data. For a clear exposition of these problems we can see for example what Fermi himself said [page 556 of F112], [F99]. In any case the forces that confined the electron in the nucleus were not clear, and there was a conviction that it was necessary to develop a new mechanics, in addition to quantum mechanics, to explain these problems.

It had been known since 1925 that the electron possessed an intrinsic angular momentum called spin. As far as the electrons bound in the atom outside the nucleus were concerned, the idea that the electron had an intrinsic angular momentum, to which a

(*) We shall report the arguments as expressed by the Authors, using their terminology.

magnetic moment was associated, certainly worked and provided a complete agreement with the spectroscopic experimental data, in particular with the separation of the spectral lines into multiplets. Moreover this idea found a place of its own in the setting of Dirac's relativistic theory of the electron, even proving to be a consequence of this theory. Moreover it was well known that the electron followed Fermi-Dirac's statistic. Therefore it was to be expected that inside the nucleus too the electron would keep these properties.

But serious difficulties immediately arose, as Goudsmit clearly pointed out in his talk.

These difficulties concerned three orders of problems: the amplitude of the separation of hyperfine structures, the values of the nuclear angular momenta, the laws of statistics. Here we will now briefly set out, point by point, the various problems in their essential contents.

On the basis of quantum mechanics, the amplitude of the separation of the hyperfine structure depended directly on the nuclear magnetic moment, which in turn depended on the magnetic moments of the electrons and the protons contained in the nucleus. It was known that the magnetic moment of the electron had a value of a Bohr magneton of $he/4\pi mc$, where m is the mass of the electron, as could actually be predicted by Dirac's equation. By analogy, the magnetic moment of the proton was at that time estimated to be equal to a nuclear magneton $he/4\pi Mc$, where M is now the mass of the proton and thus with a value about 1838 times smaller than that of the electron, taking into account the differences in mass between electron and proton in the preceding formulae. In reality, a few years later, in 1933, Otto Stern (Nobel Prize for Physics 1943) [71] found experimentally in Hamburg, by means of the molecular ray technique, a value greater by a factor of between 2 and 3. But this does not substantially change the problem.

Because of the great difference between the two magnetic moments a small separation in the hyperfine structure for atoms with an even number of electrons was expected, since their magnetic moments could have cancelled each other out, and therefore there would only be the very small effect of the magnetism of the protons. Vice versa, a very large separation of hyperfine structure for atoms with an uneven number of electrons was to be expected since, in these cases, the predominant effect would be that of the magnetism of the electron, which was of the order of a Bohr magneton, compared to that of the proton which was 1838 times smaller.

Instead the isotope ^{6}Li, which should have had a nucleus containing 6 protons and 3 electrons, and ^{14}N, which should have had a nucleus with 14 protons and 7 electrons, showed a practically non existent hyperfine structure despite the uneven number of intranuclear electrons.

As far as the values of the nuclear angular momenta were concerned, these could be estimated by using properties of the molecular spectra [166]. For example, Friedrich Hund [96] had established that, in the case of a homonuclear diatomic molecule, the ratio (R) between the intensities of the alternating strong and weak lines in a spectral band was bound to the nuclear angular momentum J by the relationship $R = (J+1)/J$. Therefore the value of J could be obtained by measuring R. With this method, applied to the hydrogen molecule, it had been found for example that the momentum of the proton had a value of 1/2.

Considering the nucleus made up of protons and electrons which had an angular momentum equal to 1/2, a half-integer nuclear moment was expected when the nucleus contained in total (between protons and electrons) an uneven number of particles and, vice versa, an integer nuclear moment was expected when the nucleus contains an even number of particles. Instead, as Goudsmit pointed out, studying the hyperfine structures of some nuclei, it was found that nuclei which were considered to be made up of an even number of particles had a half-integer nuclear moment, in particular some isotopes of cadmium (charge 58) and an isotope of lead (charge 82).

A similar situation had been found studying the molecular spectra too. In particular, the measurements by L.S. Ornstein and W.R. van Wijk taken in 1928 [125] on the spectrum of rotational bands of the molecular ion of nitrogen N_2^+ proved that the angular momentum of the nucleus of nitrogen was integer and equal to 1. A conclusion in agreement with Kronig's measurements [105] in the same year. Instead, with the hypothesis that the nucleus of nitrogen was made up of 14 protons and 7 electrons, thus by an uneven number of particles, the angular momentum should necessarily be half-integer.

On the basis of the general principles of quantum mechanics, if it was accepted that all atomic nuclei were composed of a set of protons and electrons, it could be deduced since the protons, like the electrons, obeyed Fermi-Dirac's statistic, that the nuclei should follow Fermi-Dirac's statistic, or Bose-Einstein's statistic according to whether the total number of particles that composed them was uneven or even. On the basis of this rule, the nucleus of nitrogen, which was supposed to contain an uneven number of elementary particles (14 protons and 7 electrons), should have obeyed Fermi-Dirac's statistic. Instead, as had been established experimentally, again by studying the molecular spectrum of the rotational bands of the nitrogen molecule, the nucleus of nitrogen turned out to follow Bose-Einstein's statistic, "as if the nucleus contained an even number of particles", as Goudsmit observed in his talk. Since the number of protons was even, while the number of electrons was uneven, it was as if the nuclear electrons did not contribute to the statistic of the nucleus.

It is interesting to observe that this alleged anomaly of nitrogen was first pointed out by W. Heitler and Gerhard Herzberg (1904-1999) in 1929 [94] on the basis of some experiments [133,134] carried out in the same year by Franco Rasetti on the Raman effect in the nitrogen molecule, during a study period abroad at the California Institute of Technology in Pasadena with the support of the International Education Board. Rasetti himself reported on these experiments and other later ones [135, 136] in a communication which, however, was not published in the Proceedings of the Conference as we can tell from Racah's account of the Rome Conference [132]. Rasetti's reply to Heitler and Herzberg is in [137].

It should be stressed that Rasetti's measurements, along with the later ones made in 1930 on his return to Rome in reply to some criticisms received, and with which he confirmed the previous results, are placed in a context completely unrelated to Nuclear Physics.

There would be others, above all, as we have said, Heitler and Herzberg, who would identify serious difficulties in Rasetti's results for the statistics of nuclei based on the proton and electron nuclear model.

After having discussed all these difficulties for electrons within the nucleus in relation to their spin Goudsmit concluded as follows "The present data show that the magnitude of the hyperfine structure agrees with a nuclear magnetism of the order of that of the proton, even in cases where there are an odd number of electrons in the nucleus (N, Li, Cd, Pb). It is also clear from the foregoing that the laws which hold for the outside electron configurations, can not be applied to the nucleus. The results about the nuclear moment can be remembered by saying that the electrons in the nucleus seem to lose their spin and magnetic properties and that only the protons determine the spin moment and the magnetic moment of the nucleus".

Another problem regarding the presence of electrons in the nucleus concerned the so-called continuous energy spectrum of beta rays, which we spoke of before. As Ellis and Wooster [66] had demonstrated, the continuous spectrum of beta rays showed that during a single beta decay process the electrons exited directly from the nucleus each with a different energy. This seemed to be in contradiction with the fact that, despite this difference in energy between the electrons emitted during a beta process, finally there was always a single type of radioactive product since the law of decay, for a given atomic species, was always the same. In other words all the nuclei of the same species decayed in the same way. In order to solve this problem Bohr, at the Rome Conference following a line sketched out during a Conference at the Chemical Society in London on 8 May 1930 [23], was even prepared to assume a "departure from the law of energy conservation in nuclear disintegrations". Bohr observed: "Just like the γ-ray products, all β-ray products have a well-defined rate of decay, but nevertheless for each product the energy of the emitted β-particle varies continuously within wide limits. If energy were conserved in these processes, it would imply that the individual atoms of a given radioactive product were essentially different, and it would be difficult to understand their common rate of decay. If, on the other hand, there is no energy balance, it is possible to explain the law of decay by assuming that all nuclei of the same product are essentially identical. [...] Still, we must remember, after all, that the essential stability of atoms is an implicit assumption in the whole classical description of natural phenomena, and we cannot therefore be surprised if classical concepts fail in accounting for their own foundation. Just as we have been forced to renounce the ideal of causality in the atomistic interpretation of the ordinary physical and chemical properties of matter, we may be led to further renunciations in order to account for the stability of the atomic constituents themselves".

In this respect what Marie Curie wrote to her daughter Irène, who had stayed in Paris, in a letter which briefly summed up the Conference two days after the beginning of events is significant, *So far I have little to tell you, except that Bohr insists a great deal on the impossibility of now applying mechanics inside the nucleus.* The letter is dated "Rome, mardi 13 1931", and can be found in the Curie Archive, Paris.

An alternative way to that proposed by Bohr to solve the problem of the continuous spectrum of beta rays, however, already existed and it was that based on Pauli's hypothesis of the neutrino. This hypothesis had been put forward for the first time, informally, by Wolfgang Pauli on 4 December 1930, in a letter, which then became famous, addressed to the "Radioactive Gentlemen and Ladies" ("Liebe radioaktive Damen und Herren") who had participated at the Gauverein-Tagung conference in Tübingen, and had been put forward again orally by Pauli at the 88th meeting of the American Association for the Advancement of Science, in Pasadena from 15 to 20 June 1931.

According to Pauli's idea, the nucleus also contained particles of a new type in a number equal to that of the electrons, initially called "neutrons" by Pauli, and later renamed "neutrinos" by Fermi, as we shall see.

These particles, in Pauli's concept, were neutral, with very small mass, they had an angular moment 1/2 equal to that of the electron, and obeyed Fermi-Dirac's statistic. With the hypothesis of the "neutrino" the problem of the continuous beta spectrum was immediately solved: it was enough to suppose that in the beta disintegration a neutrino was emitted every time with the electron, so that the sum of the energies distributed between these two particles was constant.

All the problems raised at an experimental level by the spin of the nuclear electron, which, as we have seen, only concerned the nuclei that were believed to be made up of an uneven number of electrons, and for which the contribution of the electrons appeared to be missing, seemed to be resolved. Indeed, since the number of neutrinos was equal to that of the electrons, the spin of these particles "neutralised" the spin of the electrons and therefore, in the case of an uneven number of electrons, it was as if the electrons were not present. On the other hand, when there was an even number of electrons, as we have seen, everything agreed and everything continued to agree even adding a neutrino for every electron present: in this case, indeed, even the neutrinos would be even numbered and so their spin would have cancelled each other out.

At the Rome Conference, Samuel Goudsmit, who had also been at the Pasadena meeting, advocated Pauli's idea of the neutrino. After pointing out the difficulties connected to the presence of electrons in the nucleus, he continued as follows: "At a meeting at Pasadena in June 1931, Pauli expressed the idea that there might exist a third type of elementary particles besides protons and electrons, namely "neutrons". These neutrons should have an angular momentum $\frac{1}{2}h/2\pi$ and also a magnetic moment, but no charge. They are kept in the nucleus by magnetic forces and are emitted together with β-rays in radioactive desintegration. This, according to Pauli, might remove present difficulties in nuclear structure and at the same time in the explanation of the β-ray spectrum, in which it seems that the law of conservation of energy is not fulfilled. If one would find experimentally that there is also no conservation of momentum, it would make it very probable that another particle is emitted at the same time with the β-particle. The mass of these neutrons has to be very much smaller than that of the proton, otherwise one would have detected the change in atomic weight after β-emission. Pauli also believes that neutrons may throw some light on the nature of cosmic rays".

Samuel Goudsmit, in a letter to Karl Darrow (1891-1982) on 15 May 1934, observed that Pauli arrived at the Conference just as he named him in his report. Indeed Pauli had disembarked at Naples on his return from a conference in the USA, and had arrived late for Goudsmit's report on 13 October.

This was the first time that the hypothesis of the neutrino was proposed and published in an international context. It should be said however, that the hypothesis of the neutrino, while on the one hand it solved the problem of the continuous spectrum of beta rays and the experimental problems raised by the spin of the nuclear electron, on the other hand, it not only completely left open the problem of the confinement of the electron in the nucleus but it extended these same problems to the neutrino, since this particle too should have been bound inside the nucleus. Not by chance the hypothesis of the neutrino did not immediately achieve success, and nor was it discussed further during the Conference For example, Goudsmit himself, after having publicly presented it, did not embrace it. Even Peter Debye, in his concluding report of the Conference, made no reference to this hypothesis while instead he stressed Bohr's position regarding the need to abandon the "law of energy conservation in nuclear disintegrations".

In the summing up in Debye's closing speech (not published in the Proceedings but preserved in the Archive of the Accademia d'Italia at the Accademia dei Lincei) we can read: "[...] Difficulties arise however when we try to analyse the properties of nuclei and of their constituent parts. The speaker touched upon all the difficulties of the study which had been presented in the course of the discussions, and finally lingered over Prof. Bohr's observations which, starting from the behaviour of the electrons inside the nucleus and from the continuous spectrum of beta rays, seem to attack the deepest foundations of physics itself, even going so far as to put in doubt the principle of conservation of energy. These speculations, however bold, seem to offer the easiest way out of the difficulties in interpreting the experimental results, which would otherwise seem insurmountable".

In any case, despite the great difficulties faced by the nascent Nuclear Physics, one important result seemed to have been reached at the Rome Conference. A clear boundary line had been drawn between the questions that it seemed possible to solve with quantum mechanics and those that instead could not hope for any help from that theory. This boundary line was the same one already drawn by Gamow at the beginning of his talk and on which all the speakers had in the end found themselves in agreement. As Gamow had observed: "We know that there are two rather differing kinds of constituent parts of atomic nuclei, we may call them *heavy* and *light* constituent parts. To the first class belong the protons and also complex constituent parts such as α-particles. For these particles we can estimate that, due to the relatively great masses, their motion may be described according to unrelativistic mechanics and the nuclear processes involving these particles only can be treated in detail by means of the present quantum theory.

On the other hand the light constituent parts, the nuclear electrons, move in the nucleus with velocities near to that of the light and relativistical treatment is necessary. This is just the point where the present means of theoretical physics fail to help us".

To conclude we can say that the Rome Conference was not just a great organisational success but also scientifically. What Ellis wrote to the President of the Reale Accademia d'Italia, on his return to Cambridge, on 22 October, 1931 (Fig. 20) was very significant. After expressing great appreciation for the hospitality he had received he went on: *Also on the scientific side I would like to say how much I personally have benefited from the*

Conference It is difficult yet to appreciate fully how much I have gained, but it is with gratitude that I thank you for the honour which you did me in inviting me to attend.

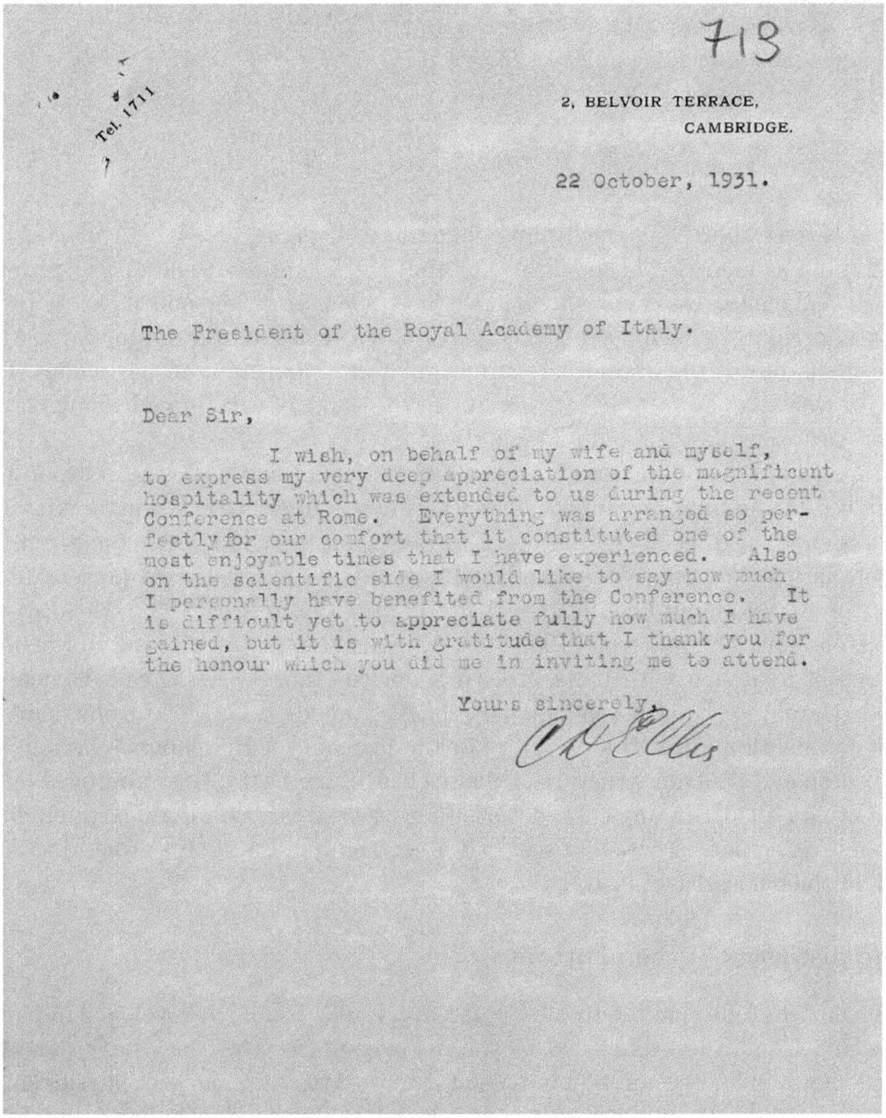

718

Tel. 1711

2, BELVOIR TERRACE,
CAMBRIDGE.

22 October, 1931.

The President of the Royal Academy of Italy.

Dear Sir,

I wish, on behalf of my wife and myself, to express my very deep appreciation of the magnificent hospitality which was extended to us during the recent Conference at Rome. Everything was arranged so perfectly for our comfort that it constituted one of the most enjoyable times that I have experienced. Also on the scientific side I would like to say how much I personally have benefited from the Conference. It is difficult yet to appreciate fully how much I have gained, but it is with gratitude that I thank you for the honour which you did me in inviting me to attend.

Yours sincerely,
C D Ellis

Fig. 20. – Letter from C. D. Ellis to the President of the Reale Accademia d'Italia (22 October 1931) - BANLC.

4 Nuclear Physics in Rome after the Rome Conference: 1931-1933

In his closing speech at the Rome Conference, kept at the Accademia dei Lincei, Corbino spoke as follows: "I therefore think that the future course of nuclear physics will be greatly influenced by this week that we have spent together and its most profound results will perhaps be seen in work that will come to fruition over many years. And this was the main aim of the organisers of the conference, and mine in particular. I thank you all for what you have brought here; but in the name of each of us I thank the others for what each of us has accrued and acquired".

Corbino's hopes and forecast about the future of Nuclear Physics would soon become true. Indeed in the following months Nuclear Physics underwent a remarkable development, it is enough to recall the discovery in that same year 1932 of the neutron (Chadwick, February 1932), the first nuclear disintegration with accelerated protons (John D. Cockroft (1897-1967) and Ernest T. S. Walton (1903-1995), April 1932), the discovery of the positron (Carl D. Anderson (1905-1991), August 1932), the discovery of deuterium, an isotope of hydrogen with mass 2, with a nucleus that was also called "deuton" or "diplon" (Harold C. Urey (1893-1981), F. C. Brickwedde e G. M. Murphy) and finally the first formulation, after the discovery of the neutron, of Heisenberg's famous theory on the proton and neutron structure of the nucleus (June 1932), later improved by Majorana (February 1933). In any case, despite the great scientific and organisational efforts, the Rome Conference, as we shall see, had a very limited impact in promoting research activity in the nuclear field in Italy.

4.1 The discovery of the neutron

In the months following the Rome Conference, intense research developed in particular in Paris and Cambridge which led to the discovery of a new elementary particle, the neutron, that would become of fundamental importance for nuclear physics and that would be at the heart of the research developed by Fermi into induced radioactivity. On 17 February 1932 a Letter to the Editor [31] by James Chadwick appeared in the prestigious journal "Nature", with the title "Possible existence of a neutron", where the discovery was announced. A lengthier article was published later in the "Proceedings of the Royal Society of London" [32].

Let us briefly run through the fundamental steps that led to the discovery. It all began with the identification by Walther Bothe and Herbert Becker [28] of a new "penetrating

F. Guerra and N. Robotti, *The Lost Notebook of ENRICO FERMI*,
https://doi.org/10.1007/978-3-319-69254-8_4

radiation", which Bothe also spoke of at the Rome Conference, which was emitted by some light nuclei (Li, Be, B, F, Al, Mg, Na) when they were bombarded with alpha particles emitted by a very intense source of polonium (3-7 mCi(*)). This radiation was interpreted as made up of gamma rays. The experimental apparatus consisted of the source of alpha rays, a lead or iron absorber in the shape of a truncated cone, and a Geiger point counter, all aligned. From the measurement of the absorption coefficient, according to the Klein-Nishina formula, which binds the frequency (and therefore the energy) of the radiation diffused by a given substance to its absorption coefficient, the energy of the radiations, which was of the same order of magnitude as that of the natural γ-radioactive elements, could be determined. In the case of beryllium and boron, the two cases in which the radiations were most penetrating, the energies were $11 \cdot 10^6$ and $1 \cdot 10^6$ eV, respectively.

From December 1931, just three months after the Rome Conference, Frédéric Joliot and Irène Curie worked on the nature of this "penetrating radiation" at the Institut du Radium in Paris. Through a series of experiments in which a much more intense polonium preparation than that used by Bothe and Becker, which they themselves had developed, was used as a source of alpha particles ($\sim$ 100 mCi) and using an ionisation chamber connected to a Hoffmann electrometer as a radiation detector instead of the Geiger counter used by Bothe and Becker, Joliot and Curiee realised that the "radiation" discovered by Bothe and Becker, and which they too considered to be electromagnetic, was even more penetrating than Bothe and Becker had estimated. In the case of boron and beryllium, the energies in play were "very high, lying between those of the most penetrating rays emitted by radioelements and those of cosmic rays".

In particular, Joliot and Curie, working in parallel, he on boron and she on beryllium, found that the energy, in the case of boron, was of the order of "$11 \cdot 10^6$ eV" [98], while in the case of beryllium it was even of the order "of 15 or 20 million electron volts" [48].

Later, as they continued their research, they came to results of strategic importance. By means of "a very fine experiment" (as Chadwick defined it in his Nobel Lecture on 12 December 1935), they found that "radiation", when it was made to pass through some "screens contaning hydrogen, such as paraffin, water, cellophane", placed at the entrance of the ionisation chamber, caused the emission by these hydrogenated substances of protons with energies respectively of the "order of $4.5 \cdot 10^6$ and of $2 \cdot 10^6$ eV". The Joliot-Curies, still believing that the "penetrating radiation" was electromagnetic in nature, interpreted this phenomenon as a sort of Compton effect in which the incident photons, here represented by "γ nuclear rays" (as they called the "penetrating radiation" of the boron and of the beryllium), in the collision with the molecules of the hydrogenated substance made a proton exit from the latter and not an electron (as happened in a common Compton effect). On the basis of this interpretation and bearing in mind the

(*) The curie (Ci) — millicurie (mCi) — was replaced by the becquerel (Bq) in the current SI system as the unit of measurement of the activity of a radionuclide, 1 Ci = 37 GBq. In this volume in places the word milliCurie has been kept in order to respect the original texts.

energy of the expelled protons, Joliot and Curie deduced, using the Compton effect formulae in which the mass of the electron was replaced by that of the proton, that the energies of the "nuclear γ rays" of the beryllium and of the boron were actually much greater than they had previously estimated on the basis of the absorption coefficients. These energies were respectively $50 \cdot 10^6$ eV and $35 \cdot 10^6$ eV, compared to the previous values of $15 - 20 \cdot 10^6$ eV and $11 \cdot 10^6$ eV. Certain, and proud, of now having discovered a new Compton-type phenomenon, they concluded as follows [49, page 275]: "It now seems to have been established by these experiments that a high frequency electromagnetic radiation is capable of releasing protons with great velocity in hydrogenoid bodies".

As they specified in a note presented at a session of the Academy of Science on 22 February 1932 (when Chadwick's Letter on the discovery of the neutron had not yet been published in "Nature") [50] "the radiation of Po-Be was able to expel from the atoms not only hydrogen nuclei, but also other heavier ones such as helium and carbon". In any case Joliot and Curie had no doubt that it was an electromagnetic radiation. At the same time however, the more the properties of this new "radiation" were investigated the more anomalous facts emerged. As the Joliot-Curies noted, if one "tried to apply to the phenomenon of the projection of the nuclei the formulae that allow diffusion absorption to be calculated, results were obtained that were incompatible with the experimental facts". Indeed, following the hypothesis that the radiation of Po-Be was $50 \cdot 10^6$ eV, the coefficient of absorption by a proton, on the basis of the Klein-Nishina formula, should be $2.04 \cdot 10^{-31}$, that is to say 10^5 times smaller compared to that of the electron, while instead from their experiments it was shown that "a proton absorbs as much as four electrons together". As a result, still clinging to their idea of the electromagnetic nature of the "radiation of Po-Be", which was the only one they had in mind, they proposed that "the absorption [of the gamma rays] corresponding to the projection of the atomic nuclei" should be considered a "new type of interaction between radiation and matter", also suggesting a possible mechanism for it.

In any case when Joliot and Curie put forward this proposal, the hypothesis of the neutron was already born, and it was actually born on the basis of results very similar to those they had found and concerning the "projection of atomic nuclei".

As he reported in his Letter in "Nature" on 17 February 1932 [31], Chadwick had carried out a series of experiments at the Cavendish Laboratory to study the properties of "radiation excited in beryllium" using a "valve counter", in which a small ionisation chamber sent its signal, appropriately amplified by an electronic circuit, to an oscillograph and its deflection signalled the sudden production of ions due to the entrance of an ionising particle. In this way he had deduced that "the radiation ejects particles from hydrogen, helium, lithium, beryllium, carbon, air, and argon". The particles expelled from hydrogen behaved with regard to range and ionising power like protons with a velocity of about $3.2 \cdot 10^9$ cm/s, while those expelled by the other elements "appear to be in each case recoil atoms of the elements". These results, according to Chadwick, could not be explained by the hypothesis that the "radiation from beryllium is a quantum of radiation", unless the laws of energy conservation and of the quantity of motion were set aside. Indeed, on the basis of these laws, if it was hypothesised, as the Joliot-Curies had held, that the "radiation from beryllium" was made up of photons of $50 \cdot 10^6$ eV, in the case of nitrogen for example, the atom rebounding from the collision with such a photon should have had an energy no greater than 400000 volts, and it therefore should produce no more than 10000 ions and in normal conditions in the air it should have a range of about 1.3 mm. Instead, as Chadwick had seen, the hit nitrogen atoms produced, in their rebounding motion, over 30000 ions and, as Chadwick had verified in collaboration with Norman Feather (1904-1978), their range in the air was 3 mm in normal conditions.

Incongruities very similar to these, and clearly incompatible with an electromagnetic hypothesis on the nature of this "radiation", were also obtained by Chadwick in the case of the other "rebounding" atoms.

In any case according to Chadwick all these difficulties disappeared if a new hypothesis was introduced regarding the nature of this penetrating radiation, assuming that it was made up of a new nuclear constituent made up "of particles with mass 1 and charge zero", called "neutrons", and if it was hypothesised that the recoil atoms produced were set in motion following the collision with these new heavy particles.

With regard to the idea of the neutron, it should be said that this was not a new idea and at the Cavendish Laboratory, where Chadwick had worked in the post of Assistant director of research since 1923, it had been around for some years. As Chadwick recalled in his Nobel Lecture [35], the first to put forward the idea of this new particle, "with the properties of the neutron we now know", had been Rutherford himself. In his Bakerian Lecture on 3 June [148], with the title "Nuclear constitution of atoms" in the paragraph devoted to the "Constitution of nuclei and isotopes" Rutherford, after having advanced the possibility that a nuclear electron could "bind to a hydrogen nucleus" (remember that it was still 1920 and at that time nuclei were considered to be composed of protons, or hydrogen nuclei, and of electrons) he continued saying that: "[...] it involves the idea of the possible existence of an atom of mass 1 which has zero nucleus charge. Such an atomic structure seems by no means impossible. [...] Such an atom would have very novel properties. Its external field would be practically zero [...] and in consequence it should be able to move freely through matter".

As Chadwick recalled in his final article on the discovery of the neutron [32], many experiments had been carried out at the Cavendish Laboratory, starting in 1920, "in these neutrons". Chadwick himself [35, page 340] "made several attempts to detect them — in discharge tubes actuated in different ways, in the disintegration of radioactive substances, and in artificial disintegrations produced by α-particles" (in this case in collaboration with Rutherford [150]), without success however. In this context, in which the search for the neutron was an open problem, it is natural that all these anomalous results which had gradually emerged from the Joliot-Curies' experiments (and that the Joliot-Curies, precisely because they had no other idea in mind, had obstinately tried to interpret in terms of "γ nuclear rays") would take on a significance and a completely different weight for Chadwick. So much so that he was driven to repeat the same experiments carried out by the Joliot-Curies, partly modifying them, with the aim however of finally finding in this way an experimental proof of the existence of the neutron, which is what happened.

Once the neutron was introduced, interpreted as a nuclear component with mass 1 and charge zero, Chadwick assumed that the production of the neutrons occurred by means of the following process of capture of the incident alpha particle by the $^{9}_{4}$Be nucleus, formation of the nucleus of $^{12}_{6}$C and emission of a neutron

$$^{9}_{4}\mathrm{Be} + ^{4}_{2}\alpha \rightarrow ^{12}_{6}\mathrm{C} + ^{1}_{0}n.$$

From the energy balance of this nuclear reaction, the velocity of the neutron emitted in the forward direction was about $3{\cdot}10^9$ cm/s. At this point, referring to the elastic collision of a neutron with this velocity with the various "atoms of matter" studied experimentally, in his Letter to "Nature" he concluded as follows: "The collision of this neutron with the atoms through which it passes gives rise to the recoil atoms, and the observed energies of the recoil atoms are in far agreement with this view". Moreover with the neutron hypothesis even a new fact observed by Chadwick had a "simple explanation", in other words that "the protons ejected from hydrogen by the radiation emitted in the opposite direction to that of the exciting α-particle appear to have a much smaller range than those ejected by the forward radiation". Indeed it was enough to refer to the laws that govern the elastic collision between two massive particles, such as the neutron and the proton, to explain this behaviour.

Chadwick's results are of fundamental importance. The neutron entered into nuclear physics, and was destined to produce a profound upheaval of all the ideas prevailing until then, and to form the basis of decisive future developments.

Just after the publication of Chadwick's Letter in "Nature", in which the discovery of the neutron was announced, Franco Rasetti immediately took an interest in the problem. From October 1931 to July 1932, using a grant from the Rockefeller Foundation, he was at the Kaiser Wilhelm Institut für Chemie at Berlin-Dahlem to study experimental nuclear techniques in the laboratory directed by Lise Meitner. It should be noted that Meitner, one of the greatest leading figures in this area, was present at the Rome Conference. It was an exceptional opportunity: in Berlin Rasetti had available the most advanced instrumentation in the nuclear field.

In any case the topic he decided to tackle was certainly not a pioneering one at that particular moment, in which the properties and the very nature of this new particle were all still to be analysed in depth, together with the relative consequences for nuclear structure. Indeed Rasetti, after having observed the evidence of the new particle, applied himself to research which tended in a certain sense to "retrieve" the old interpretation, first put forward by Bothe and Becker and later supported by the Joliot-Curies, according to which "penetrating radiation of beryllium" contained γ rays. To do this Rasetti used Geiger-Müller counters (with walls 0.5 mm thick) and a strong preparation of polonium made available to him by Lise Meitner [138, pages 252-253] as a source of α particles to bombard the beryllium and thus obtain "the penetrating radiation". The "penetrating radiation", after being filtered through a 2 cm thickness of lead, was made to pass through two Geiger-Müller counters placed in coincidence. In these conditions a large number of coincidences were produced which disappeared however if a sheet of aluminium 5 mm thick was placed between the two counters. Since, as was already known, the penetrating power of the neutrons was much greater than this thickness (indeed they passed through 10-20 cm of lead), the coincidences could not be attributed to these particles. Therefore, according to Rasetti, it had to be concluded that the "penetrating radiation" was not composed exclusively of neutrons but it also contained a gamma radiation component and that the coincidences were due to Compton electrons produced by this gamma radiation in the collision with the aluminium sheet.

Rasetti's result was arrived at independently, and almost at the same time, both by Bothe and Becker [18,19], and by the Joliot-Curies [51, 52]. Bothe and Becker, whose results preceded Rasetti's even if their publication of them was slightly later, used two Geiger-Müller counters in coincidence, as Bothe had already announced at the Rome Conference and as Rasetti had done. Instead the Joliot-Curies studied the absorption of this "penetrating radiation" with an ionisation chamber through various thicknesses of lead.

It should be noted that both Bothe and Becker and Joliot and Curie had been proponents of the electromagnetic nature of the "penetrating radiation of beryllium", and that therefore this new result could have shown that they too, from a certain point of view, were right. It should be said however that this was true only in the case of Bothe and Becke who, having already used a Geiger counter since their first measurements in 1930, had only been able to detect the gamma radiation component, and in no way could they have identified the neutron since this particle, being neutral, could not be revealed with a Geiger counter. The case was different for the Joliot-Curies who, despite having seen the effects of the collision of the neutron against light nuclei, which were the same effects then seen by Chadwick, did not understand the existence of this new nuclear constituent and persisted in believing in an exclusively electromagnetic nature of the "penetrating radiation of beryllium". Anyway the Joliot-Curies, on the basis of these new results and bearing in mind the masses and the energies involved, in June 1932 [53] managed to establish that the "neutrons and the photons", that constituted the radiation of beryllium, were "emitted simultaneously according to the equation $^{8}\mathrm{Be} + \alpha \rightarrow {}^{12}\mathrm{C} + n + h\nu$".

The remarkable situation created immediately after the publication of the article by the Joliot-Curies [49], in which the emission of protons at very great velocity by hydrogenated substances under the influence "of very penetrating γ rays" was recognised, and before the publication of Chadwick's article on the "possible existence of a neutron" [31], is neatly described in the well-known apologue according to which Ettore Majorana in Rome realised that the Joliot-Curies' results implied the existence of the "neutral proton", a new particle responsible for the emission of the protons by simple elastic collision, an emission that is clearly impossible by γ rays with the usual plausible energies. This episode was told by various sources, even ones as authoritative as Edoardo Amaldi [9], Emilio Segrè [157], Bruno Pontecorvo [130], and was repeated by Leonardo Sciascia in his well-known "philosophical novel" about Ettore Majorana [156], summed up in the sentence that it is claimed Majorana addressed to the Joliot-Curies: "What fools, they have discovered the neutral proton, and they didn't notice". The adjective used to describe the Joliot and Curie spouses' lack of perspicacity varies from one source to another, and even became decidedly vulgar in the CERN-Courier on 27 July 2004, but that is the essence.

Without going into the value of the historical validity of the episode, which is based solely on the memory of those present, but which could be a reconstruction *a posteriori*, we observe that if it really happened, and Majorana therefore understood everything before Chadwick, then the research group in Rome missed their chance to secure the credit for the experimental demonstration of the existence of the neutron. Indeed, as we have mentioned, Franco Rasetti was in Berlin at that time, with full access to the most advanced instrumentation for nuclear research and extremely strong sources of polonium. Just as he was able to verify the existence of the neutron experimentally immediately *after* Chadwick's announcement, so he would have been able to demonstrate the existence of the neutron *before* Chadwick if Fermi had informed him of Majorana's idea in a timely fashion. But evidently history follows its own inscrutable path.

4.2 Fermi at the Paris Conference in 1932

As was to be expected, the discovery of the neutron raised great interest in the international scientific community. Much research was undertaken with the aim of studying the characteristics of this new particle and many review articles were published in the various international journals. In Italy too the discovery of the neutron met with great excitement.

Already in the 15-31 May 1932 issue of "La Ricerca Scientifica", in the section "Notizie Varie" (various news), a two page note appeared prepared by the editorial staff [142] devoted to "The neutron", that concluded as follows (perhaps not very prophetically): "For now it does not seem that the discovery of the neutron leads to any remarkable practical application. Considered instead from the scientific point of view it is of very great importance since one can already glimpse that the neutron will have a considerable part to play in future investigations into the structure of atomic nuclei" [142, page 599].

This same journal, to celebrate the discovery of "this new constituent of matter, the neutron", and at the same time to stress the fact that this discovery had been "made in the Cavendish Laboratory directed by Lord Rutherford", and therefore to underline "the importance [...] that the name of Lord Rutherford has today in the history of physics", published a note [141] actually in the same issue on "The scientific life of Lord Rutherford", recognised as "the founder of nuclear physics, the dominant figure in this new order of studies".

The scientific importance of the discovery of the neutron was also highlighted by the journal "Scientia" that same year 1932, with a "Critical Note" by Gleb Wataghin [164] in which this discovery was actually defined as "one of the most important successes in nuclear physics", and by "Il Nuovo Cimento" in 1933 with a long note by Zaira Ollano on "Il neutrone" (The neutron) [124]. The journal "L'Elettricista" even published promptly a brief announcement by "Dr. Giuseppe Occhialini" [122].

Despite the significance given to Chadwick's discovery it nevertheless seems that in Rome the neutron almost passed unnoticed. In July 1932, about five months after the publication of Chadwick's Letter in "Nature", Fermi attended, as Italian representative, the "Congrès International d'électricité" organised in Paris from 5 to 12 July, speaking in particular at the "Première Section: Science de l'électricité et du magnétisme, Théories générales; Isolants, Conducteurs; Radioactivité; Corps magnétiques", with a report on Nuclear Physics read on 7 July, and corresponding to the "Rapport no 22 (Italie)", with the title "La physique du noyau atomique" [F99].

Fermi's report, although it occurred about a year after the Rome Conference, reflected the situation of Nuclear Physics as it was just after the Conference without any mention of later developments that had happened in the meantime. Indeed there was no reference, for example, to the achievement by Cockroft and Walton [38] on 14 April 1932 of the first artificial disintegration of a nucleus (of lithium) by bombardment with a beam of accelerated protons and, above all, there was no reference to the discovery of the neutron which, as we have already said, was announced by Chadwick in a Letter in "Nature" on 27 February 1932 [31]. Only at the end of his report did Fermi very briefly cite these two discoveries saying: "After this manuscript was written important experiments in nuclear physics were carried out. To mention only the most significant work I quote the following [...]". And he then continued giving a brief summary of the results obtained by Chadwick and by Cockroft and Walton.

With regard to Fermi's position, it should be observed that while the experiments by Cockroft and Walton were fairly recent, Chadwick's discovery dated back to long before, to the end of February. Moreover, as we have said, Rasetti in Berlin had also been working on the problem of the neutron and at that time he was certainly in contact with Rome. So this failure to mention the discovery of the neutron in the report prepared by Fermi seems strange. It appears even stranger if you consider that at the same Conference Madame Curie (Curie and Fermi were the only speakers to talk about Nuclear Physics) had spoken at length about the neutron [47] and had also quoted Rasetti's experiments on the gamma radiation that accompanied the neutron, as well of course as the experiments, also on gamma radiation, carried out by her daughter Irène together with Frédéric Joliot. Fermi's attitude, underestimating the neutron, was further reiterated in a review article of his published in August in "La Ricerca Scientifica" with the title "Lo stato corrente della fisica del nucleo atomico" (The current state of the physics of the atomic nucleus) [F100], and at the same time in "L'Elettricista" [F101].

Even if Fermi's report had been written long before the Paris Conference, clearly the discovery of the neutron had initially been underrated. One reason for Fermi's attitude may be due to the fact that when it came down to it the neutron, even if it represented a new development, was a heavy constituent of the nucleus for which, as for the proton, quantum mechanics could always be used. Instead for Fermi the problems that needed to be solved concerned the light components of the nucleus, the "nuclear electrons". It is not by chance that Fermi's report concentrated on "focussing attention above all on the phenomena that represented the greatest difficulty in interpretation for current corpuscular mechanics" and that, as had been widely discussed at the Rome Conference and as Fermi reiterated, were linked to the presence of electrons in the nucleus. Fermi wrote: "These difficulties in interpreting the nuclear structure are very serious [...] on the other hand one must never forget that the very fundamental ideas of quantum mechanics can most likely no longer be applied to the study of the interior of the atomic nucleus, above all as far as the behaviour of the electrons is concerned".

And returning to the great problem of the confinement of electrons inside the nucleus, he reaffirmed: "This seems to indicate yet again that the concepts of ordinary quantum mechanics are not applicable to the study of the dynamics of the electrons of atomic nuclei".

This position however did not prevent Fermi from acting as spokesman for the hypothesis of the neutrino which until then, as we have already said with regard to the Rome Conference in 1931, had received almost no credence in the international scientific community and which, according to Fermi, could be a solution to a series of problems which seemed insuperable such as the continuous spectrum of beta rays and the spin of the "nuclear electron". Fermi called the neutrino "neutron", just as Pauli had initially called it, without worrying that in the meantime this name had been used by Chadwick to indicate the new heavy particle that he had discovered, following a tradition at the Cavendish Laboratory. So in the discussion after Fermi's talk misunderstandings developed, for example with L. Wertenstein, which forced Fermi to clarify: "These neutrons are not those that have been discovered but would have a much lower mass". Moreover it is clear that the name "neutrino" had not yet been devised by Fermi.

In his report Fermi wrote: "One might think for example, according to a suggestion from Pauli, that in the atomic nucleus there are "neutrons" that are emitted at the same time as beta particles. They could pass through great thicknesses of matter only losing a small part of their energy and thus practically escape observation. The existence of the "neutron" could no doubt provide a very simple explanation for certain phenomena that are currently not very comprehensible".

This position, completely accepting the neutrino hypothesis, together with an attitude of extreme awareness of the problem of the presence of electrons in the nucleus, would help ensure that Fermi, just over a year later as we shall see, would tackle the problem of beta decay, soon arriving at a solution. However at the time of the Paris Conference, almost a year after the Rome Conference, a systematic nuclear physics programme was still far from being launched in Rome.

In reality, as can be reconstructed on the basis of a statement by Rasetti [6, page 548], partly also confirmed in a letter from Fermi to Segrè on 30 September 1932 which we will talk about later, an attempt had already been made to start an experimental type of activity in Rome in Nuclear Physics too; there were scant results however. Immediately after the Rome Conference, according to Rasetti, Amaldi and Fermi had actually built a Wilson chamber which worked badly however and could therefore not be used.

The atmosphere of profound indecision in Rome at that time is described well in a letter that Fermi wrote on 30 September 1932, while he was on holiday in Rignano sull'Arno, sent to his young colleague Emilio Segrè who at the time was in Hamburg at Stern's laboratory, thanks to a study grant, and was working on spin flipping (Umklappung) in molecular beams. Fermi wrote: *Dear Emilio, [...] I have devoted this summer to finishing tidying up the book on "molecules and crystals". I finished it today! Now I just have to prepare a bit of bibliography, which of course I cannot do here in the country. I will return to Rome on 9 October for the Progress of Science meeting, and I shall stay there, except perhaps for a few trips for a couple of days. I have absolutely no work plans for the next year: I don't even know if I will go back to Wilson-chambering, or whether I will go back to being a theoretician again. When I was in Rome Franco was at Postumia; so I haven't seen him and I don't really know what his ideas are. Certainly the problem of equipping the Institute to work on nuclei is becoming ever more urgent if we don't want to lapse too far into an intellectual doze.*

The reference is to Franco Rasetti, who then had returned to Rome after his stay at Meitner's laboratory in Berlin.

In his letter Fermi continued, referring to the future move of the Physics Institute from Via Panisperna where it was situated to its new location, which now houses the Department of Physics of University of Rome "La Sapienza": *Solid progress on the plans for the construction of a new institute in the university city is now being made. The outstanding characteristic, as you will perhaps have heard from Edoardo [Amaldi], will be being cut off from higher Physics; however there will be plenty of space. Let's hope that there will also be something to put inside it!*

It is a fact that between the two options of returning to being a theoretical physicist or returning to being an experimental physicist in the months that immediately followed between the end of 1932 and the early months of 1933 Fermi chose the first alternative, picking up again his work on the hyperfine structures of spectral lines and presenting a definitive paper, actually together with Segrè, at the Meeting of the Accademia d'Italia on 10 March, with the title "Sulla teoria della struttura iperfine" (On the theory of hyperfine structures) [F107].

Over the same period he also worked for the first time on cosmic rays. In a paper in collaboration with Bruno Rossi, presented at the Accademia dei Lincei on 5 February 1933, he studied in particular the effect that the Earth's magnetic field ought to produce on cosmic rays, at various latitudes and heights, on the hypothesis that they were "constituted primarily of electrical corpuscles (electrons and protons) originating from cosmic space" [F105]. This was the only paper published by Fermi on cosmic rays during his period of activity in Italy. His close scientific contact with Bruno Rossi in a field like that of cosmic rays in which the Geiger-Müller counter was the fundamental detection instrument must certainly have played a major role in Fermi's later research. Bruno Rossi was a great expert on counters and electronic circuits. When Fermi would later need Geiger-Müller counters to detect any neutron induced beta radiation Rossi's advice and assistance would be decisive.

4.3 Orso Mario Corbino and the particle accelerators

Corbino was also well aware that the problem of successfully launching important experimental activity in Nuclear Physics was one of the priorities not only for the Physics Institute in Rome but for all of Italian physics. Even before the Rome Conference he had been the foremost supporter of this line, above all with regard to the field of nuclear disintegrations which he considered to be the most promising field of enquiry. At the Joint Meeting of the XXV General Assembly of the Italian Physics Society and the XXI Annual Meeting of the Italian Society for the Progress of Science, which took place in Rome from 9 to 15 October 1932 and where Fermi also participated with a communication [F104] on atomic physics, on the rotation spectrum and oscillation of ammonia, Corbino, who had been invited to deliver the general report, tackled, not by chance, the topic "Le nuove esperienze sulla disintegrazione degli atomi " (The new experiments on the disintegration of atoms) [43]. At the heart of Corbino's talk was the new apparatus developed by Cockroft and Walton "in Rutherford's laboratory", and the apparatus designed at the same time by Ernest O. Lawrence (1901-1958) and M. Stanley Livingston (1905-1986) in California, and with which they had succeeded in producing accelerated proton beams capable of causing nuclear disintegrations. This type of experiment [37] had started immediately after the formulation, in 1928, of Gamow's theory on the passage of alpha particles through a barrier of nuclear potential [75, page 514]. Indeed, on the basis of this theory, it was immediately realised that the capacity shown by alpha particles to produce nuclear disintegrations could also be achieved in the case of protons, without having to impose on them an energy equal to the energy of alpha particles, for example of the order of three and a half million electronvolts, but by imposing on them much lower energies, about one tenth of those of alpha particles, and therefore easier to reach. On this point Corbino observed: "This represents one of the not infrequent cases when theoretical research can provide the impulse and extremely efficient guide for experimental enquiry".

The formula obtained by Gamow for the penetration probability W of a particle with charge $Z'e$, mass m and energy E inside a nucleus with atomic number Z and with radius r_0 was:

$$W = \exp\left(-\frac{e^2}{\hbar}\sqrt{2m}\frac{ZZ'}{\sqrt{E}}(2u_0 - \sin 2u_0)\right),$$

where $u_0 = \cos^{-1}\sqrt{r_0 E/Ze^2}$ and $\hbar = h/2\pi$.

As a result, in the case of the proton, to have, for example, the same penetrating power as the alpha particle it was sufficient, because of its half charge and its approximately quarter mass, to impose on it an energy only one tenth that of the alpha particles, and so it was sufficient to create accelerator electrical fields with potential differences of the order of a few hundred thousand volts. For example, already for a proton of 300 000 eV a probability of penetrating the nucleus of beryllium was predicted equal to $6.2 \cdot 10^{-3}$. Moreover, once an energy equal to that of alpha particles had been successfully imposed on a proton, it would have had a probability of penetrating inside a nucleus much greater than an alpha particle and would thus be shown to be a much more powerful projectile.

The method developed by Cockroft and Walton to accelerate a charged particle, in particular protons, was based on an "ingenious combination" of large capacitors and thermoionic valves which, from an alternating voltage of 200 000 V, allowed a continuous voltage of 800 000 V to be obtained and that, in consequence, allowed protons to be produced with an energy equal to 800 000 eV. Lawrence and Livingston's method (which is the idea behind the cyclotron) consisted instead of submitting a beam of protons, created in a common discharge tube, to the action of an alternating electric field acting only on some sections of the trajectory and of a constant magnetic field. This resulted in spiralling trajectories with constant energy. The apparatus "with only 4000 volt from the alternating accelerator field" allowed a beam of protons to be obtained, all with the same energy, of the order of 1 000 000 eV. This method, even if it could produce protons with a greater velocity than Cockroft and Walton's method, as Corbino pointed out, had the disadvantage hat "the number of protons put in motion was about a thousand times fewer than that obtained by the British physicists".

In any case, regardless of the advantages and disadvantages of the two methods, it is true that, with these new technological innovations, research in the field of nuclear disintegrations was further bolstered since, for the first time, the possibility of artificially producing a new kind of projectile in order to bombard nuclei had materialised. Moreover, this new type of projectile presented a series of advantages compared to the alpha particles of natural radioactive substances. In the first place the beams of protons achieved had a much greater intensity than the usual sources of alpha particles, since "a microampere of positive ions is equivalent to about 180 grammes of radium, as far as the number of particles emitted is concerned" [37, page 477]). Moreover these beams were not accompanied by penetrating radiations such as beta rays and gamma rays, which could have disturbed the measurements, and, finally, their velocity could be varied at will, obviously within certain limits.

As we indicated before, the first artificial disintegration of a nucleus by a proton was achieved by Cockroft and Walton on 14 April 1932 [38], bombarding lithium with protons with energy of 800 kV. Subsequent experiments [39, 40] extended to other light atoms (^{7}Li, ^{19}F, ^{23}Na) had confirmed the great efficiency of these new projectiles at causing nuclear disintegrations and had shown that these latter occurred with the absorption of the proton by the target nucleus and with the subsequent emission of an alpha particle, therefore with a nuclear reaction of the (p, α) type.

The ballistic efficiency of the proton, as Corbino pointed out, had been confirmed by Cockroft and Walton in the case of uranium too and it had proved to be even greater than that of alpha particles (even if in reality their energy was lower). Until that moment, indeed, "no physical agent had ever proved to be capable of accelerating or slowing down the natural rhythm of breakdown of uranium; not even collision with faster alpha rays". The protons, instead, seemed able to do so. Bombarding uranium with protons even of only 300 eV, Cockroft and Walton had established that the emission of alpha particles had increased. This result, which quite rightly had fascinated Corbino so much, would however be retracted by Cockcroft himself at the Solvay Conference in 1933, on the basis of "a precise study of this effect" undertaken by Rutherford and Mark Oliphant (1901-2000) in which it was shown [123] that the increased emission of alpha particles was not produced by the action of the protons on uranium but was due to the presence during the measurements of boron impurities which, under the action of the protons, underwent nuclear disintegrations, also emitting alpha particles.

In his speech Corbino highlighted the fact that, as well as the accelerated proton, another projectile useful for nuclear disintegrations had recently been made available to the scientific community: the neutron. Obviously a projectile such as the neutron, with no charge, could not be accelerated by an electrical field but the advantages of using it were considerable. Corbino expressed himself as follows: "[the neutron] when it is in motion penetrates matter deeply because it does not undergo the braking action due to the electronic atmosphere of atoms; and so, proceeding undisturbed in its motion, it will end up meeting some nucleus on its way. Approaching the nucleus, the collision will no longer be obstructed by mutual electrical repulsion, as happens with the proton or with an alpha particle, and so it is to be expected that a jet of fast neutrons will be very effective in breaking up the matter it propagates in".

In fact indications of the efficiency of the neutron as a new projectile had already emerged. A few days after Chadwick's discovery of the neutron Norman Feather, one of Chadwick's colleagues at the Cavendish Laboratory, had tried to obtain the disintegration of nitrogen using a neutron source consisting of a small sheet of beryllium bombarded by alpha particles derived from polonium, thus a (Po+Be)-type neutron source, the same used in "Dr. Chadwick's" experiments with positive results [67].

In particular, observing the reaction in a Wilson chamber he succeeded in establishing that the neutron was absorbed by nitrogen, leading to a nucleus of boron and to the expulsion of an alpha particle according to the (n, α) reaction

$$^{14}\mathrm{N} + {}^{1}n \rightarrow {}^{11}\mathrm{B} + {}^{4}\mathrm{He},$$

which is the inverse of the reaction that produced the neutron.

Corbino observed, referring to these and later experiments: "[...] from some recent experiments by Feather it emerged that a flow of neutrons is capable of breaking down nitrogen and oxygen by mechanical collision, and not one every hundred million as for the proton but at least one in two gave rise to a useful effect".

Neutrons, despite their efficiency as projectiles to produce nuclear disintegrations, had a major limitation however according to Corbino: the fact that they originated from a secondary emission "when a nucleus explodes because bombarded with alpha particles", and therefore they were produced in very limited numbers. As a result, for the moment, according to Corbino it was necessary to concentrate on the proton as projectile, and "to insist on the procedures described above" to produce accelerated protons. Corbino was well aware of the difficulties and of the high costs that would be incurred if a decision was taken to install a plant of the type designed by Cockroftt and Walton, which by the way he favoured over the cyclotron created in Berkley. According to Corbino however this effort should be made at all costs if Italy was to be guaranteed a primary scientific role. Corbino ended his speech thus: "I hope that I will be able to speak before you again at the second ten year anniversary [*of the Fascist march on Rome*]; and that I will be able to describe more interesting results and that they will have been obtained by Italians".

Corbino's chance to speak of the great successes achieved by Italian physics in the nuclear field, and in particular in Rome, was not long delayed as we shall see. However these results were not achieved through the construction of large accelerator machines, such as those Corbino had envisaged to produce ever more powerful proton beams in order to bombard nuclei, but followed a different course, theoretical at first and then experimental, undertaken independently in Rome by Fermi at the end of 1933, where the neutron played the leading role, and that, within a few months, led to the discovery of radioactivity induced by this particle.

Anyway Corbino's dream of building a particle accelerator in Italy of the type designed by Cockroft and Walton was not immediately followed up. Indeed it would be built in 1939 at the Istituto Superiore di Sanità (Higher Institute of Public Health) [7], by which time Corbino had died two years earlier and Fermi had left for the USA, when the aim was no longer to launch Italian physics in the field of nuclear disintegrations but to boost nuclear instrumentation in order to keep pace with the largest foreign laboratories.

What did happen in the months following Corbino's speech in 1932 was that in Italy too experimental research in Nuclear Physics finally began to take off.

4.4 The official beginning of Nuclear Physics in Rome: March 1933

The official beginning of experimental activity in the field of Nuclear Physics in Italy, and in particular in Rome, is represented by the "programme of experimental research" proposed for the year 1933 by the Committee for Physics, Applied Mathematics and Astronomy of the C.N.d.R., and approved in the Plenary Session on 7-8-9 March 1933-XI [26](*). Here we find in first place, "a group for enquiry into nuclear physics", divided between "three Institutes appropriately equipped for this purpose". These Institutes were the Physics Institute of the University of Rome, the Physics Institute of the University of Florence and the Physics Institute of the University of Padua. Each Institute was assigned a specific research topic, with no overlap between the three places. The neutron was one of the subjects included and so this new particle, in 1933, began to become part of the official research programmes of Italian physics. However Florence, and not Rome, as was to be expected on the basis of later developments in this field, was the location proposed for this kind of study and where neutron physics in Italy ought to have been born. Indeed in the report with which Amedeo Giannini (1886-1960), vice chairman delegate of the C.N.d.R, illustrated the research plan approved by the C.N.d.R. for the year 1933, we find written: "the Physics Institute of the University of Florence will concern itself with excitation of neutrons in various elements with alpha particles with various energies, and with the disintegrations produced by neutrons when passing through matter". This programme, as we can see from the research plan proposed by the Committee for Physics, Applied Mathematics and Astronomy [26], "mainly involved" Gilberto Bernardini, Daria Bocciarellii, Lorenzo Emo Capodilista (1909-1973). Thanks to a study grant from the C.N.d.R. Bernardini immediately went to Berlin-Dahlem, to the Kaiser Wilhelm Institut für Chemie where Meitner was working and where Rasetti had already been for a year, and there he concerned himself with the excitation of neutrons

---

(*) As was normal for official documents at that time the Roman numerals after the dates indicate the year of the Fascist Era that began on 29 September 1922 [Translator's Note].

in beryllium. In particular he studied the excitation curve of beryllium as a function of the energy of the incident alpha particles. On 9 August 1933, a few months after his arrival, he sent an article on this subject to the journal "Zeitschrift für Physik" where it was then published [20]. These measurements by Bernardini were later confirmed by Emo Capodilista during a stay of his still at the Berlin-Dahlem laboratory.

As a matter of fact all the experiments on the neutron carried out by Italian physicists in this first period, Bernardini and Emo Capodilista, and previously Rasetti, were carried out outside Italy and using neutron sources of the Po-Be type made available to the Italian physicists by a foreign laboratory, the laboratory in Berlin-Dahlem.

The second research group financed by the C.N.d.R. mainly concerned "cosmic rays", which at that time, as we already saw with regard to the Rome Conference, were considered part of the field of Nuclear Physics. The Institute designated for this research was the one in Padua where Bruno Rossi, Italy's greatest expert in this field, had recently transferred [25]. "Research into the artificial disintegration of nuclei with alpha particles and artificially accelerated protons" was also envisaged.

The last research group planned for the year 1933 concerned gamma ray spectroscopy. The Institute involved was the one in Rome. "Corbino, Fermi and Rasetti" participated in this research. Clearly the subject of this research was not due to a choice by the C.N.d.R but derived from a proposal by the Roman participants themselves. So the fact that it was decided to study gamma rays in Rome, leaving to others the task of dealing with the neutron, despite the experience in this sector that Rasetti had already acquired at Meitner's laboratories and despite the attention Corbino had paid to the potential of this nuclear component as a projectile in nuclear disintegrations, indicates to us how much the importance of this particle, still in early 1933, was not fully understood in Rome and instead how much the tradition of experimental studies in spectroscopy was still firmly entrenched, carried on first of all by Rasetti since the late '20s.

The Roman group should have worked, in particular, on the "study of nuclear energy levels both by studying ways for their artificial excitation and by perfecting techniques for gamma ray spectrometry". They should also have studied "the problem of the diffusion of gamma rays especially relative to the diffused radiation of the nucleus". This programme led to Fermi and Rasetti creating a bizarre bismuth crystal spectrograph for gamma rays, as reported in the article published by them in the 15 November 1933 issue of "La Ricerca Scientifica" [F111].

In order to test this new spectrograph Fermi and Rasetti exploited the well known property that in radon (or radium emanation) decay, a strong gamma radiation was present together with the emission of alpha particles and therefore they used, as a source of gamma rays, "glass capillary tubes, 15 mm long and with a diameter of 0.2-0.3 mm, filled with radium emanation". The sources available had an intensity that varied between 100 and 150 mCi and were "prepared expressly by Prof. G. C. Trabacchi of the Radium Office". This office [115] was a section of the Physical Laboratory of Public Health and was located in Rome in the same building on via Panisperna which housed the Physics Institute. Fermi and Rasetti expressed themselves as follows, already on this first occasion, with regard to the support given by the Physics Laboratory of Public

Fig. 21. – Apparatus for extracting radon in the Laboratorio Fisico - Physics Laboratory of the Istituto della Sanità Pubblica - Institute of Public Health ("Ufficio del Radio - Radium Office") in via Panisperna - ISP.

Health, and in particular by Giulio Cesare Trabacchi (its Director for the long period 1922-1958) in launching a line of research in Nuclear Physics in Rome: "Without the means that this Institute possesses and without the continuous and keen collaboration of Prof. Trabacchi this research would not have been possible, and therefore we wish to express to him our gratitude here as well". The Radium Office had the institutional task of supervising the possession of radioactive substances in Italy and of administering the distribution of radium, bought by the State, between the various medical treatment centres. Moreover, analogously to foreign Radium Institutes, it was equipped with apparatus to extract radon which developed continuously from the deposition of radium salts and to purify it (Fig. 21, 22). At the end of the process the radon, liquefied with liquid air, was collected in a capillary tube, sealed by flame, separated from the equipment and divided into smaller tubes which were also sealed. In the ordinary extraction, which usually happened every fortnight, the capillary tube was "usually split into 32 pieces, each 1 cm long and containing 16 milliCurie of radon".

It is interesting to observe that this type of radon sources, here used by Fermi for the first time as a gamma ray source, were used again by Fermi in the spring of 1934 when

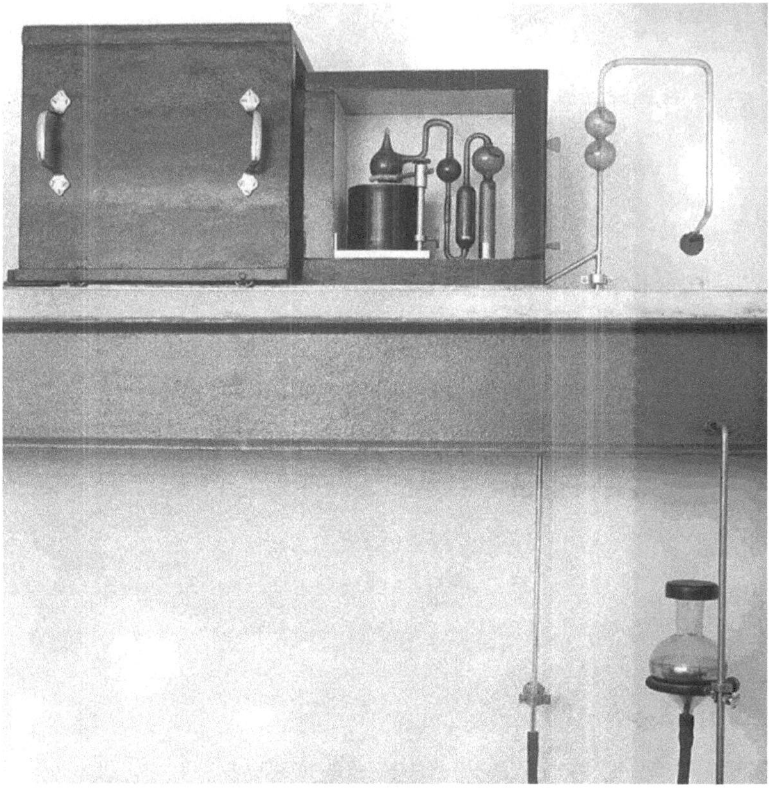

Fig. 22. – The safe in the Ufficio del Radio of the Istituto della Sanità Pubblica (Institute of Public Health), where the Radium preparation was stored - ISP.

he started to research neutron induced radiation, adding inside them beryllium powder, along with the radon, and thus transforming them into neutron sources to be used to bombard various substances.

Again, as part of the C.N.d.R research programme cited above, Rasetti had extracted from old solutions of radium salts ($RaBr_2$), which had been made available by the "Direzione Generale della Sanità Pubblica" (General Directorate for Public Health), yet again due to Trabacchi's intervention, a preparation of 110 mCi of radium D, from which every six months polonium or radium F [139] could be extracted. In fact radium D (which has a very long mean lifetime of 22 years and, in consequence, accumulates over time in radium salts) decays into radium E (mean lifetime 7 days), and this decays into radium F or polonium (mean lifetime 200 days). In Fig. 23 we can see the products of the decay of the uranium-radium series.

The operation of separating radium D from radium salts was far from simple. With this project in view on the first of July 1933 Rasetti, using a "grant [given to him] by the Reale Accademia d'Italia" (remember that Fermi was an influential member of

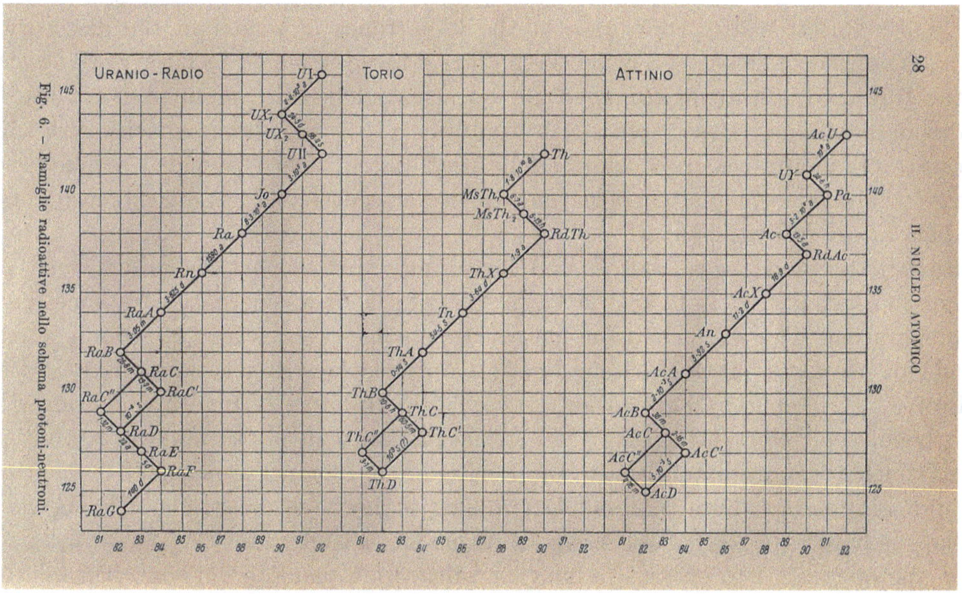

Fig. 23. – The radioactive series - in [140, page 28] (NZE).

this Academy), returned for a month to the Kaiser Wilhem Institut in Berlin-Dahlem, "because, of all of them, that laboratory is the best specialised in the chemistry of radioactive elements". He took with him "from Rome a small tube containing 4 mg of thirty year old $RaBr_2$, with the aim of carrying out on this small quantity of substance the operations that should then be applied on a larger scale". On the importance of being able to have a strong quantity of polonium available in Rome Rasetti expressed himself as follows:

"This element is perhaps the most precious for physicists of all the radioactive elements given its property of being an intense source of alpha particles without beta and gamma rays being emitted at the same time. For this reason, almost all research in the fascinating field of the artificial disintegrations of nuclei were carried out with polonium. Amongst other things, thanks to polonium the famous disintegration of beryllium was possible, which led to the discovery of the neutron, and finally it provided the first artificial source of positrons. One can therefore claim that possessing strong polonium preparations was a necessary condition for carrying out research in nuclear physics in our Institute".

When he spoke of an "artificial source of positrons" Rasetti alluded to the experiments by Joliot and Curie, begun in the first months of 1933, according to which some light elements, in particular aluminium, if bombarded with the alpha particles of polonium, emitted positrons. As we shall see later these experiments would then lead in January 1934 to the discovery of alpha-particle–induced radioactivity.

Despite Rasetti's firm conviction of the importance of polonium the discovery of neutron-induced radioactivity, which Fermi arrived at a few months later working alone at the Physics Institute in Rome, could be achieved because Fermi did not use the polonium obtained by Rasetti with such trouble as a source of alpha particles with which to create his neutron sources. Instead he used radon, which provided a much more intense production of alpha particles and which was continuously and easily available at the Laboratory of Public Health.

Rasetti continued in his article in 1933: "Since this element (polonium), just like radium D which derives from it, is not available for sale, it was necessary to proceed with its preparation. The writer had the idea of using some preparations of radium recently bought by the Directorate of Public Health. For this the writer is particularly grateful to Prof. G. C. Trabacchi, Director of the Physical laboratory of Health who granted his most benevolent intervention [...]. Thus strong preparations of polonium were obtained, one of which was immediately sent to Dr. Bernardini in Florence for research on neutrons. The preparation of radium D, about 110 millicuries, is one of the strongest in the world, perhaps only inferior to that possessed by M.me Curie's laboratory".

In the operations to extract radium D, which took place throughout the month of November 1933, Rasetti was assisted by the chemist Oscar D'Agostino who had formerly collaborated with Nicola Parravano, Director of the Chemical Institute of the University of Rome and Chairman of the Committee for Chemistry of the Consiglio Nazionale delle Ricerche. For this activity D'Agostino benefitted from a C.N.d.R. research grant that had been conferred on him from 1 November 1933 for "research on radium", and that he then used, from the end of the month of January 1934, for a stay in Paris at the "Radium Laboratory of M.me Curie" where he went "to gain experience in the manipulations of radioactive chemistry", as can be read in a "proposal from Prof. Rasetti in agreement with His Excellency Prof. Parravano", approved by the Directorate of the C.N.d.R on 22 November 1933 (D'Agostino Archive - Avellino).

With regard to D'Agostino's imminent stay in Paris, in the Curie Archive we found two letters of introduction written on the same day 26 January 1934 by Fermi to Madame Curie and by Rasetti to Frédéric Joliot respectively, and that D'Agostino probably took with him. In them it was stated that D'Agostino was going to Paris "to enhance his knowledge of the manipulations of Radioactive Chemistry" (E.F.), and "to enhance his knowledge of the chemistry of radioactive elements" (F.R.), respectively. In the case of Rasetti's letter, specific research topics were suggested, complementing the activity previously carried out in Rome with Rasetti on the separation of radium D, and linked to a possible creation of new sources of alpha particles starting from the products of the decay of the radium chain.

By pure coincidence D'Agostino's arrival at the Institut du Radium in Paris and the arrival with him of two letters of introduction took place about ten days after the discovery at that same institute by Frédéric Joliot and Irène Curie of alpha-particle–induced radioactivity. In any case neither Fermi nor Rasetti was yet aware of this discovery when they wrote the two letters, much less did D'Agostino become involved on this subject during his stay in Paris.

In the Archive in Avellino many documents are kept bearing witness to D'Agostino's stay in Paris including, for example, a declaration by A. Debierne, Director of the Curie Laboratory, which states that D'Agostino had "informed himself of the methods and techniques employed" in that laboratory and in which D'Agostino was amply praised for his active participation.

At the Institut du Radium, as he wrote in a letter to the Chairman of the National Committee for Physics of the C.N.d.R. on 25 June 1934 (Fig. 24), D'Agostino followed "the courses held by M.me Curie, M. and M.me Joliot and M. Debierne", and "with the authorisation of M.me P. Curie(*) and under the guidance of M.me Joliot and Dr. Haissinsky" he followed "in the various laboratories of the Institute the procedures used there with regard to the chemistry of radioactive substances, carrying out many practical exercises and beginning, in collaboration with Dr. Haissinsky, research into some electrolytic methods of separating products of the Th family". This Parisian experience, as we shall see, would prove fundamental for D'Agostino when, on returning to Rome for the Easter holidays, he would be "detained" there by Fermi (Fig. 25) and involved in the research into neutron induced radioactivity, thus becoming Fermi's first collaborator.

Returning to the research project in the field of Nuclear Physics which over the year 1933 it was attempted to launch at the Physics Institute in Rome, as well as the installation of a bismuth crystal spectrograph and the preparation of a source of radium D which we have just mentioned, the purchase towards the end of the year of two Wilson chambers and a small motor for one of them should also be pointed out. We found the details in the Financial Archive of the Physics Institute of the Royal University of Rome. Indeed in the "Registro inventariale dei beni mobili" (Inventory ledger of goods), in the section "VI. Dynamic electricity" they are listed in succession: (with "Acquisition date 9/12/1933", "Acquisition number 599", "Acquisition slip 1260") a "pneumatically operated Wilson chamber, made up of cylinder, floating piston, three way tap with spring mechanism and three bronze rings", Supplier A. Contini, "Purchase price 2500", followed by (with "Acquisition date 9/12/1933", "Acquisition number 600", "Acquisition slip 1260") an "experimental Wilson Chamber with steel springs, with piston cylinder cast in bronze, complete with spring loading mechanisms", Supplier A. Contini, "Purchase price 5700", and finally (with "Acquisition date 9/12/1933", "Acquisition number 601", "Acquisition slip 1261") a "Small electric motor 190 Volt (s.n. 15) speed reduction speed 1/50", Supplier Rastelli, "Purchase price 170", under "Observations" "Applied to Wilson chamber" was added in pencil.

So it seems that, after Amaldi and Fermi's failed attempts to construct a Wilson chamber by themselves, it was decided to move on and to simply proceed with the purchase of this type of apparatus (two chambers no less) directly from a supplier of scientific instrumentation (A. Contini & Son). The two Wilson chambers, as reported

(*) Here D'Agostino follows the French usage calling Marie Curie by the name Madame Pierre Curie.

Istituto di Fisica
della R. Università
di Roma

Roma 25 giugno 1934/XII

Illmo. Sig. Presidente del Com. Nazionale per la Fisica
Cons. Nazionale delle Ricerche
Roma

In seguito al conferimento fattomi di una borsa di studio da parte del Cons. Nazionale dell Ricerche dopo avere coadiuvato il Prof. Rasetti nella estrazione di Ra D da un vecchio preparato di Ra, mi sono recato alla fine di gennaio del corrente anno all'Institut du Radium a Parigiv

Ivi ho frequentato i corsi tenuti da Mme. Curie, M. e Mme. Joliot e M. Debierne. Con l'autorizzazione di Mme. P. Curie e sotto la guida di Mme. Joliot e del Dr. Haissinsky ho seguito nei vari laboratori dell'Istituto i procedimenti ivi in uso nei riguardi della chimica delle sostanze radioattive, eseguendo praticamente molte esercitazioni ed iniziando, in collaborazione col Dr. Haissinsky una ricerca su alcuni metodi elettrolitici di separazione dei prodotti della famiglia del Th.

Ritornato per le vacanze Pasquali a Roma, sono stato trattenuto come lo sono tuttora presso l'Istituto fisico della R. Università dal prof. Fermi per le ricerche ivi in corso sulla radioattività artificiale.

Nel riferire alla S. V. sulla mia attività La prego di voler disporre affinchè mi sia corrisposta la seconda rata della borsa.

Fig. 24. – Letter from D'Agostino to the C.N.d.R., 25 June 1934 - ACS.

ISTITUTO DI FISICA
DELLA
R. UNIVERSITÀ DI ROMA
VIA PANISPERNA N. 89-A

Roma 26 giugno 1934/XII

Caro Bordoni,

il Dr; O. D'Agostino che ha una borsa di studio del C.N.R. per Parigi, è , come ti accennai, stato trattenuto da me per lavorare qui alle nuove radioattività artificiali.

So che ti ha scritto perchè gli venga pagata la seconda rata della borsa di studio. Vedi se è possibile accontentarlo.

Coi più cordiali saluti e ringraziamenti

Enrico Fermi

Fig. 25. – Letter from Fermi to the C.N.d.R., 26 June 1934 - ACS.

in the "Table of variations with additions or reductions, of items present on 31 October 1933-XII in the Physics Institute of the Royal University of Rome from 1 November 1933-XII to 31 October 1934-XIII", were only removed from the inventory in 1966.

The "Acquisition slip 1260" for the two Wilson chambers, at the beginning indicates, again referring to the same company A. Contini & Son: "Slit for gamma ray spectrograph, consisting of four right angle bronze pieces, two cm. 20 × 20 and two 20 × 40. Cast iron plate plus base and one piece in elektron applicable to the centre (Value to be recorded in Ledger 1250)". So it would therefore seem that in December 1933, there was still considerable interest in gamma ray spectrometry, enough to continue to invest a considerable sum of money in it. On the top of this acquisition slip and also on the slip for the small motor "fondo Rasetti. Min. Ed. Naz.le" is written in pencil. We can deduce that the cost was attributed to a different source of funding than that allocated by the C.N.d.R. On the slip for the motor "Radium room" seems to be written in pencil, which perhaps indicates the location of the equipment for nuclear research in a specific laboratory.

5 New particles and new theories: 1932-1933

While, as we have just seen, experimental activity was beginning to take shape in the Nuclear Physics field in Rome, and more generally in Italy, at an international level already from early 1932 great progress was being made in this field, both experimentally and theoretically. This however does not seem to have affected research carried out in Rome apart from a theoretical study on the positron carried out by Fermi during a visit to the United States, which we will discuss in the next paragraph.

5.1 A new particle: the positron or "anti-electron"

On the experimental side, as we discussed in chapter 3, Chadwick discovered the neutron at the Cavendish Laboratory in Cambridge in February 1932 while in April of the same year Cockroft and Walton built the first proton accelerator and achieved the first nuclear disintegration of light atoms with these particles.

In August that same year, at the California Institute of Technology (Caltech) in Pasadena, Carl D. Anderson discovered the positron experimentally. This discovery arose from a new line of research on cosmic rays, opened up a little before by Millikan in his own laboratory at Caltech and concerning the study of these rays with a cloud chamber. In the spring of 1932 as part of this research programme Anderson, who two years before in 1930 had been awarded a Ph.D. in Physics under the guidance of Millikan specialising in the use of the cloud chamber, studied the high-energy corpuscles "associated with cosmic rays" using a vertical cloud chamber, designed by him with Millikan, and placed in a uniform magnetic field, with a maximum attainable value of 21 000 G. As he stated in his first paper on the subject [10] out of 3000 photographs taken, 62 showed tracks of cosmic rays sufficiently long that, on the basis of their radius of curvature, it was possible to take "energy measurements". On the hypothesis that the particles that produced the tracks travelled downwards through the chamber Anderson, on the basis of the direction of curvature, identified tracks that could be attributed both to negatively and positively charged particles. The former, also taking into account their specific ionisation, were identified with the electrons, while the latter, since as Anderson observed "the specific ionization along the tracks showing positives is in most instances *not much greater* [the italics are ours] than that for the electrons ", were at first identified with protons which were the lightest positive particles known at that time. The important point is that Anderson, already in this first paper, also studied the "scattering of cosmic particles" through matter, introducing a lead plate 6.00 mm thick horizontally

F. Guerra and N. Robotti, *The Lost Notebook of ENRICO FERMI*,
https://doi.org/10.1007/978-3-319-69254-8_5

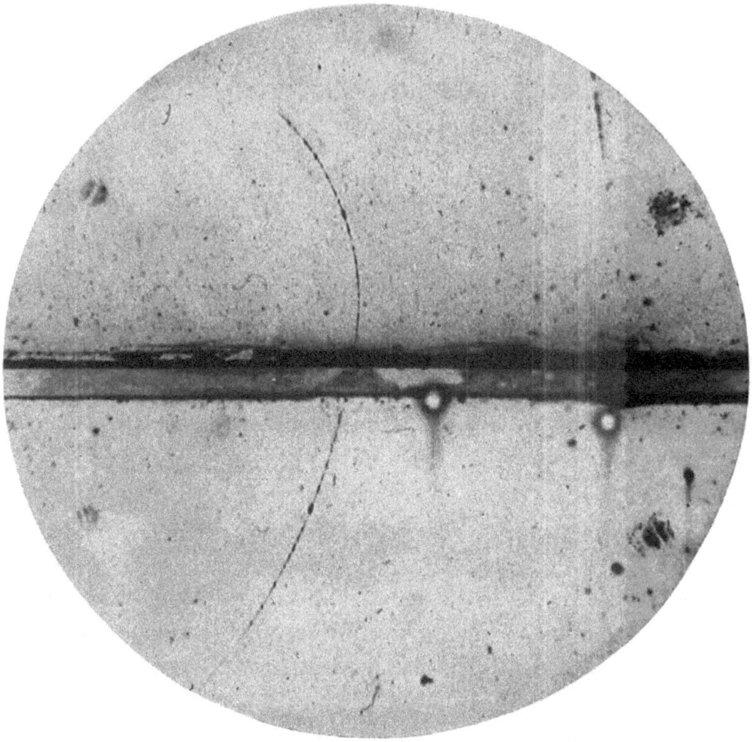

Fig. 26. – The first photograph in a cloud chamber of a positron passing upwards through a 6 mm lead plate (C. D. Anderson, 2 August 1932) - in [12] (PR).

through the chamber. For example in photograph 22 of [10, page 417], in the centre the lead plate is clearly visible and the rectilinear track of a particle crossing it almost perpendicularly. In this case, since the curvature could not be measured, because it was clearly too energetic a particle to be deflected by the magnetic field applied to the cloud chamber, the sign of the charge was uncertain. According to Anderson, it could have been a proton, in which case its energy was greater than $200 \cdot 10^6$ eV, or an electron, in which case the energy was higher than $600 \cdot 10^6$ eV.

A luckier photograph of the scattering of these "cosmic particles" through a lead plate was taken by Anderson on 2 August 1932. It is the famous first photograph of the positron, which then became its symbol, in which the track of a particle passing through a lead plate is visible and which, unlike the cases in the earlier photographs, is deflected by the magnetic field. Therefore in this case it was a less energetic particle which in particular is not very curved under the plate, and thus had a large radius of curvature, while it is very curved above the plate and thus had a smaller radius of curvature (Fig. 26).

Taking into account that the radius of curvature, according to the expression of

Lorentz force, is directly proportional to velocity, and assuming that the particle loses energy passing through the plate, it was clearly a particle that entered from the bottom of the chamber (where its track was almost rectilinear) to emerge then above the plate (where its track was very curved), towards the top of the chamber. At this point, on the basis of the direction of the curvature and taking into account the polarity of the field, it was possible to establish that it was a positively charged particle. In any case, unlike the previous cases dealt with, this particle could not be identified with a proton and much less with a heavier nucleus since, as Anderson observed, if it had had the mass of a proton the energy with which it exited the lead plate, calculated on the basis of its radius of curvature, would have been 300 000 V and thus it should have had, "according to well established and universally accepted determinations" a total range in the air of about 5 mm, while instead its actually visible range was at least ten times greater, even more than 5 cm, before being curved. Moreover its specific ionisation proved to be "close to that for an electron with the same curvature" [11]. After excluding as improbable the other possible alternatives according to which it was, instead of a positive particle moving from the bottom to the top, an electron moving from top to bottom — but in that case the electron before entering the plate would have had less energy (20 million volts) than that with which it exited the plate (60 million volts) — or that there were two independent electrons that in the same instant had produced two tracks placed in such a way as to give the impression of a single particle passing through the lead plate, Anderson arrived at the conclusion that it was a particle bearing a positive charge but which had a mass of the same order of magnitude as that normally possessed by a free negative electron. Anderson gave this new particle the name positive electron or positron [11, 12].

The existence of this new particle was confirmed a few months later by P. M. S. Blackett and G. P. S. Occhialini at the Cavendish Laboratory in Cambridge by means of a series of studies carried out, in this case too, on cosmic rays. They used a new type of cloud chamber, which they themselves had created, the counter-controlled cloud chamber [22]. This apparatus consisted of a traditional cloud chamber with the addition of a new control system, which meant that the cloud chamber would automatically start working every time one or more ionising particles passed through. This control system was made up of two Geiger-Müller counters put in coincidence, one above the chamber and the other below. When a particle had also passed through the second counter the chamber automatically started up and the photograph was taken. Since the time interval between the discharge of the counters and the end of the expansion was very short (it had been reduced even to 1/100 of a second), in this time the ions produced in the chamber by an ionising particle had moved very little with respect to the position where they had been formed and, as a result, the consequent tracks were still very clear, so as "to allow very accurate measurements". Clearly with this control system the moment of the expansion of the chamber, and therefore the moment of observation with the taking of the photograph, was no longer left to a random choice by the operator but was driven by the particle itself. To use an expression of Blackett and Occhialini's, it was "a method of making particles of high energy take their own cloud photographs". Blackett, in a letter on 15 September 1932 sent to Augusto (1878-1951), Occhialini's father and professor of Experimental Physics at the University of Genoa (A. Occhialini Archive at Arcetri) described their machine and their way of taking measurements as follows: *The apparatus is really very successful and a most fascinating instrument to use. The waiting with dark, with all set ready, for the cosmic ray to arrive to set all the mechanism in motion is rather thrilling.* The control method constructed by Blackett and Occhialini fully reflected the specific expertise that characterised the two physicists: Blackett, a researcher at the Cavendish Laboratory in Cambridge since 1921, was in fact one of the greatest experts in the field of cloud chambers while Occhialini, who from August 1932 had been benefitting from a study grant from the C.N.d.R. at the Cavendish Laboratory, had gained great experience in the field of Geiger-Müller counters and coincidence circuits as a collaborator of Bruno Rossi in Florence [107].

The cloud chamber used by Blackett and Occhialini was of the vertical type, like that used by Anderson, and the magnetic field which, unlike Anderson's apparatus, had to be applied for long periods of time, had a value of 3000 G compared to Anderson's 15000 G and was applied uniformly throughout the chamber. This device guaranteed an 80% success rate for every photograph taken. Indeed out of 700 photographs no less than 500 contained high-energy particle tracks while with Anderson's device only one photograph in 50 was significant. Blackett and Occhialini succeeded in obtaining large numbers of images not only of single particles (electrons and positrons), often emitted in pairs from a single point in the chamber, but also "showers", made up of groups of these particles (positive and negative) emitted in cascade. In these experiments Blackett and Occhialini, in order to study the interaction of these high-energy particles with matter, often introduced a plate of the various materials under examination horizontally, in the centre of the chamber, as Anderson had done.

In any case Blackett and Occhialini did not restrict themselves to confirming experimentally the existence of the positron but tried to give it a theoretical interpretation, connecting it to Dirac's theory of the electron. As we said before, Dirac's equation admitted negative-energy solutions for the electron as well as positive-energy solutions. To justify these solutions Dirac, in 1930, had proposed that all negative-energy states be considered occupied, according to the Pauli exclusion principle. If one of these states, in "exceptional" cases, was not occupied, it would show itself as "a kind of hole" and would have appeared as a particle with a positive energy and a positive charge. Dirac was first led to identify this hole with the proton and later, after the introduction of the symmetry under charge conjugation by Herman Weyl (1885-1955), with a new particle "which is not expected to be found in nature" and that had the same mass as the electron but opposite charge, called the "anti-electron" by Dirac. This new particle, according to Dirac's interpretation, could be created, together with an electron, by a highly energetic gamma ray, with an energy of at least $2mc^2$, so as to allow an electron that was in a negative-energy state to pass to a positive energy state leaving a negative-energy "hole", for example in the proximity of a nucleus that would guarantee the conservation of the momentum in the course of the process. Vice versa, the "hole" in the presence of electrons could be destroyed by an electron that then occupied the negative-energy state, giving rise to two photons with energy of at least $2mc^2$ [65].

Blackett and Occhialini in the article in which they reported their experimental results in favour of the existence of the positron, directly recalling Dirac's interpretation, identified this new particle with Dirac's "anti-electron" and, still in the context of Dirac's view, tried to give an explanation of its origin in terms of the creation of electron–anti-electron pairs "during the disintegration of light nuclei" [22, page 713]. At the time however they did not specify the agent responsible for the creation of this pair, either "particles or photons of high-energy" associated with cosmic rays [22, page 711]. Moreover, referring to the process of annihilation between a positive electron and a negative one, also predicted by Dirac, they clarified the reason why this new particle could only be discovered when cosmic rays were studied with a cloud chamber. In these theoretical aspects Blackett and Occhialini had been supported by Dirac himself, who was repeatedly thanked in the article "for most valuable discussions". It should be noted that at that time Dirac was professor at Cambridge, at St. John's College, and therefore in the same town as the Cavendish Laboratory which encouraged collaboration. In particular Dirac, as Blackett and Occhialini stressed reporting his original calculation, had calculated the probability of an annihilation process per unit time between a positive electron and a negative one and had obtained "a time of life for the positive electron that is long enough for it to be observed in the cloud chamber but short enough to explain why it had not been discovered by other methods", thus finding the reason for the cloud chamber's success in identifying this new particle [22, page 716].

Dirac's involvement in Blackett and Occhialini's research into this new particle was such that at a certain point he even reviewed the experimental part. In the Occhialini-Dilworth Archive at the Department of Physics of the University of Milan, amongst the papers regarding this period that Occhialini spent at the Cavendish Laboratory in Cambridge, there is a photograph in which Dirac is pictured, bent over, tinkering with a device (Fig. 27). As Occhialini wrote in pencil and in Italian on the back of the photograph, it showed "Dirac redesigning the stereoscopies of the positive electron".

Dirac's new interest in the experimental aspects of Physics is also borne witness to by a letter, which we will return to later, that Lord Rutherford, at that time Director of the Cavendish Laboratory, wrote to Fermi on 23 April 1934 in which he informed him that "Professor Dirac also is doing some experiments". So Dirac's involvement at an experimental level does not appear so odd considering that in the laboratory he could finally see and study his anti-electron close up, with all its properties that until then had only been predicted on paper.

5.2 The positron in the laboratory

On the basis of this first experimental research by Anderson and by Blackett and Occhialini, the existence of the positron seemed to be connected exclusively to processes that involved cosmic rays (even if these processes, in order to occur, seemed to require the presence of heavy nuclei), and did not seem to be "associated with matter under normal conditions" [22, page 714].

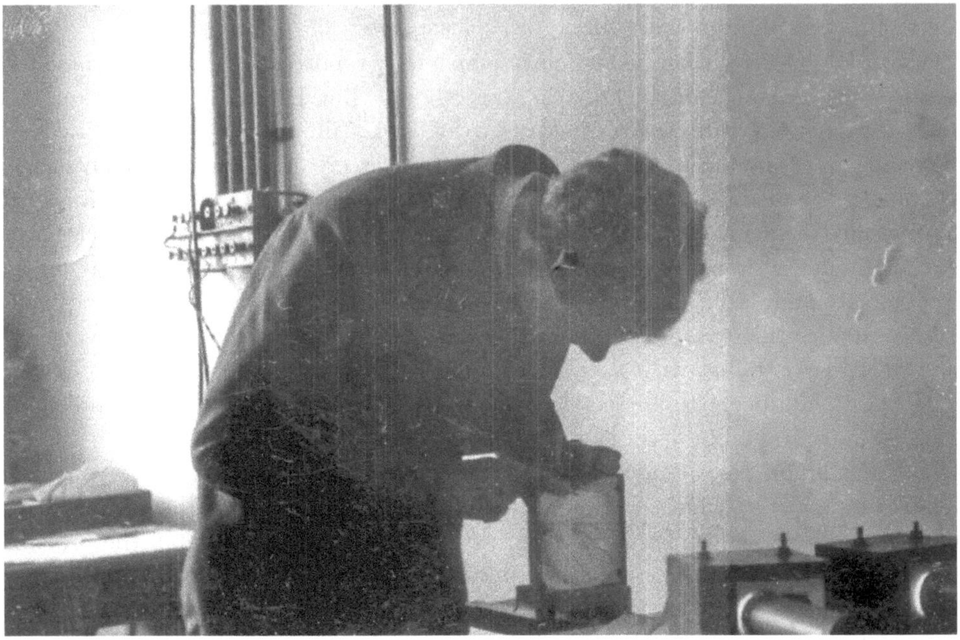

Fig. 27. – Photograph of "Dirac redesigning the stereoscopies of the positive electron" (1933) - OD.

In any case a change in this regard very soon happened. On 27 March 1933 Blackett and Occhialini again, this time together with Chadwick who was probably involved because of his expertise with regard to the neutron, discovered [33, page 473] that positrons could be produced completely by laboratory means, without recourse to cosmic rays, simply by making the "penetrating radiation" of Po-Be, which by now was known to be composed of neutrons and gamma radiation, impinge on a small lead plate. A few days later (10 April 1933), Irène Curie and Frédéric Joliot demonstrated that, of the two components (neutrons and gamma rays) that constituted this radiation, the one that was able to produce positrons was represented by the gamma rays [55]. Moreover, as the Joliot-Curies demonstrated on 22 May 1933 [56], and in parallel Lise Meitner and Kurt Philipp in Berlin [114], and Carl D. Anderson and Seth H. Neddermeyer (1907-1988) in Pasadena [13], the gamma rays emitted by Th C, when impinging on a lead plate, were also able to produce positrons. Not only that, but in both cases electrons were emitted as well as positrons. In many of the photographs, as Blackett and Occhialini had already observed, the two trajectories of the electrons, the positive one and the negative one, seemed to come from the same point in space. Hence the idea, in agreement with Dirac's prediction, even if Dirac was not quoted, that "a gamma photon with great energy, meeting a heavy nucleus will be transformed into two electrons with contrary signs". These electrons, both positive and negative, were called "materialisation electrons" by Joliot and Curie, as suggested by Marie Curie, in order to specify their origin. At the Solvay

Conference in 1933 Joliot and Curie would observe: "If this hypothesis is confirmed, we are witnessing for the first time the transformation of electromagnetic radiation into matter".

The discovery of the positron and above all the processes of creating and destructing pairs of electrons and positrons, raised great interest around the world. Fermi too tackled the subject from the theoretical point of view. In a study with Uhlenbeck [F109, F110], carried out during his stay in August 1933 at the University of Michigan, he worked on the process of annihilation of an electron and a positron. In particular he tried to confirm whether a certain spectral line, identified by Louis H. Gray (1905-1965) and Gerald T. P. Tarrant (1906-1965) in the "scattering of hard γ rays" [80], could be attributed, as Blackett and Occhialini had suggested, to an annihilation process but he found that, on the basis of Dirac's theory, it did not seem possible.

The positron, as studied in accordance with Dirac's theory, did not however seem capable of an origin of its own and of an autonomous existence, independently of that of the negative electron. It only seemed to exist because it had been created by the collision of the hard gamma ray with a heavy nucleus, an electron-positron pair. In any case new research carried out by the Joliot-Curies would soon open up a completely new scenario for the appearance of the positron. As they announced in the 19 June 1933 session of the Academy of Sciences in Paris [57], positrons could in fact be emitted in a nuclear transmutation produced by alpha particles in some light elements, such as aluminium and boron, without electrons and without the action of γ rays.

Placing a source of alpha particles of 20 mCi of Po, covered by a sheet of aluminium, over the orifice of a cloud chamber to which a magnetic field of 400 G was applied, Joliot and Curie had indeed observed tracks of positrons, all derived from the source. Since they were not present when the aluminium was replaced by a sheet of silver or of paraffin, Joliot and Curie concluded that the positrons were emitted by the aluminium under the action of the alpha particles of the source. Boron too had the same behaviour as aluminium.

These observations represented the first step towards the discovery of alpha-particle–induced radioactivity. For these positrons, in order to stress their nuclear origin and at the same time to differentiate them from those previously discovered and called "materialisation positrons", Joliot and Curie proposed the name "positive transmutation electrons". In a page of their laboratory notebook, in early June 1933, we can read "Distribution énergie électrons transmutation Po+Al".

The hypothesis of "transmutation positrons", even if, as we shall see, it raised serious doubts in the international scientific community, completely changed the picture of nuclear transmutations as they had been conceived until that moment. Joliot and Curie observed in a later paper in July 1933 [58]: "The transmutations so far known, caused by γ rays, by fast protons or by neutrons occur with the emission of protons, of neutrons or of α particles, in these phenomena the emission of nuclear electrons has never been observed".

With regard to the nuclear transmutations known in 1933 and the type of reactions produced, please see the table (Fig. 28) included in a report by A. S. Russell (1888-1972), published early in 1934 in the "Annual Reports of the Chemical Society" [146], in which only reactions of the (α, p), (α, n), (p, α), (n, α) type appear, in accordance with what Joliot and Curie maintained.

Curiously, Rasetti was a supporter of the idea of the "transmutation positrons". As we saw before, when Rasetti together with Oscar D'Agostino succeeded in obtaining a preparation of polonium he was mainly interested in the fact that this preparation would have made a positron source available in Rome. Indeed, according to Joliot-Curies' claims, it would be enough to bombard aluminium with alpha particles of polonium and a source of positrons would be obtained.

The discovery of "transmutation positrons", beyond its intrinsic value, was very important for Joliot and Curie because it allowed them to settle a rather problematic result with regard to aluminium that they had obtained a few months earlier (presented by J. Perrin in the session of 6 February 1933 [54]) during some research on the emission of neutrons by light elements bombarded by alpha particles.

This is what it was about. Joliot and Curie, using an ionising chamber filled with methane, and in some cases butane, "in order to increase the effect of the neutrons", had succeeded in demonstrating that, as well as beryllium and boron, the elements fluorine, aluminium, sodium if bombarded with alpha particles also emitted neutrons. A confirmation of Joliot-Curies' results with regard to aluminium was provided a few days later by Pierre V. Auger (1899-1993) and Gabriel Monod-Herzen, in a note presented by J. Perrin in the session on 20 February 1933 of the Académie des Sciences [14]. The emission of neutrons by aluminium must have appeared somewhat strange for those who, like Joliot and Curie, were well acquainted with Chemistry. In fact the initial nucleus had necessarily to be the isotope ${}^{27}_{13}\mathrm{Al}$, since aluminium had no other isotopes. As a result the emission of a neutron necessarily led to the hypothesis of the following nuclear (α, n) transmutation:

$$
{}^{27}_{13}\mathrm{Al} + {}^{4}_{2}\mathrm{He} \rightarrow {}^{30}_{15}\mathrm{P} + {}^{1}_{0}n,
$$

where the final product of the transmutation, ${}^{30}_{15}\mathrm{P}$, was an "unknown" isotope of phosphorus [99, page 141], never seen before and therefore probably non-existent. Only in early January 1934 did they realise that the previous nuclear reaction was in reality the beginning of the process of α-particle–induced radioactivity on aluminium.

In any case, at the time they did not realise the importance of this result but tried to adjust things in such a way as to make this "unknown" isotope "disappear". To do this they exploited the newly discovered property that aluminium, under bombardment by α particles, emits positrons as well as neutrons. So they hypothesised that the transmutation of aluminium took place with the simultaneous emission of a neutron and a positron, even if these particles had been seen separately and with different instruments: the neutrons with an ionisation chamber by means of the hit protons and the positrons with a cloud chamber. In the supposed reaction aluminium, after capturing an alpha particle, was transformed into the stable isotope of silicon ${}^{30}_{14}\mathrm{Si}$:

$$^{27}_{13}\mathrm{Al} + ^{4}_{2}\mathrm{He} \rightarrow ^{30}_{14}\mathrm{Si} + ^{1}_{0}n + e^{+}.$$

In this way, by compacting it all into a single nuclear reaction, Joliot and Curie managed to insert the transmutation of aluminium into the general setting of the nuclear transmutations known at that time, according to which the final product of a nuclear transmutation was a stable nucleus. Even if the Joliot-Curies' idea of the simultaneous emission of the neutron and of the positron would eventually prove to be wrong, it did represent a very important step towards the discovery of alpha-particle–induced radioactivity because it was by working on this idea that, in the end, it could be completely overcome and therefore the correct solution could be arrived at. Moreover, as Joliot and Curie pointed out and as had been known from the first transmutations achieved in the early '20s, aluminium under bombardment by alpha particles emitted protons, transforming into the stable isotope of silicon according to the following (α, p) reaction:

$$^{27}_{13}\mathrm{Al} + ^{4}_{2}\mathrm{He} \rightarrow ^{30}_{14}\mathrm{Si} + ^{1}_{1}p.$$

As a consequence, on the basis of Joliot and Curie's new results, which seemed to show that aluminium when bombarded by alpha particles could also emit a neutron and a positron, the stable isotope of silicon $^{30}_{14}\mathrm{Si}$, it had to be assumed that the aluminium nucleus — and therefore a same nuclear species (let us remember that aluminium has no isotopes) — could be transformed in two different ways, either by the emission of protons, or by the emission of neutrons and positrons. Joliot and Curie observed: "We believe that sometimes a neutron and a positive electron are emitted instead of a proton".

A similar behaviour was also found by Joliot and Curie in the case of some other light elements such as boron. Joliot and Curie concluded as follows: "These considerations lead us to admit that the proton is complex and arises from the union of a neutron and of a positive electron". When Joliot and Curie made these observations they were far from thinking that, behind the nuclear reactions they were studying, alpha-particle–induced radioactivity lay hidden, and with it a new type of beta decay, positive decay, where there was the transformation of a proton into a neutron in the nucleus with the emission of a positron and a neutrino.

5.3 The Heisenberg-Majorana theory of the nucleus

Returning to the development of Nuclear Physics, starting in early 1932, it should be pointed out that on the theoretical side, in particular with regard to modelling the nucleus, with the discovery of the neutron, a new heavy nuclear constituent like the proton but without charge, new development prospects were able to open up and, by means of Heisenberg's theory in 1932-33, it was possible to arrive at the formulation of a first stable nuclear model.

Indeed, immediately after the discovery of the neutron, Heisenberg had the idea of trying to construct a theory of the nucleus, no longer considering electrons as an integral part of the nucleus but considering exclusively protons and neutrons as "fundamental independent" components [90, 91]. In this way all the problems raised by the presence of electrons in the nucleus, and for which it had not been possible so far to construct a stable nucleus model, in particular the fact that in accordance with known laws electrons could not be confined inside the nucleus, with Heisenberg's idea were suddenly set aside. Not only that. Now only having to deal with heavy particles, and therefore with particles for which the laws of quantum mechanics were valid, finally the possibility of nuclear modelling, using this theory fully, was guaranteed.

Heisenberg's idea of doing without electrons, understood as "fundamental independent" components of the nucleus, clearly was not feasible before the discovery of the neutron, when, if it was wished to construct a model of the nucleus, it was possible to count only on protons and electrons. Clearly Heisenberg's way of operating revolutionised the constitution of the nucleus, as it had been conceived until that moment. Indeed a nucleus with atomic weight A and atomic number Z, while it was considered before to be composed of A protons and $A - Z$ electrons, now had to be considered as composed of Z protons and $A - Z$ neutrons. Heisenberg, in order to bind these two types of heavy particles, protons and neutrons, together in the nucleus introduced a complex system of forces, including an exchange force, analogous to that which operates in the case of molecules. In particular he hypothesised a very close analogy between the mutual proton-neutron action and the action that occurred in the ionised hydrogen molecule H_2^+ where the atom of hydrogen H and the atom of ionised hydrogen H^+ exchanged an electron. In the case of the proton-neutron action, according to Heisenberg's idea, what the particles exchanged, as well as position and spin, was the charge. This however did not mean that the charge was associated with the electron. On this point Heisenberg was explicit: there was "no movement of electrons". With regard to the exchange integral $J(r)$ which he introduced to characterise the interaction between proton and nucleon, and which he compared to the exchange integral "in molecular theory", Heisenberg observed: "In any case it is more correct to interpret the exchange integral $J(r)$ as a fundamental property of the neutron-proton pair, without reducing it to the movement of electrons". In giving a mathematical representation of the exchange action, Heisenberg considered the proton and the neutron to be two different quantum states of a same particle, characterised by a new variable ρ which took the value $+1$ or -1 according to whether the particle was a neutron or a proton, in this way anticipating a formalism that would then become that of isotopic spin. All in all therefore, as Heisenberg observed, "each particle in the nucleus is characterised by 5 quantities: the three coordinates of position (x, y, z), the spin σ_z along the z-axis and a fifth number ρ which can be ± 1". A complex of forces was assumed between the particles of the nucleus, which included the Coulomb force between the protons and an appropriate exchange force between protons and neutrons. In this way, Heisenberg, in June 1932, arrived at the formulation of a nuclear theory which led to a stable nucleus model, demonstrating that it was possible, within the framework of quantum mechanics, to treat the nucleus as made up exclusively of protons and neutrons from the theoretical point of view.

Clearly the path followed by Heisenberg left open the problem of the origin of the electrons emitted by some radioactive isotopes in beta decay: indeed in order to explain this phenomenon successfully, as Heisenberg himself admitted, it seemed necessary to assume that the electrons were also somehow contained in the nucleus. In any case now, with the development of Heisenberg's theory, the presence of electrons in the nucleus was no longer of any consequence in the problem of the stability of the nucleus since this problem had now been solved. Not only. Since now electrons had lost their connotation of "fundamental" constituents of the nucleus, the problem of their presence inside the nucleus became a separate problem, completely detached from the problem of the structure of the nucleus. For example, in order to explain beta decay, according to Heisenberg, it could be considered, following an idea first put forward by Iwanenko in a Letter sent to "Nature" on 21 April 1932, and published the following 28 May [97], that electrons were "all packed" inside the neutrons, that is to say that neutrons were compound structures, made up of a proton and an electron which, "in favourable circumstances", broke down into a proton and an electron "violating the laws of energy conservation and of the quantity of motion". Heisenberg, at the moment that he assumed this violation of the laws of conservation for beta decay, clearly did not adopt the hypothesis of the neutrino but aligned himself with Bohr's line of thought, which we have already amply discussed when speaking of the Rome Conference and which at that time did in fact seem the most feasible one.

Naturally with this hypothesis of the "compound neutron", as we shall see better in the next chapter, the electrons that are now to be found "packed in the neutrons", analogously to what had already happened to the "nuclear electrons", faced insuperable difficulties from the point of view of spin and statistics if one wished to guarantee that both the spin of the neutron and that of the proton continued to have the value 1/2. Not only that, the problem of their confinement in the nucleus still persisted, rather it was further exacerbated, since now it had to be understood how they could be contained in structures even smaller than the nucleus itself, such as neutrons. In any case, for now Heisenberg took his time on these problems: in the short term he was interested in succeeding in completely explaining the structures of nuclei and their properties. As he wrote in a letter sent to Bohr on 20 June 1932 (kept in the Bohr Archives in Copenhagen): *the basic idea is to unload all the fundamental difficulties onto the neutron and to let quantum mechanics penetrate into the nucleus.*

With regard to the possibilities of Heisenberg's theory, a fundamental contribution was made by Ettore Majorana at the end of February 1933, during a study visit in Leipzig, at Heisenberg's Institute. We have already spoken of Majorana before with regard to the early developments in Nuclear Physics in Rome and we said that his *laurea* thesis represented the first groundbreaking study carried out in Rome in this field. Strangely, in the years after his *laurea*, Majorana no longer worked on Nuclear Physics but on questions relating to atomic Physics and problems connected to relativistic wave equations for elementary particles. Only at the end of 1932 did he really turn his attention to this field of research, asking the C.N.d.R., through Fermi, for a study grant to use with "Herr

Prof. Heisenberg" in Leipzig as well as for a brief stay with Niels Bohr in Copenhagen to devote himself to "theoretical research mainly concerned with the structure of nuclei and the relativistic formulation of the new theory of quanta", as seen in his precise study programme presented in the letter to the C.N.d.R. on 9 January 1933, reported in [82, Fig. 35].

Clearly Majorana's decision to go to Leipzig was motivated by the publication, a few months earlier, of Heisenberg's first two papers on the structure of the nucleus. In Rome Majorana was the first to understand the importance and the possibilities of these papers by Heisenberg and, with them, the fundamental role of the neutron in interpreting the constitution of the nucleus. According to some interpretations, advanced by a very authoritative source [9], immediately after the announcement of the discovery of the neutron in the spring of 1932 Majorana had actually constructed a theory of the nucleus based on exchange forces, completely independently of Heisenberg. This interpretation, although plausible on the basis of Majorana's previous experience on exchange forces at the molecular level and on his profound phenomenological intuition, is sadly not supported by any objective documentation.

In any case Majorana's decision, immediately supported financially by the C.N.d.R., to take an interest in the theory of the nucleus and to work in particular in close contact with Heisenberg, was an outstanding choice. After arriving in Leipzig at the end of January 1933,within a few weeks he in fact succeeded in making some fundamental corrections to Heisenberg's theory, reducing the interactions between protons and nucleons solely to appropriately modified "exchange forces". Majorana's modifications, which as Majorana himself observed were "substantial" and which had been made "to re-establish agreement with experiment", concerned both the sign of the forces which was changed from negative to positive and the spin, which was not exchanged.

With these corrections, as Majorana observed, "the great stability of the alpha particle" was immediately guaranteed and "the general trend of the mass defect curve and of nuclear volumes" successfully found a simple explanation.

These results, which were briefly announced in "La Ricerca Scientifica" [108], were immediately published by Majorana in the prestigious journal "Zeitschrift für Physik" [109], the same journal in which Heisenberg's first two papers appeared [90,91], and also a third [92], made available to Majorana before publication. To satisfy a request from the C.N.d.R. an expanded summary was also published in "La Ricerca Scientifica" [110].

These results improved Heisenberg's earlier theory substantially, and earned Majorana not only Heisenberg's immediate appreciation in national conferences and at the Solvay Conference in October 1933 but also unanimous international recognition, so much so that "Heisenberg's theory" was from then on called the "Heisenberg-Majorana theory" and "Heisenberg's exchange forces" became the "Heisenberg-Majorana exchange forces". Majorana's international fame in the nuclear field is further documented in a handwritten letter of 21 April 1935, addressed to "Signor Majorana", kept in the Family Archive, in which he was invited to attend an international conference on Nuclear Physics planned for the following 20-30 September at the Physical Technical Institute in Leningrad. The letter is signed by M. Bronstein (Matvei Petrovich Bronstein (1906-1938)), Secretary of the Conference. "Signor Majorana" immediately answered in French accepting the invitation.

Unfortunately the conference was not held until two years later, in 1937 and in Moscow, without Majorana's participation. It is significant that the other two Italians invited to the 1935 conference were Enrico Fermi and Bruno Rossi. So Fermi, Majorana and Rossi were the reference points for Italian Nuclear Physics: a fine acknowledgement for Majorana. The letter is transcribed in the appendix and reproduced in Figs. 29-32.

In an interval during his stay in Leipzig Majorana also spent a month and a half at Bohr's Institute in Copenhagen, another centre of excellence for the New Physics. His visit was prepared with the intervention of George Placzek (1905-1955), who in a letter from Copenhagen dated 23 February 1933 (Fig. 33), and kept in the Majorana Family Archive, wrote: *Illustrious Inquisitor. On my arrival in Copenhagen I found your letter here. I immediately went to Bohr informing him of your praiseworthy intention. As the Most Holy inquisition may perhaps already have predicted Bohr sends word that that he will be very glad to see you here.* The letter is transcribed in full and translated in the Appendix. Placzek showed a very expressive use of the Italian language and one must remember that Majorana was called "Great Inquisitor" in the Physics Institute in Rome because of his profound critical spirit. Enrico Fermi of course was the "Pope".

Despite the great success achieved in Leipzig, when Majorana returned to Rome in early August 1933 strangely he ceased all official science activity, not publishing any more papers until the spring of 1937.

It should be noted that between 1931 and the end of 1932, before his departure for Germany, Majorana produced no less than 10 publications [83], including two Letters to "La Ricerca Scientifica", in one of which he announced the future publication in an extended version of a piece of research of his, as if to guarantee priority of it for himself,

TABLE III.

Process.	Reference.	Process.	Reference.
$H+Li^7 \longrightarrow 2He^4$	42, 43, 44,	$He^4+B^{11} \longrightarrow n+N^{14}$	48, 56, 58
	45, 50, 51	$He^4+N^{14} \longrightarrow H+O^{17}$	67
$H+Be^9 \longrightarrow He^4+$?	45	$He^4+F^{19} \longrightarrow H+Ne^{22}$	67
$H+B^{11} \longrightarrow 3He^4$	42, 44, 45	$He^4+F^{19} \longrightarrow n+$?	56, 57
$H+F^{19} \longrightarrow He^4+$?	45	$He^4+Ne \longrightarrow H+Na$	67
$D+Li^6 \longrightarrow 2He^4$	47, 50, 51	$He^4+Na^{23} \longrightarrow H+Mg^{26}$	67
$D+Li^7 \longrightarrow n+2He^4$	50, 52	$He^4+Na^{23} \longrightarrow n+$?	56, 57
$D+Be^9 \longrightarrow n+B^{10}$	47, 52	$He^4+Mg \longrightarrow H+Al$	67
$D+Be^9 \longrightarrow He^4+$?	47	$He^4+Al^{27} \longrightarrow H+Si^{30}$	67
$D+N^{14} \longrightarrow He^4+$?	47	$He^4+Al^{27} \longrightarrow n+$?	56, 58, 60
$n+N^{14} \longrightarrow He^4+B^{11}$	52, 59, 65	$He^4+Al^{27} \longrightarrow e^+ +$?	62
$n+O^{16} \longrightarrow He^4+C^{13}$	52	$He^4+Si \longrightarrow H+P$	67
$n+F^{19} \longrightarrow He^4+$?	54	$He^4+P^{31} \longrightarrow H+S^{34}$	67
$n+Ne^{20} \longrightarrow He^4+O^{17}$	54	$He^4+S \longrightarrow H+Cl$	67
$He^4+Li^7 \longrightarrow n+B^{10}$	56, 58	$He^4+Cl \longrightarrow H+A$	67
$He^4+Be^9 \longrightarrow n+C^{12}$	48, 56, 58	$He^4+A \longrightarrow H+K$	67
$He^4+B^{10} \longrightarrow H+C^{13}$	67	$He^4+K \longrightarrow H+Ca$	67
$He^4+B^{10} \longrightarrow n+e^+ +C^{13}$	62		

Fig. 28. – Table of the disintegrations of the elements produced by neutrons, known in 1933 - [146, page 355] (CS)

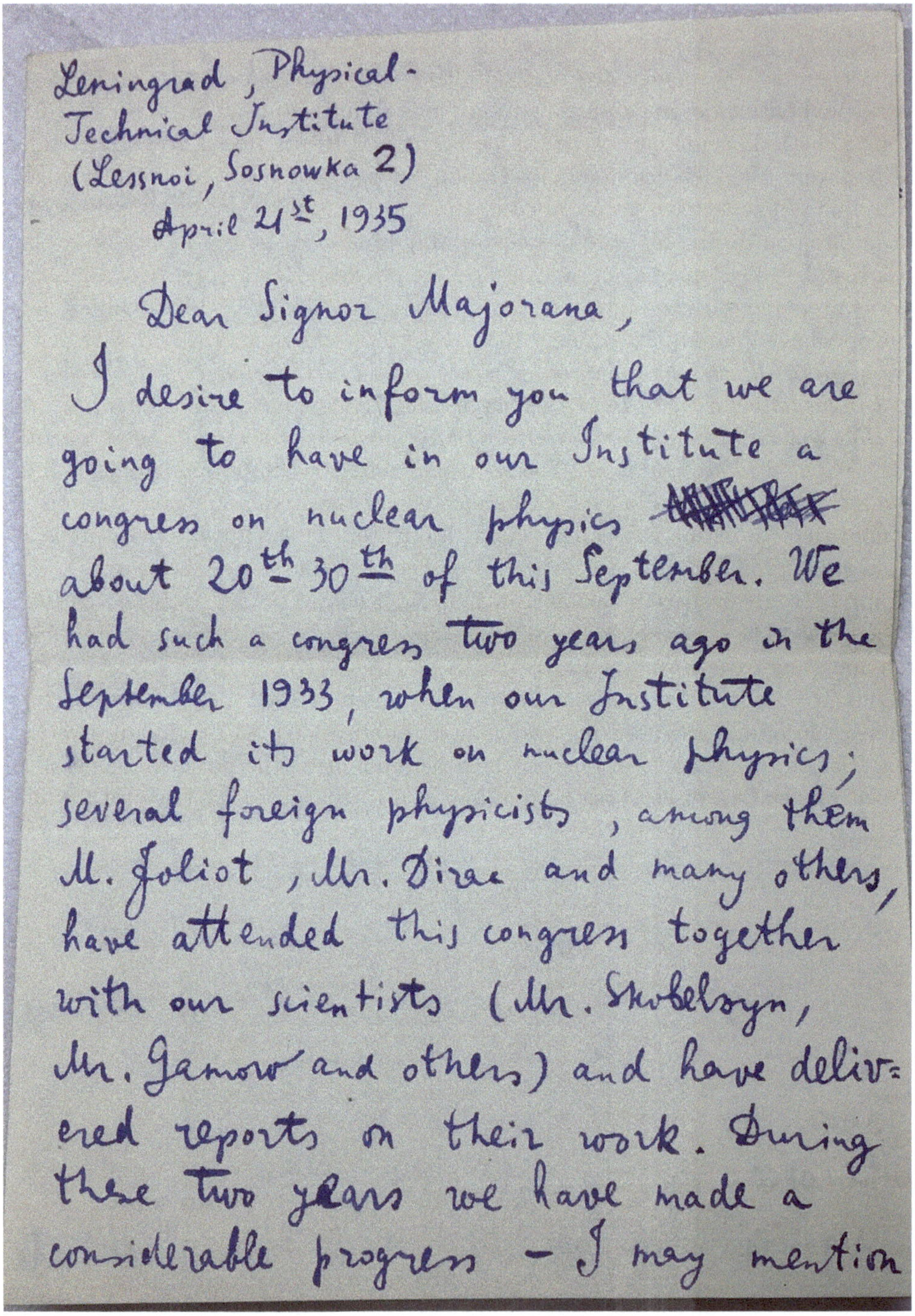

Leningrad, Physical-
Technical Institute
(Lessnoi, Sosnowka 2)
April 21st, 1935

Dear Signor Majorana,

I desire to inform you that we are going to have in our Institute a congress on nuclear physics about 20th–30th of this September. We had such a congress two years ago in the September 1933, when our Institute started its work on nuclear physics; several foreign physicists, among them M. Joliot, Mr. Dirac and many others, have attended this congress together with our scientists (Mr. Skobelzyn, Mr. Gamow and others) and have deliv-ered reports on their work. During these two years we have made a considerable progress – I may mention

Fig. 29. – Invitation letter from Bronstein to Majorana (21 April 1935), page 1 - AFM.

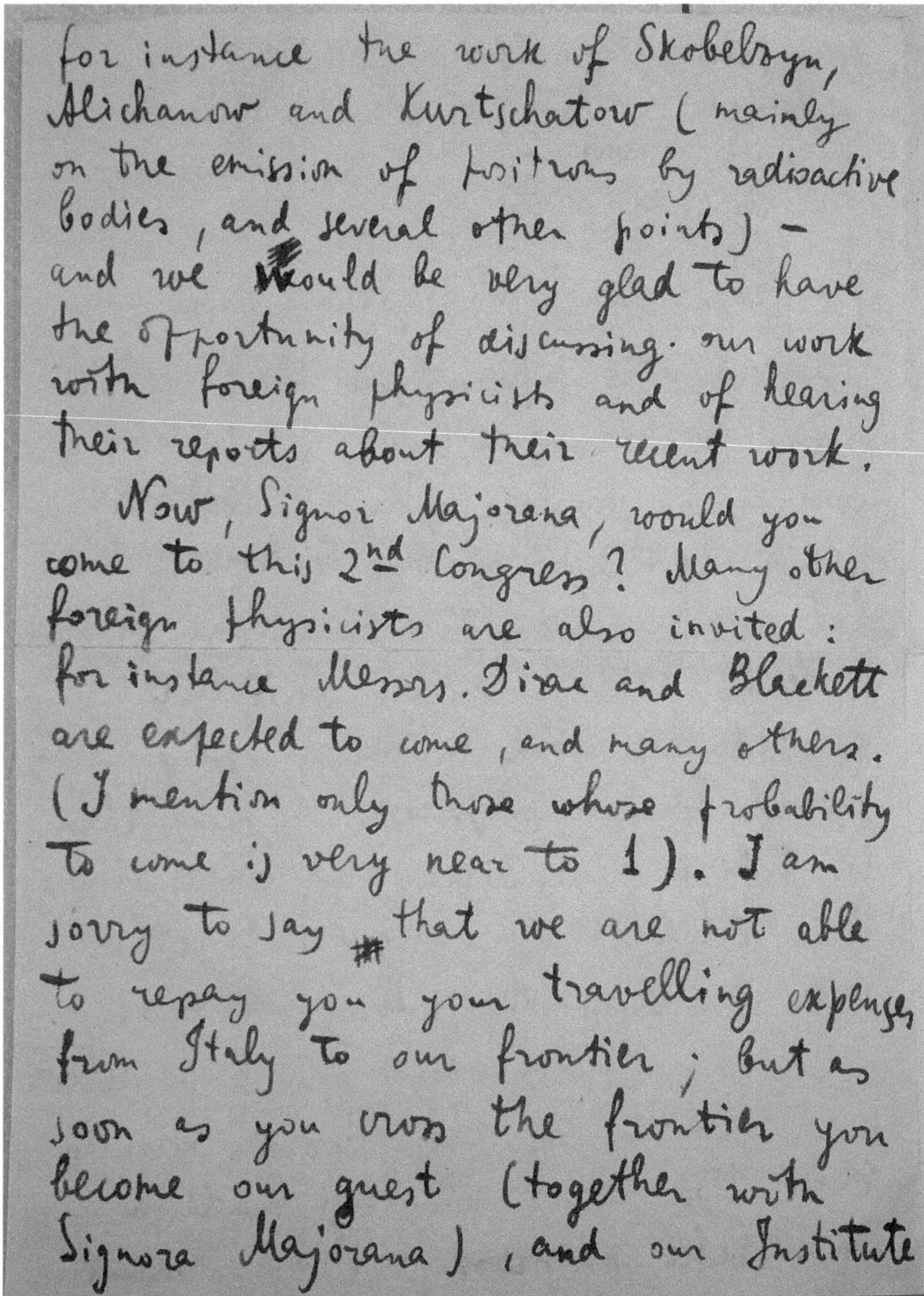

for instance the work of Skobelzyn,
Alichanow and Kurtschatow (mainly
on the emission of positrons by radioactive
bodies, and several other points) –
and we would be very glad to have
the opportunity of discussing our work
with foreign physicists and of hearing
their reports about their recent work.
Now, Signor Majorana, would you
come to this 2nd Congress? Many other
foreign physicists are also invited:
for instance Messrs. Dirac and Blackett
are expected to come, and many others.
(I mention only those whose probability
to come is very near to 1). I am
sorry to say that we are not able
to repay you your travelling expenses
from Italy to our frontier; but as
soon as you cross the frontier you
become our guest (together with
Signora Majorana), and our Institute

Fig. 30. – Invitation letter from Bronstein to Majorana (21 April 1935), page 2 - AFM.

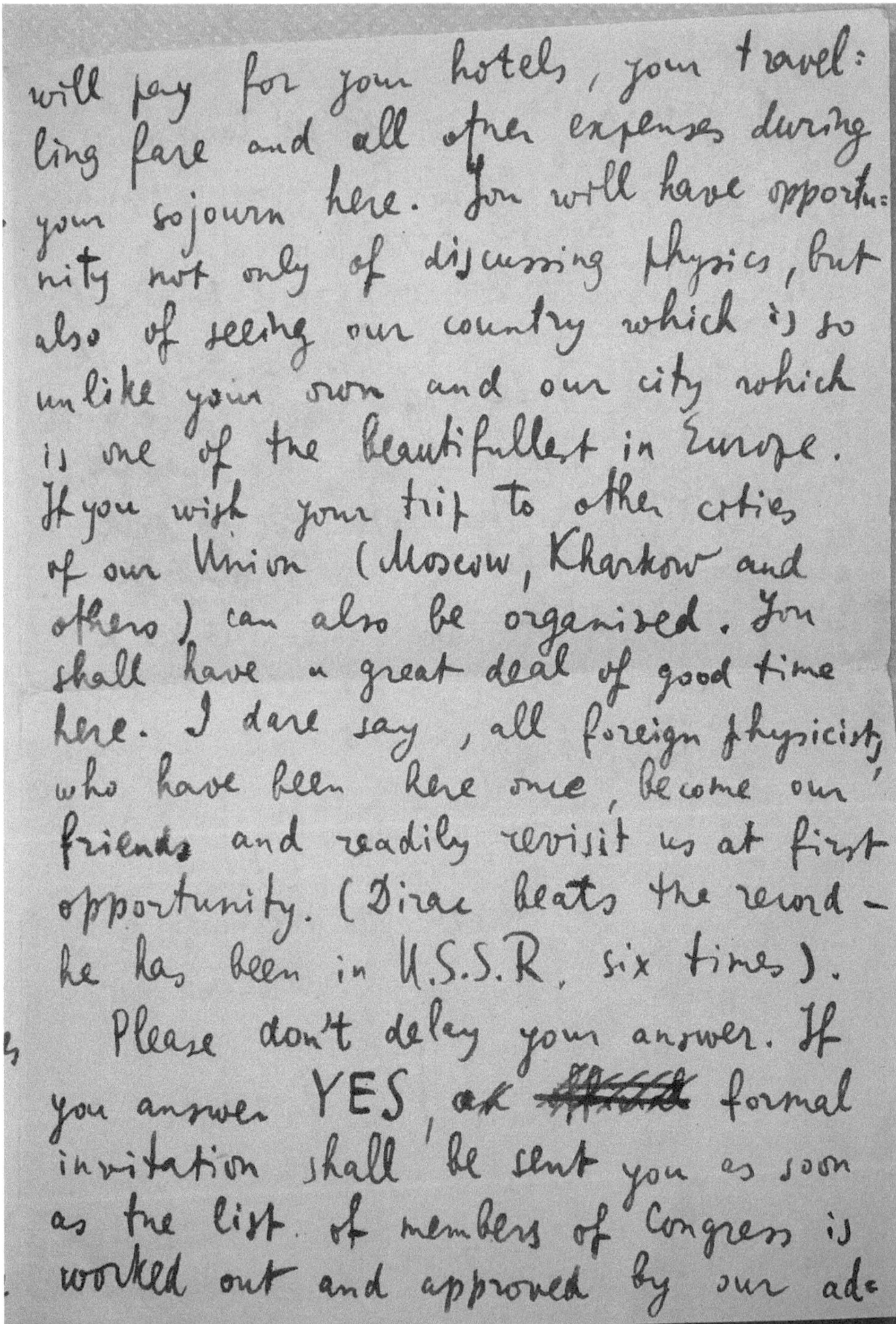

will pay for your hotels, your travel-
ling fare and all other expenses during
your sojourn here. You will have opportu-
nity not only of discussing physics, but
also of seeing our country which is so
unlike your own and our city which
is one of the beautifullest in Europe.
If you wish your trip to other cities
of our Union (Moscow, Kharkow and
others) can also be organised. You
shall have a great deal of good time
here. I dare say, all foreign physicists,
who have been here once, become our
friends and readily revisit us at first
opportunity. (Dirac beats the record -
he has been in U.S.S.R. six times).
Please don't delay your answer. If
you answer YES, a formal
invitation shall be sent you as soon
as the list of members of Congress is
worked out and approved by our ad-

Fig. 31. – Invitation letter from Bronstein to Majorana (21 April 1935), page 3 - AFM.

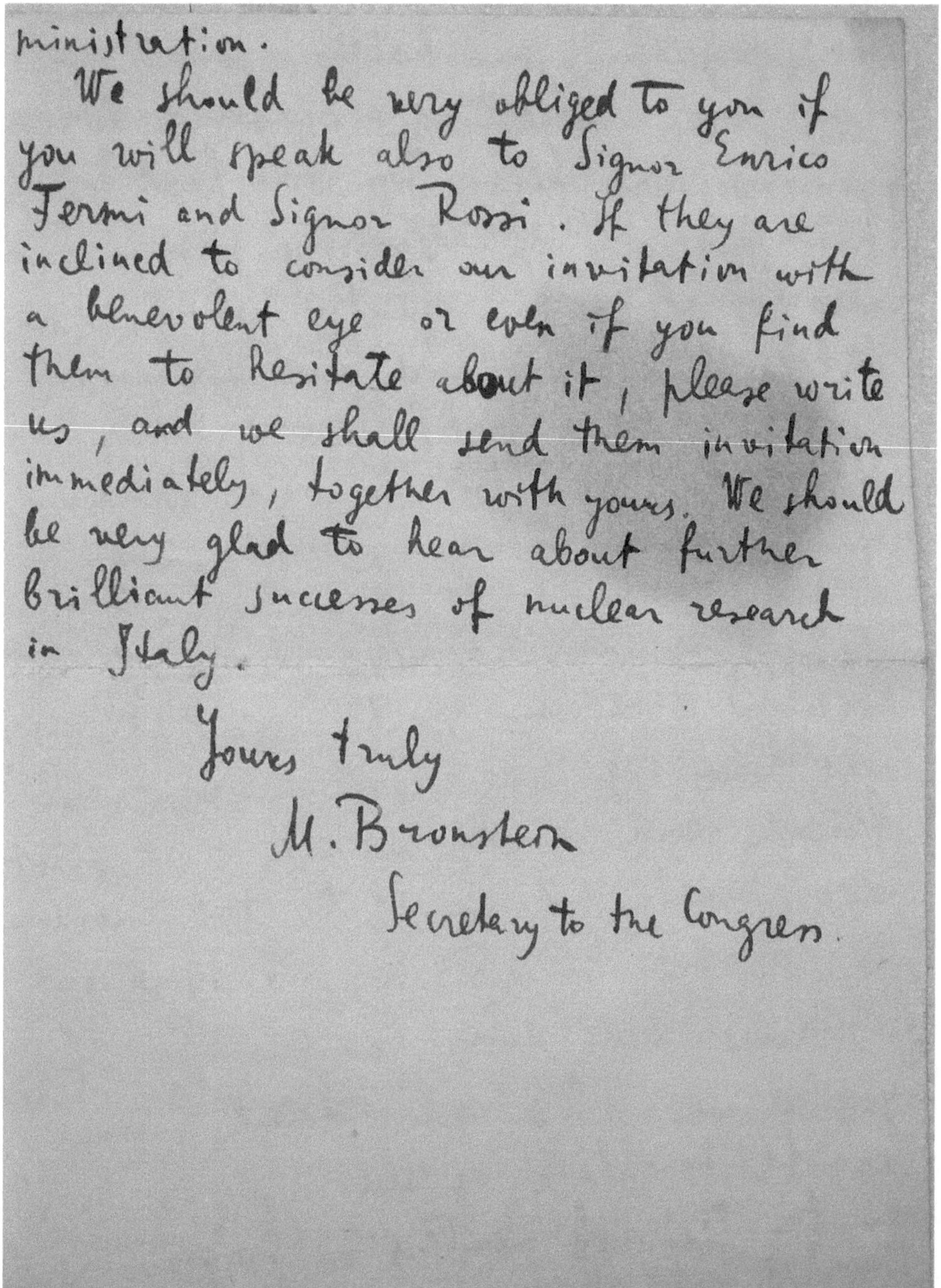

ministration.

We should be very obliged to you if you will speak also to Signor Enrico Fermi and Signor Rossi. If they are inclined to consider our invitation with a benevolent eye or even if you find them to hesitate about it, please write us, and we shall send them invitation immediately, together with yours. We should be very glad to hear about further brilliant successes of nuclear research in Italy.

Yours truly

M. Bronstein

Secretary to the Congress.

Fig. 32. – Invitation Letter from Bronstein to Majorana (21 April 1935), page 4 - AFM.

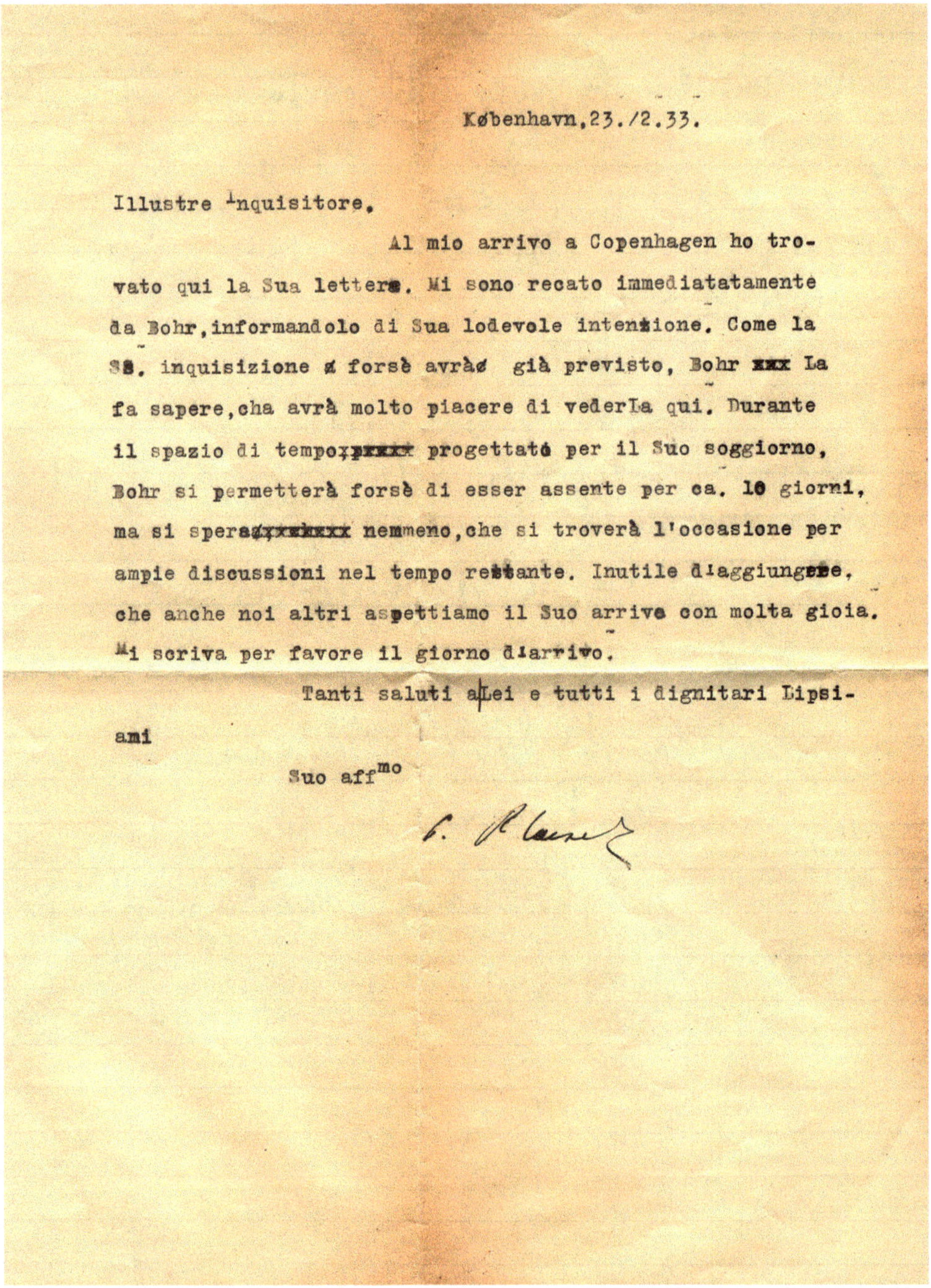

København, 23./2.33.

Illustre Inquisitore,

Al mio arrivo a Copenhagen ho trovato qui la Sua lettera. Mi sono recato immediatatamente da Bohr, informandolo di Sua lodevole intenzione. Come la SS. inquisizione forsè avrà già previsto, Bohr La fa sapere, cha avrà molto piacere di vederLa qui. Durante il spazio di tempo progettato per il Suo soggiorno, Bohr si permetterà forsè di esser assente per ca. 10 giorni, ma si spera nemmeno, che si troverà l'occasione per ampie discussioni nel tempo restante. Inutile d'aggiungere, che anche noi altri aspettiamo il Suo arrivo con molta gioia. Mi scriva per favore il giorno d'arrivo.

Tanti saluti a Lei e tutti i dignitari Lipsiami

Suo aff^{mo}

G. Placzek

Fig. 33. – Letter from Placzek to Majorana (23 February 1933) - AFM.

and he presented a brief summary of it. This must be said in order to demonstrate how careful Majorana was with this aspect of his professional activity, and how he did not need to be pushed to publish his results as has too often been told.

This total absence from the international scientific world of Ettore Majorana, whose name had now become officially associated with Heisenberg's and with the theory of the nucleus, appears even stranger if you consider that it coincides with the precise moment of the launching and development in Rome of intense activity in the Nuclear Physics field, above all by Enrico Fermi who, as we shall soon see, within less than a year, starting in October 1933 immediately after the Solvay Conference and less than four months after Majorana's return to Rome, arrived at the formulation of the theory of beta decay (December 1933), at the discovery of neutron-induced radioactivity (March 1934), and at the discovery of the neutron slowing down effect (October 1934).

Majorana, despite having become a few months earlier a recognised leader in the field, did not participate in these developments in Nuclear Physics in Rome. After his return to Rome Majorana fell into a long official scientific silence which can be considered as Majorana's first "disappearance", anticipating his final disappearance in 1938. This silence was momentarily interrupted only in 1937 by a single and last article of his concerning the formulation of the symmetrical theory of the electron and the positron in the framework of second quantisation [111]. After that nothing, apart from his appointment to the chair in Theoretical Physics at the Royal University of Naples for his "great reputation for remarkable expertise", a post he filled for less than three months and, finally, his disappearance at the end of March 1938 in circumstances still wrapped in mystery. The scientific, academic, personal, affair of Ettore Majorana is described in our publications [82, 83, 85].

At the beginning of his stay in Leipzig Majorana received, amongst other things, a postcard from Emilio Segrè dated 3 February 1933 (Fig. 34) and a letter from Enrico Fermi dated 11 February (Fig. 35, 36). This previously unpublished correspondence, together with the two letters from Bronstein and Placzek mentioned before, is kept in the Family Archive and has been made available to researchers thanks to the kind helpfulness of Ettore Majorana Jr. Firstly the correspondence bears witness to the friendly relations that existed at the time between Majorana on the one hand and Segrè and Fermi on the other but it also provides important information on scientific activity in Rome at the time. For example Segrè wrote: *Dear Ettorre, [...] There are no great changes here Fermi and I are writing on hyperfine structures. Ado [Amaldi] is making short waves or trying to, Rasetti Bi crystals and the others entsprechend.* Segrè used in friendly terms the epic-poetic version of the name "Ettorre", instead of "Ettore". The name of the famous Trojan hero appears about a hundred times as "Ettorre" because of metrical requirements in the translation of Homer's Iliad by the poet Vincenzo Monti, as exemplified in the famous line: "Tal fu d'Ettorre il favellar superbo" (Such of Hector was the proud speech).

Fermi in his letter expressed himself with great familiarity and affection stating among other things: *Dear Majorana, [...] In a few days I will send you a copy of the manuscript on hyperfine structures that Segrè and I are laboriously refining these days. Nothing*

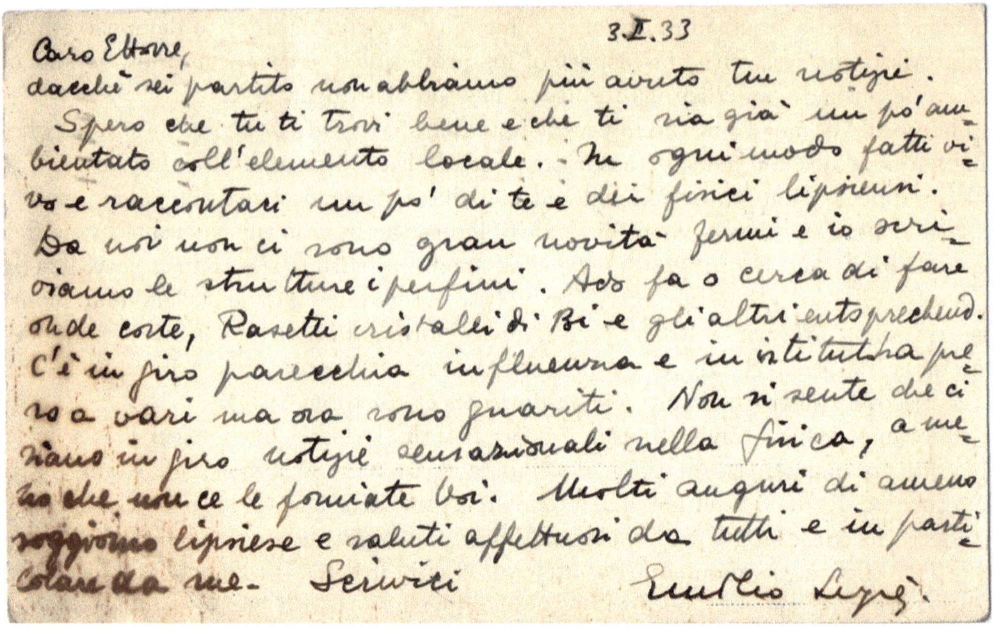
3.II.33

Caro Ettore,
dacché sei partito non abbiamo più avuto tue notizie. Spero che tu ti trovi bene e che ti sia già un po' ambientato coll'elemento locale. In ogni modo fatti vivo e raccontaci un po' di te e dei fisici lipsiensi. Da noi non ci sono gran novità. Fermi e io scriviamo le strutture iperfini. Ado fa o cerca di fare onde corte, Rasetti cristalli di Bi e gli altri [illegible]. C'è in giro parecchia influenza e in istituto ha preso vari ma ora sono guariti. Non si sente che ci siano in giro notizie sensazionali nella fisica, a meno che non ce le forniate voi. Molti auguri di ameno soggiorno lipsiese e saluti affettuosi da tutti e in particolare da me. Scrivici

Emilio Segrè

Fig. 34. – Postcard sent by Emilio Segrè to Ettore Majorana in Leipzig (3 February 1933) - AFM.

sensational is happening here. Not finding anything sensible to do in theory I have started working experimentally, and together with Rasetti we are trying out various methods to make existing techniques for radioactive measurements worse; there is no denying that we have had some success in this direction.

The postcard and the letter are transcribed in full and translated in the Appendix.

After Majorana's return to Rome there were no further direct contacts between Fermi and Majorana, as Majorana himself declared in an official document presented when competing for a university chair in 1937 [83, Fig. 63], indeed Majorana's relations with the whole via Panisperna group had deteriorated badly for reasons that are not easy to analyse because of the lack of documentary evidence. We cite a very significant episode.

On 9 November 1933 the Swedish Academy of Science announced that the 1932 Nobel Prize for Physics had been bestowed on Werner Heisenberg. Immediately all representatives of world culture sent their congratulations in the most various forms. Heisenberg kept all the messages he received in a file in his personal archive. Amongst them a telegram from Rome (Fig. 37), transmitted by the Deutsche Reichspost and dated 11-XI-33 and written in German in a very formal and cold style, stands out. It read: "HERZLICHSTE GRATULATIONEN CORBINO FERMI RASETTI SEGRE AMALDI WICK" (most heartfelt congratulations ...). We can note that the order of the signatures rigorously follows the degree of academic seniority at the Physics Institute in Rome. Ettore Majorana who was in a Rome at that time is not included in the list, not even in last

ISTITUTO DI FISICA
DELLA
R. UNIVERSITÀ DI ROMA
VIA PANISPERNA N. 89-A

Roma 11 febbraio 1933

Caro Majorana,

ti ringrazio della tua lettera, da cui mi pare di capire che ti sei già discretamente ambientato nell'ambiente lipsiese.

Ti manderò tra qualche giorno una copia del manoscritto sulle strutture iperfine, che Segrè e io stiamo faticosamente elaborando in questi giorni.

Qui non sta succedendo nulla di sensazionale. Non trovando nulla di sensato da fare nella teoria io sto mettendomi a lavorare sperimentalmente, e insieme a Rasetti cerchiamo vari metodi per peggiorare le tecniche esistenti sulle misure radioattive; in questo indirizzo non si può negare che abbiamo qualche successo.

Attendiamo da Peierls notizie sopra il convegno che avete in questi giorni.

Dovresti dire da parte mia a Bloch, se non glielo ha già detto Peierls, che non c'è nessuna difficoltà perchè egli venga qui dal primo ottobre, perchè io non ci sono di solito ancora, ma l'istituto è aperto. Che lo sconsiglio

Fig. 35. – Letter from Fermi to Majorana in Leipzig (11 February, 1933) - first page - AFM.

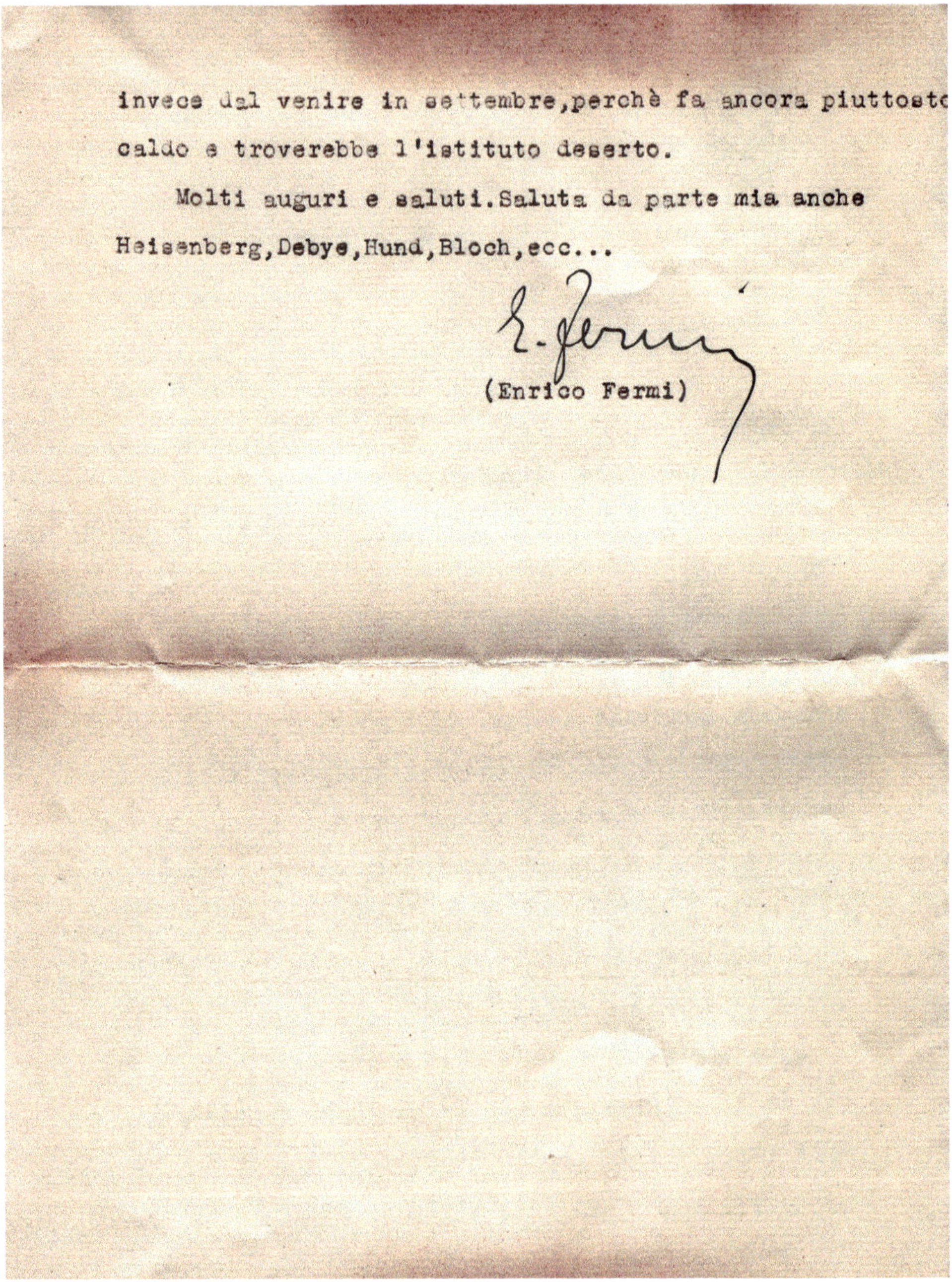

invece dal venire in settembre, perchè fa ancora piuttosto caldo e troverebbe l'istituto deserto.

Molti auguri e saluti. Saluta da parte mia anche Heisenberg, Debye, Hund, Bloch, ecc...

E. Fermi

(Enrico Fermi)

Fig. 36. – Letter from Fermi to Majorana in Leipzig (11 February, 1933) - second page - AFM.

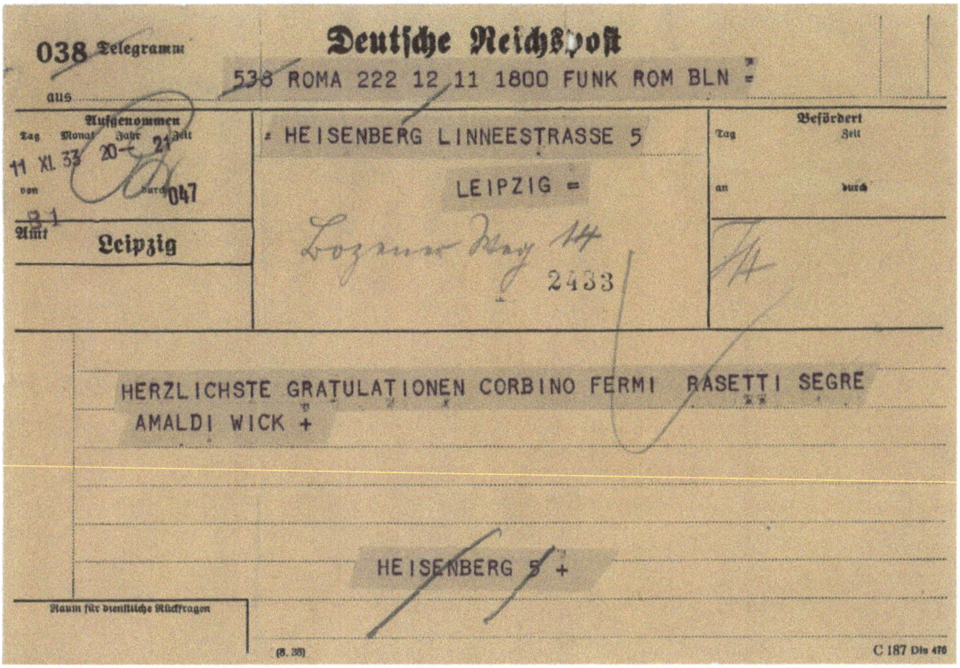
038 Telegramm
Deutsche Reichspost
538 ROMA 222 12 11 1800 FUNK ROM BLN =
aus
Aufgenommen
Tag Monat Jahr Zeit
11 XI 33 20- 21
von durch 047
Amt Leipzig
= HEISENBERG LINNEESTRASSE 5
LEIPZIG =
2433
Befördert
Tag Zeit
an durch
HERZLICHSTE GRATULATIONEN CORBINO FERMI RASETTI SEGRE
AMALDI WICK +
HEISENBERG 5 +
Raum für dienstliche Rückfragen
C 187 DIN 476

Fig. 37. – Telegram of congratulations for being awarded the Nobel Prize for Physics sent to Heisenberg from Rome on 11-XI-33 and signed "Corbino, Fermi, Rasetti, Segrè, Amaldi, Wick" - HA.

place. But Majorana, who had left Leipzig at the beginning of August three months before, nevertheless also sent his own "Gratulationen" on the same date, of course in his own style, as we have already reported in [83]. His very intense message (Fig. 38, 39) is written in very touching Italian on the two sides of a very small personal card where in the printed heading the title "Dr." was crossed out by a pen stroke. This is the text: *Dear Professor, Allow me (if you haven't forgotten me!) to express to you my ardent best wishes on the occasion of the new solemn acknowledgement of your prodigious work. With deep admiration Your Ettore Majorana.*

Despite the deterioration in his personal relations with Majorana, Fermi nevertheless was certainly fully convinced that the approach followed by Heisenberg and Majorana in tackling the problem of nuclear structure, according to which only protons and neutrons were considered fundamental constituents of the nucleus, excluding electrons, was physically sound. This is shown for example by a report he made to the SIPS in October 1933 [F112]. Naturally the question of beta decay remained open. On this point Fermi at first embraced the idea adopted by Heisenberg according to which the neutron was a complex structure, composed of a proton and an electron, without further analysing the matter. In fact at that time it seemed to Fermi that the problem of beta decay could only be solved by the arrival a new theory, different from ordinary quantum mechanics, according

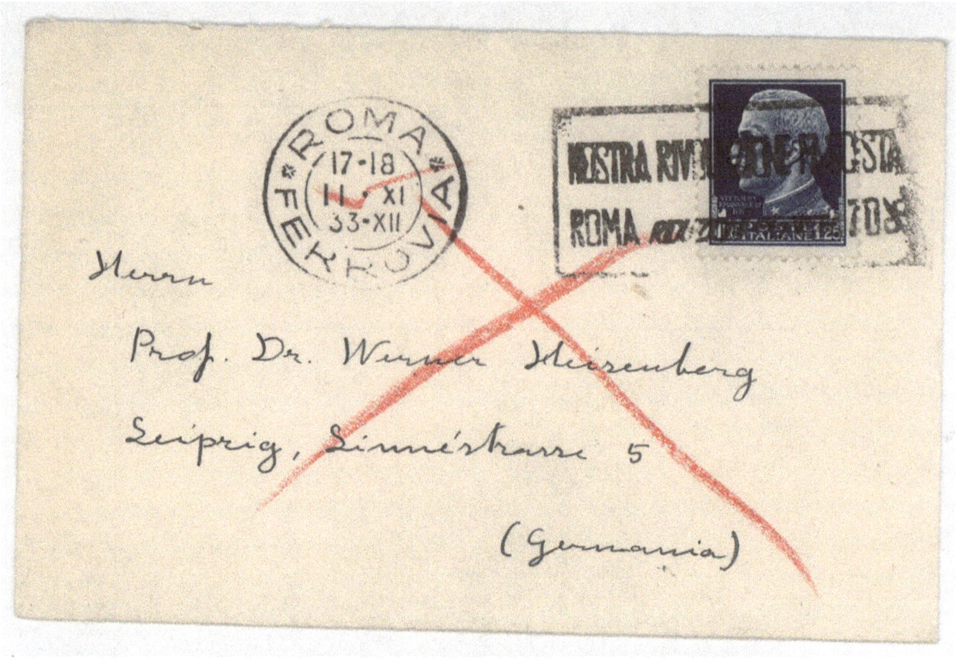

Fig. 38. – Business card, dated Rome 11.11.1933, sent by Majorana to Heisenberg when he was awarded the Nobel Prize for Physics, envelope - HA

to a position that in any case already existed and was commonly held amongst the various participants at the Rome Conference in 1931, and so at the moment it could not really be addressed. Indeed Fermi observed in the report to the SIPS in 1933 [F112], contrasting Heisenberg and Majorana's new ideas to the now superseded ones which envisaged the nucleus composed of protons and electrons: "Instead, according to the theories of Heisenberg and Majorana protons and neutrons must be considered fundamental elements of nuclear structure. Neutrons could be thought of as being in their turn constituted of the union of a proton with an electron. This is probably necessary to explain the possibility of a disintegration with emission of beta rays. It would in any case be an aggregation occurring according to laws that are *different* from those of ordinary quantum mechanics [our italics]: just as ordinary mechanics ceases to be valid to describe the behaviour of electrons in the change of scale from ordinary mechanics to atomic mechanics, so the new change of scale from atomic phenomena to nuclear phenomena would make a new change necessary. The details of course are unknown to us for now : instead it seems that ordinary quantum mechanics is adequate to describe the behaviour of relatively heavy corpuscles such as protons and neutrons, even inside the nucleus. Indeed, due to their considerable mass, they have fairly small velocities, even in nuclear orbits, compared to that of light, so that relativistic corrections of their motions always remain of secondary importance".

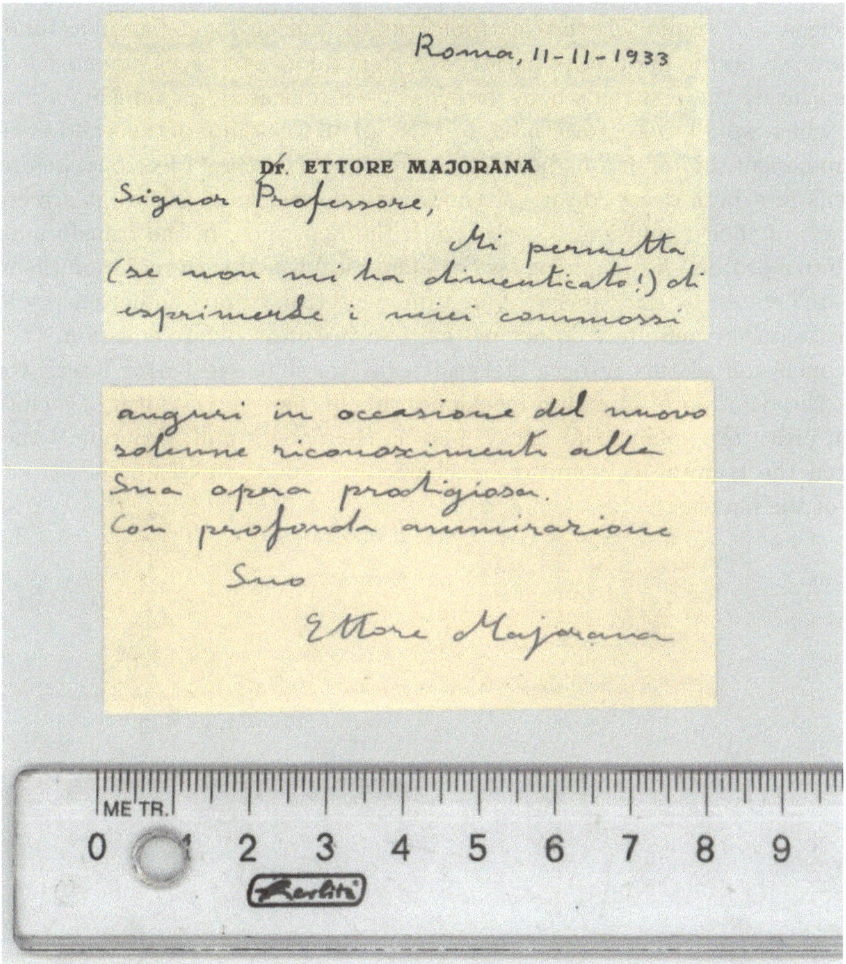

Roma, 11-11-1933

Dr. ETTORE MAJORANA

Signor Professore,

Mi permetta (se non mi ha dimenticato!) di esprimerle i miei commossi auguri in occasione del nuovo solenne riconoscimento alla Sua opera prodigiosa.

Con profonda ammirazione

Suo

Ettore Majorana

Fig. 39. – Business card, dated Rome 11.11.1933, sent by Majorana to Heisenberg when he was awarded the Nobel Prize for Physics, front and back.

This was Fermi's position with regard to beta decay a few days before taking part in the VIII Solvay Conference which took place in Brussels from 22 to 29 October 1933. He was convinced that the behaviour of electrons inside the nucleus, and in consequence beta decay as well, was governed by laws totally different from those of quantum mechanics and that therefore a solution to this could not be found within this theory but should be sought in a new theoretical setting.

Instead, immediately after the Solvay Conference, Fermi's position was completely different. Indeed when in December 1933 Fermi tackled and solved the problem of beta decay he did not seek a new theory alternative to quantum mechanics with which to explain this phenomenon. Just as Heisenberg and Majorana, in order to construct a theory of the nucleus, had “removed” the electrons from the nucleus, understood as fundamental components, so Fermi, in order to construct a theory for beta decay, “removed” the electrons from inside the neutron where they had been relegated, introducing a remarkable new idea which would then prove to be the basis of all weak interactions and in general of all the fundamental interactions between elementary particles. Electrons, before exiting the nucleus in a beta decay, do not exist inside the nucleus as such but are created at the moment of their expulsion, together with the neutrinos, in the transformation of a neutron into a proton. A full analogy is therefore established with radiation theory where photons are created by the passage of an atomic electron from a higher-energy level to a lower one, and they definitely do not pre-exist in the atom before emission.

The conclusion of this project of Fermi's, as we shall see better later, would also represent the triumph of quantum mechanics, in this case in the form of quantum field theory, in order to explain beta decay, just as Heisenberg and Majorana's theory had represented the triumph of quantum mechanics in order to explain the structure and stability of the nucleus.

6 The Solvay Conference and Fermi's theory of beta decay: October-December 1933

6.1 The Seventh Solvay Conference

Between 22 and 29 October 1933 the Institut International Solvay organised in Brussels the "Septième Conseil de Physique" or, as it would later be remembered, the "Septième Solvay Conference". In the tradition of these Conferences the theme under discussion, on which the greatest experts in the field were summoned to report, had to be a topical subject, of great scientific interest and with big problems still open. The first Conference dated back to 1911 and significantly it was dedicated to the "Theory of radiation and quanta", a field in which the first developments were beginning to be seen in those very years.

The aim of the Solvay Conferences was to take stock of the situation in a specific field, showing both the successes achieved and the difficulties still present, in the hope that the state of the discipline chosen as the theme of the Conference could make successful progress by means of direct discussions between the various participants. It could be said that, all told, the Rome Conference in 1931 had been organised in the same spirit.

In 1933 the general theme of the Solvay conference was identified as "Structure and properties of atomic nuclei" ("Structure et propriétés des noyaux atomiques"). Even if this Conference took place a little under two years after the Conference in Rome, the propulsive drive which Nuclear Physics enjoyed in those years was such as to justify such a choice. As Paul Langevin (1872-1946), Chairman of the Scientific Committee, recalled in the opening speech of the Conference, the decision "to restart the Rome Conference two years later" had been taken earlier, "on the basis of the interest of the subject and counting on the fact that eighteen months later, that is to say today, the subject would have been sufficiently renewed for us to have new things to tell each other, taking as read the things that had been the object of the meetings in Rome".

History would prove this choice to be absolutely right. Indeed in that two year interval not only, as Langevin observed, had "the two latest offspring of nuclear and corpuscular Physics been seen to appear in succession: the neutron and the positive electron", but the bases were also laid for new breakthroughs in Nuclear Physics which came to light precisely following this Solvay Conference.

Fourteen Nations were represented at the Conference (Enrico Fermi was present for Italy) for a total of 35 participants. As was observed in the opening speech, the par-

F. Guerra and N. Robotti, *The Lost Notebook of ENRICO FERMI*,
https://doi.org/10.1007/978-3-319-69254-8_6

ticipants were distributed "in equal parts between experimentalists and theoreticians", because "it was necessary to compare very closely the efforts of the former and the latter".

The subjects of the six invited papers and the names of their speakers were particularly significant. Here below we give the list of the talks in the order in which they were given:

J. D. Cockcroft – "La Désintégration des éléments par des protons accélérés" (The disintegration of the elements by accelerated protons),

J. Chadwick – "Diffusion anomale des particules α. Transmutation des éléments par des particules α. Le neutron" (Anomalous diffusion of α particles. Transmutation of elements by α particles. The neutron),

I. Curie and F. Joliot – "Rayonnement pénétrant des atomes sous l'action des rayons α" (Penetrating radiation of atoms under the action of α rays),

P.A.M. Dirac – "Théorie du positron" (Theory of the positron),

G. Gamow – "L'origine des rayons α et les niveaux d'énergie nucléaires" (The origin of α rays and nuclear energy levels),

W. Heisenberg – "Considérations théoriques générales sur la structure du noyau" (General theoretical considerations on the structure of the nucleus).

As can be seen from the titles of the talks, the subjects dealt with completely covered the various sectors in which Nuclear Physics had developed in recent years, in particular since the Rome Conference. The discussions, which took place after the talks and were reported printed in the Proceedings of the Conference, were extremely significant.

Of the various questions dealt with during the Conference, both in the invited papers and in the discussions, we will restrict ourselves to only recalling those that represented a sort of backdrop for later and more immediate developments in Nuclear Physics, in particular for the formulation of Fermi's theory on beta decay (December 1933), for the discovery of alpha-particle–induced radioactivity by F. Joliot and I. Curie (January 1934) and finally for Fermi's discovery of neutron induced radioactivity (March 1934).

First we would like to point out the atmosphere of complete trust in the Heisenberg-Majorana theory and the broad consensus it achieved. Immediately after Heisenberg's talk, during which enormous credit was attributed to Majorana, so much so that in reality Heisenberg directly expounded Majorana's formulation of nuclear structure and exchange forces, the debate that followed did not touch in any way the principles of the theory, on which there was total agreement, but turned to other problems left open by this theory, such as for example beta decay. Pauli intervened explicitly on this point [127] personally reproposing the hypothesis of the neutrino as a solution for the continuous energy spectrum presented by the electrons emitted in beta decay, and recalling that he had put forward this hypothesis already in June 1931 on the occasion of a Conference in Pasadena, and that for the new particle "Mister Fermi has proposed the name «neutrino»", to distinguish it from "heavy neutrons".

In Pauli's comments the word «neutrino» attributed to Fermi appeared in print for the first time. He must have coined it starting from the term neutron, initially used by Pauli, and modifying it, according to Italian usage, in the diminutive form, after Chadwick had demonstrated the existence of the neutron, a particle which instead belonged to the tradition of the Cavendish laboratory.

The proposal of the neutrino, which at the Rome Conference, as we saw, did not receive any consideration even if it remained noted in the Proceedings, here instead opened up a brief debate on the possible experimental methods with which to try to identify this hypothetical particle, aware however of the intrinsic difficulties of this type of experiment.

As Chadwick observed: "'It is certain that the neutrino, if it exists, will be extremely difficult to identify".

History, as is well known, would prove Chadwick right: even today the true nature of the neutrino is an open problem and the subject of intense research. In any case it is a fact that throughout the Conference, in the absence of any experimental verification, the neutrino was not considered by those who spoke on the problem of beta decay to be one of the possible "key players" in this decay and was almost ignored.

Another aspect to emerge from the Solvay Conference and which deserves to be mentioned is the scepticism shown with regard to the proposal of considering the neutron as a compound structure, made up of a proton and an electron, in order to explain the origin of the electrons present in beta decay, as was suggested for the first time by Iwanenko. The clearly alternative proposal of considering the proton as a compound structure, made up of a neutron and a positive electron, as argued by Joliot and Curie to explain some nuclear transmutations they had identified produced by alpha particles on light elements (Al, B). was also greeted with the same attitude. As we have said before, in these transmutations a same nucleus seemed to have two different transmutations, one with the emission of protons and one with the emission of neutrons and positrons.

In both alternatives, either a compound neutron or a compound proton, difficulties immediately arose as was be observed by many.

The hypothesis of the compound neutron, as well as seeming to be incompatible with the values of the masses involved and the mass defects then known [158, page 299], faced insuperable difficulties from the point of view of spin and statistics. As Chadwick commented for example [158, page 103], both the spin of the neutron and the spin of the proton had a value 1/2, in $h/2\pi$ units, as obtained from the law that governs the spin of light nuclei. Therefore, if the hypothesis of the compound neutron was to be maintained, spin zero would have to be attributed to the electron and Bose's statistics applied to it, "which is contrary to the way it behaves in a free state". Clearly this objection would have fallen if it had been hypothesised that the neutron also contained, as well as a proton and an electron, a neutrino with a spin, presumed to be 1/2, which could combine vectorially with the spin of the electron, thus resolving the matter: but no one made that observation publicly, not even Fermi or Pauli himself, although both were present. Heisenberg too was well aware of the difficulties pointed out by Chadwick. Referring to the hypothesis of compound neutron, that he himself had once maintained, he observed during his talk: "It is difficult to give a precise meaning to the assertion that a neutron is made up of an electron and a proton, because, interpreting it literally, one would be led to incorrect conclusions with regard to the spin and statistics".

The hypothesis, maintained by the Joliot-Curies, of the compound proton made up of a neutron and a positive electron also met with the same type of difficulty. In this case even if, on the basis of the values then known, the mass of the proton was smaller than the sum of the mass of the neutron and positron, the bond energy between the neutron and the positron was too small for these two particles together to be able to form a stable structure. Moreover, as was pointed out again by Chadwick, in order to guarantee both for the proton and the neutron a spin equal to 1/2 it was necessary to "admit that the spin of the positive electron will be either 0 or 1", with the consequence that "the reciprocal distinction of the positive electron and of the negative electron would become difficult to understand". Here too, as for the compound neutron, this objection linked to the spin would have been answered by resorting to the neutrino hypothesis. Instead, as Chadwick commented: "It therefore seems that all the facts suggest that the neutron and the proton are both elementary particles".

To further complicate the picture of the complexity, or non-complexity, of the proton or the neutron there was also the position adopted by Francis Perrin (1901-1992). Perrin's reasoning was of a general nature and was based on the possibility of the materialisation of a gamma ray in an electron-positron pair and vice versa. According to Perrin, if it was maintained that the neutron was composed of a proton and a negative electron, it was also necessary to maintain that the proton was made up of a neutron and a positive electron, since the two reactions:

$$ {}^{1}_{0}n \leftrightarrow {}^{1}_{1}p + e^{-}, \quad {}^{1}_{1}p \leftrightarrow {}^{1}_{0}n + e^{+} $$

"necessarily are both possible or both impossible". "Without doubt there is therefore a complete symmetry from the point of view of the complexity between the neutron and the proton, since these particles were either both elementary and independent, or both complex".

So the problem of understanding the ultimate nature of the neutron and the proton and the link between these two particles still seemed completely open, and with it also the problem of the presence of electrons in beta decay and of "transmutation positrons" in the nuclear reactions seen by Joliot and Curie. In any case, while in the case of beta decay it seemed obvious that the origin of the electrons had to be connected to process that occurred inside the nucleus, the situation for the Joliot-Curies' "transmutation positrons" was different. Indeed there seemed to be a strong objection to the hypothesis that they came from a nuclear transformation, which Meitner had raised immediately after the Joliot-Curies' talk, based on some research into transmutations of aluminium and fluorine after bombardment with alpha particles that she had carried out using a Wilson chamber.

Although this research had confirmed the Joliot-Curies' result according to which positive electrons were emitted and that these reached a far higher energy than that of the negative electrons that accompanied them, if the number of positive electrons emitted was compared with the number of recoil protons produced at the same time by any neutrons present (remember that, in a Wilson chamber, the neutrons, not having a charge, were identified by the protons that they put into motion in the collision with hydrogenated substances) it is shown that only positive electrons were present. Meitner stated that "it was not possible to identify any neutrons", not even in the case of aluminium where the number of positrons detected was four times greater than in the case of fluorine. Meitner commented: "This is not favourable to the idea that, in this case, the emission of the neutron occurs concurrently with that of the positive electron". With this observation, Meitner certainly was not thinking of a delay process, inside the nucleus, between the emission of neutrons and the emission of positrons, as would be suggested by Francis Perrin, as the discussion continued, and as would soon be discovered by the Joliot-Curies following another route. On the contrary. For Meitner the fact that, in none of the nuclear transmutations observed had the simultaneous emission of a neutron and a positron been seen, meant that in reality positrons, unlike neutrons, did not originate directly from the nucleus, as the Joliot-Curies had instead maintained with their idea of "transmutation positrons". For example it could be supposed, as Blackett had suggested during a discussion, that these positrons were created at the moment and that therefore they came, for example, from the "internal conversion" of a γ ray in an electron-positron pair.

At a certain point this hypothesis was also put forward, almost as a "fallback" position, by Irène Curie who, however, pointed out that with this hypothesis it could not be explained why the number of negative electrons exiting the aluminium was smaller than the number of positive electrons as observed experimentally. Also, if the idea that the positrons did not come directly from inside the nucleus but from the materialisation of a gamma ray was accepted, there still remained the problem that Joliot and Curie had tried to overcome precisely by means of the "transmutation positrons" hypothesis, according to which a nucleus, such as that of aluminium, emitting only a neutron would be transformed into an "unknown" element.

It should be said that this hypothesis on the origin of positrons as due to the materialisation of a gamma ray would be considered, immediately after the Solvay Conference, by Leo Nedelsky (1903-2006) and J. Robert Oppenheimer (1904-1967) in a letter to the Editors of "Physical Review" on 18 November 1933 and completely demolished [118]. After finding a general expression for the probability of the creation of an electron-positron pair, starting from a nuclear gamma ray and applying it to the "observations of Curie and Joliot" concerning aluminium, Nedelsky and Oppenheimer concluded as follows: "The positives observed in aluminum are, from the point of view of present theory, altogether unexplained"'. This fact clearly restricted the margins for manoeuvre for a possible interpretation of the origin of positrons. Indeed, dropping the "materialisation" hypothesis, if one wished to explain the origin of positrons attention should be concentrated exclusively on the heavy particles that compose the nucleus (neutrons and protons), which is what Fermi would do, the following month, formulating his theory of beta decay.

In any case, as indicated before, a new hypothesis for the mechanism leading to the emission of positrons had emerged during the discussion opened up at the Solvay Conference by the Joliot-Curies' talk. It could be assumed, despite Meitner's opposition, that positrons not only originated from the nucleus but that their emission derived from a phenomenon of radioactive decay, of the same type as natural beta decay, with the difference that in this case it concerned the emission of positive electrons. This idea was suggested by Francis Perrin, immediately after Meitner's comments, without however identifying in the observations expounded by Meitner on the "non-simultaneity" between the emission of the neutron and that of the positron, a possible support for this idea of his. Perrin started from the observation that, on the basis of the results of Joliot-Curie, the energy of the emitted positrons presented a continuous energy spectrum, similar to that of the natural beta spectrum. According to Perrin, in order to interpret the energy differences between the various positrons it could be supposed, following the Joliot-Curies' idea, that the positrons were emitted simultaneously with the neutrons and that the overall energy was distributed between the two particles. This hypothesis, however, as Perrin noted dealing directly with the case of aluminium, could not stand up because the neutrons always had an energy less than 1 MeV which was thus too small to be able to compensate the energy differences between the positrons, which were less than or equal to 2.6 MeV, with the energy differences between the neutrons. Therefore, according to Perrin, as reported in the Proceedings: "It therefore seems reasonable to suppose that the mechanism proposed by Mr Joliot breaks down into two successive emissions, first of a neutron, and then of a positive electron, with the intermediate formation of an unstable nucleus (in the case of aluminium); this nucleus will thus present a radioactivity due to positive electrons and it will no longer be surprising that in this case one finds a continuous spectrum as for the β rays in natural radioactivities".

After having observed that also in some materialisation processes of a gamma ray into an electron and positron pair an energy deficit had been found, Perrin concluded his comments as follows, as reported in the Proceedings: "We are therefore led to suppose that the phenomena of radioactivity with the emission of positive electrons and perhaps the phenomena of materialisation of pairs of electrons bring into play energy mechanisms analogous to those observed in natural beta radioactivity" [158, page 180].

We found the original transcript of Perrin's comments in the "Archives Joliot-Curie, Boite JC4", typed on tissue paper, and sent, probably copied by Langevin to the Joliot-Curies, together with the transcripts of other comments made after their talk. On this text (delivered on 1 February and sent back on 8 February 1934) a series of changes, cancellations and clarifications appear, handwritten in pencil by Perrin, with a view to the publication of the Proceedings, obviously after the discovery of alpha-particle–induced radioactivity, and which were all included in the printed text. The real difference between the original text and the one published later is that in the published text the chemical symbol of the new radioactive element which should be produced is explicitly added, ${}^{30}_{15}P$, as it was later published by the Joliot-Curies on 15 January 1934 in the article in which they would communicate the discovery of alpha particle induced radioactivity.

Pauli and Bohr immediately commented on a possible analogy between the emission of positrons and natural beta decay suggested by Perrin, with rather different opinions.

While Pauli assumed a very doubtful stance, based on the conviction supported by Meitner, that positrons, unlike beta electrons, did not emerge from the nucleus, Bohr proved to be more open to the possibility. Pauli observed: "The conclusion according to which there is a close analogy between the emission of positrons due to artificial disintegration and the spontaneous emission of β rays does not strike me as being very sound. In the first case one can say that the particles are produced at the periphery or the gates of the nucleus strictly speaking, while the β rays seem to come from the nucleus itself". Instead Bohr commented: "It is a problem of prime importance to know whether the energy is conserved or not in the processes in which aluminium is bombarded by α particles. Without doubt, the observation that positrons do not all have the same velocity is not in itself an argument against energy conservation, since we do not yet know how the emission of positrons is produced. If, as Mr Joliot supposes, the positrons really came from inside the nucleus, the circumstances would be very similar to those of β rays".

So the origin of Joliot and Curie's positrons, like that of the negative electrons emitted in natural beta decay, represented a thorny problem to solve. Anyway, looking hard at these results, it seems that most of the premises necessary to arrive at the formulation of Fermi's theory of beta decay were already developed. The fact is that after the Solvay Conference it was clear to Fermi that the nucleus is made up exclusively of protons and neutrons, according to Heisenberg and Majorana's theory. So from this perspective for Fermi the electrons present in beta decay are created at the moment of their expulsion together with the neutrinos since they cannot be inside the nucleus. Another idea amply discussed at the Solvay Conference that Fermi would take into careful consideration and that he would try to incorporate into his theory of beta decay, is the idea of "transmutation positrons", understood as the product of a nuclear reaction. Perrin's suggestion, according to which the positron could be the product of a positive beta decay, must also have influenced Fermi while he was developing his theory on beta decay.

Indeed this theory, as we shall see, as well as interpreting negative beta decay, was potentially also able to explain a possible positive beta decay, predicting in fact a symmetrical position between the neutron and the proton and thus the possibility of transforming a neutron into a proton and vice versa.

6.2 Fermi's theory of beta decay

As we have revealed several times, immediately after the Solvay Conference Fermi tackled and solved the problem of beta decay.

The solution he arrived at was published in a first short version in "La Ricerca Scientifica" in 1933, towards the end of December [F115], and in an extended form in "Il Nuovo Cimento" in January 1934 [F116] and in the same month in "Zeitschrift für Physik", where 15 January 1934 appears as date of arrival [F117]. The difference between the three versions lies in the fact that all the mathematical passages were given only in the last two publications, which were essentially identical apart from the addition of the number "I" in the title of the German version, to indicate that it was only the first part

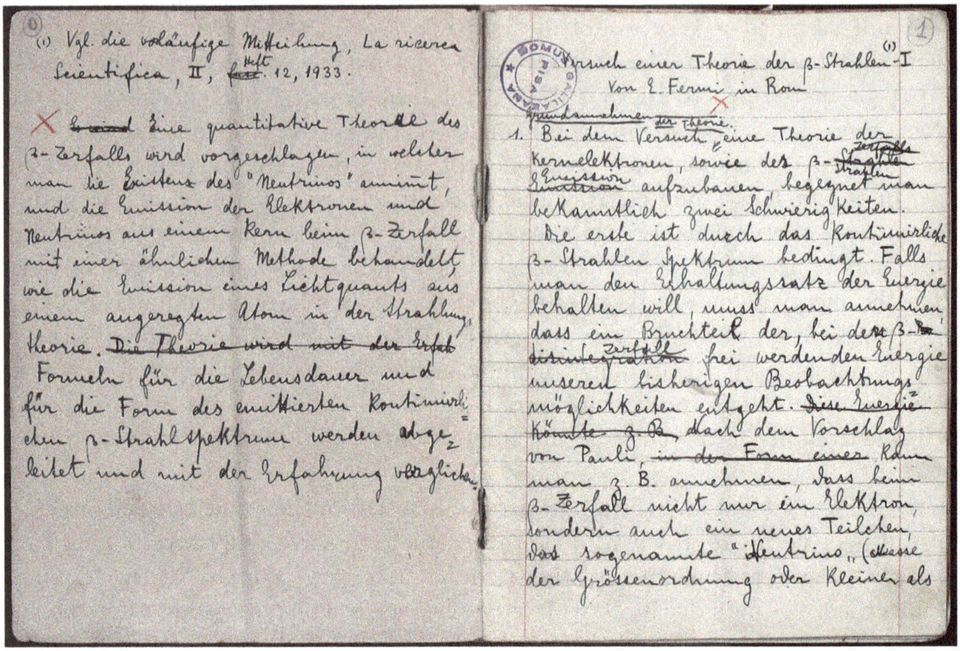

Fig. 40. – First page of the manuscript of the article "Versuch einer Theorie der β-Strahlen. I" [F117], in notebook 17 (1934) - FDG.

of a much larger study, which however was not developed further by Fermi. Moreover in the Italian version, along with Heisenberg's first paper on the theory of nuclei [90], Majorana's article in Zeitschrift [109] was also cited while in the German version this latter paper was inexplicably expunged, in practice depriving Majorana of an important acknowledgement of his contribution. Finally, another difference concerned the value of the "proportionality constant g" introduced by Fermi, which would later be called the "Fermi constant", which in the first communication was estimated as equal to $5 \cdot 10^{-50}\text{cm}^3 \cdot \text{erg}$, while in the other two articles it was estimated as equal to $4 \cdot 10^{-50}\text{cm}^3 \cdot \text{erg}$, in the units of measure then used.

Fermi's original manuscript of the article in German (Fig. 40) is kept in the Domus Galilaeana in Pisa, while a typescript of the same article with formulae inserted by hand by Fermi, but with the two first pages missing, can strangely be found amongst the papers of Giovanni Gentile Jr at the Department of Physics of the Sapienza University in Rome.

As Fermi stressed starting from his first paper on the subject published in "La Ricerca Scientifica", one of the most troubling problems raised by beta decay was the fact that this type of decay, unlike alpha decay, presented a continuous energy spectrum. The electrons exited directly from the nucleus, all with different energies which varied with continuity within very broad limits.

To solve this problem in formulating his theory Fermi, as a starting point, adopted the hypothesis of the neutrino, embracing Pauli's idea which we have already mentioned, according to which in β emission a new particle was emitted at the same time as the electron. This particle (later called neutrino), was electrically neutral and with a mass of the order of magnitude of the mass of the electron or smaller. In this way, as Fermi stressed, following in full Pauli's suggestion, it could be assumed that "the energy liberated in the process would be shared in any case between the two corpuscles so that the energy of the electron could take all values from zero up to a certain maximum" [F116], as was seen in experiments. With regard to any objections about the fact that this new particle had never been revealed experimentally Fermi expressed himself as follows: "The neutrino on the other hand, because of its electrical neutrality and its very small mass, would have such a high penetrating power as to escape practically any current method of observation".

Despite Fermi's strong conviction, it should be observed however, in general, that acceptance of the hypothesis of the neutrino could certainly not be taken for granted at that time, as we have already seen with regard to the Solvay Conference. For example in a note that appeared in "Nature" on 23 December 1933 [16] G. Beck, tackling the problem of beta decay while he was in Copenhagen at Bohr's institute wrote about the neutrino: "There is, however, at present no need to assume the real existence of a neutrino, and the assumption of its existence would even be an unnecessary complication of the description of the β-decay process". M. N. Saha and D. S. Kothari also dispensed with the hypothesis of the neutrino when in November 1933 [152], and later in January 1934 [153], they tried to find an interpretation of β decay in terms of the creation, by a gamma ray, of positive electron - negative electron pairs, produced as an alpha particle passed from an excited level to a lower level. In those years Fermi was a pioneer in supporting the idea of the neutrino.

The second question that Fermi faced in developing his theory of beta decay concerned the origin of the electrons that were emitted during the β decay. Since during beta decay the electrons were emitted by the nucleus, it was natural to suppose that they pre-existed in the nucleus itself. However, as Fermi pointed out and as had now been known for some time (see the chapter describing the Rome Conference in 1931), quantum mechanics, and in particular "the current theories on light particles", were not able to explain "in a satisfactory way" how these light particles could be "bound in a stable or almost stable way inside a nucleus given its small volume". Remember that before the Solvay Conference, as we have seen, this impossibility of confining the electron within the nucleus had been the fundamental reason that had driven Fermi to invoke a new mechanics with which to explain beta decay, and more generally the behaviour of electrons within the nucleus.

To overcome this "essential difficulty" (as Fermi defined it), Fermi adopted totally the Heisenberg-Majorana nuclear theory which considered the nucleus to be made up only of heavy particles, protons and neutrons, excluding electrons from the fundamental and independent components of the nucleus. Therefore he drastically removed the electrons from inside the nucleus and introduced a remarkable idea, developed "by analogy to radiation theory", and which would then prove to be the basis of all weak interactions, that is to say that: "electrons do not exist as such in the nucleus before beta emission but that they (together with neutrinos) acquire existence so to speak in the very instant in which they are emitted; in the same way as a quantum of light emitted by an atom in a quantum jump cannot in any way be considered to pre-exist in the atom before the emission process".

As far as the process by which beta emission occurred was concerned, this was also developed by Fermi "by analogy with the theory of the emission of a quantum of light by an atom excited in the ordinary process of radiation". In particular Fermi assumed that a neutron could be transformed into a proton and that "every transition from neutron to proton" was "of necessity connected to the creation of an electron that is observed as a β particle, and a neutrino, just as in radiation theory the emission of a quantum of light is connected to a certain quantum jump of the atom". In modern terminology the particle that Fermi named the "neutrino" takes the name (electron) antineutrino and so, in modern notation, the reaction assumed by Fermi is presented as:

$$n \to p + e^- + \bar{\nu}$$

where with the symbols n, p, e^-, $\bar{\nu}$ we have indicated, respectively, the neutron, the proton, the electron and Fermi's neutrino (antineutrino in current terminology).

At this point we wish to make two observations. Fermi's idea that electrons were created at the moment they were emitted was not really new. In fact it had already been put forward by G. Beck and K. Sitte [17] and later, in November 1933, by Saha and Kothari [152] when, while developing a theory for beta decay, they had hypothesised that electrons did not pre-exist within the nucleus but were created during the process of materialisation of a gamma ray in an electron-positron pair, of which the positron was absorbed into the nucleus. Blackett too, as we have seen, during the Solvay Conference had resorted to this materialisation process to explain the presence of positrons in Joliot and Curie's experiments.

Instead, what differentiated Fermi's approach, and what made it extremely innovative, was the idea of the mechanism underlying the creation of these light particles. Indeed for Fermi this mechanism had nothing to do with the transformation of radiation into matter but represented a new type of interaction which involved the transformation of a type of heavy particle into another type of heavy particle with the simultaneous creation of two light particles. As Fermi stressed: "Electrons (or neutrinos) can be created or destroyed. Moreover this possibility has no analogy with the possibility of the creation or destruction of an electron-positron pair; if indeed a positron is interpreted as a Dirac "hole" this latter process can simply be considered as a quantum jump of an electron from a state of negative energy to a state of positive energy, conserving the (infinitely large) total number of electrons".

As far as the neutrino is concerned it should be observed that F. Perrin had also arrived at similar conclusions to Fermi's in a note communicated to the Académie des

Sciences on 18 December 1933, just a few days before the publication of Fermi's theory, and devoted to the "Possibilité d'émission de particules neutres de masse intrinsèque nulle dans les radioactivités β" [128]. It should be noted that already in the title Perrin talked about β radioactivities in the plural, as if to highlight the fact, in accordance with what had been suggested at the Solvay Conference, that two types of beta decay existed, negative and positive (still to be verified experimentally!). In this communication Perrin also expressed some ideas on the reciprocal transformation between neutron and proton, again considerably anticipating Fermi's thoughts. Perrin observed in the conclusion: "If the neutrino has an intrinsic zero mass, one must also think that it does not pre-exist in atomic nuclei, and that it is created, as is a photon, at the moment of emission. Finally it seems that one must attribute to it a 1/2 spin in order that one can have the conservation of the spin in the β radioactivities and more generally in possible transformations of neutrons into protons (or vice versa) with the emission and absorption of electrons and neutrinos".

We can see that an acceptance of the solution of beta decay was in the air. In any case, even if Perrin had understood some crucial points of the mechanism that underlay beta decay, it is a fact that Fermi was the first to formulate the right theory. It should be noted that Fermi would cite Perrin's article both in his second paper on beta decay and in his third, both written in January 1934. Instead this citation does not appear in his first paper sent to "La Ricerca Scientifica" in December 1933 and therefore, presumably, at the same time as the appearance of Perrin's article.

In constructing a theory for beta decay on the basis outlined above Fermi referred to the theory of the nucleus developed by Heisenberg, which we touched upon before, and in particular to the idea of considering "the heavy particles, neutron and proton, as two quantum states connected to two possible values of an internal coordinate ρ of the heavy particle" with $\rho = +1$ if the particle is a neutron, $\rho = -1$ if the particle is a proton, and he hypothesised that the transition from neutron to proton corresponded to the transition of the coordinate ρ from the value $+1$ to the value -1.

Still acting in analogy with radiation theory Fermi, symmetrically, considered the inverse process that therefore a proton could be transformed into a neutron, and that every transition of a proton into a neutron, where the coordinate ρ passed from the value -1 to the value $+1$, was accompanied by the "disappearance" of an electron and a neutrino, "in the same way as the absorption of a quantum is connected to the quantum jump opposite [to the jump corresponding to the emission of a quantum]". It should be noted that in this way, in both cases of transformation (either neutron into proton or proton into neutron), conservation of the electric charge was assured.

Certainly Fermi, expert as he was on Dirac's theory (remember his studies into quantum electrodynamics carried out between 1928 and 1932 and his work [F110] developed with Uhlenbeck in the framework of Dirac's theory, in that very summer of 1933, on the recombination between an electron and a positron), was aware that on the basis of this theory the "disappearance of an electron", if it had negative kinetic energy, corresponded to the creation of a positive electron, and that therefore the transformation of a proton into a neutron led to the creation of a positive electron, as well as of an antineutrino

(neutrino in current terminology), that is to say, again in modern notation:

$$p \rightarrow n + e^+ + \nu ,$$

where by e^+ and ν we have indicated, respectively, the positive electron and the neutrino. Fermi, in any case, did not make this relationship explicit, but merely declared the symmetry between the neutron-proton transition (with the emission of an electron and a neutrino), and the proton-neutron transition (with the destruction of an electron and a neutrino).

Fermi's aim at that time was in fact to succeed in explaining natural beta decay, where emission of negative electrons takes place, which was the only radioactive decay known until then for which an explanation had not been found. Remember that alpha decay, where a heavy particle exits the nucleus, had already been brilliantly explained by Gamow in 1928 on the basis of quantum mechanics, making reference to the tunnel effect, and that positive beta type radioactive decay, with the emission of positrons, that is to say the very type of decay predicted by the previous relationship, had not yet been discovered experimentally even if, as we said at the beginning, it had been hypothesised by Perrin at the Solvay Conference in 1934 with regard to Joliot and Curie's experiments on the emission of positrons. Probably it was this very idea of Perrin's that must have guided Fermi in his search for a symmetrical theory for beta decay, capable of also explaining a possible positive beta decay, which is in fact what he did. Amongst the various expressions of the interaction energy between the light particles (electrons, neutrinos) and the heavy ones (protons and neutrons), that is to say between the possible Hamiltonians of the total system (light particles and heavy particles) capable of guaranteeing the various conditions imposed, Fermi chose the "simplest" one, justifying this choice a posteriori with the fact that "since the consequences of this choice are shown to be in good agreement with the experimental facts it is not necessary for now to resort to more complicated expressions". As was typical in Fermi's way of proceeding, here too a sort of simplicity principle was put to work.

To construct a theory in which the total number of particles, "just like the total number of quanta of light in the theory of radiation", was not conserved (there could indeed be processes of creation and destruction of the light particles, electrons and neutrinos), Fermi turned to the formalism of second quantisation developed by Dirac [64], Jordan and Oskar B. Klein (1894-1977) [104], and others and which was familiar to him from his previous studies in quantum electrodynamics but that until then had been mainly used only for photons.

In particular, Fermi introduced the transformation operators from neutron state to proton state and vice versa, and the creation and destruction operators for the electron and for the neutrino, in this way bringing about the possibility, implicit in Dirac-Jordan-Klein's method, of creating and destroying any type of particle. With these operators he wrote the interaction Hamiltonian which involved a small coupling constant, called the Fermi constant, and he calculated the probability of perturbative first order decay, through some drastic simplifications, so that in the end the probability distribution of the energy of the β particle emitted depended practically only on the extension of the phase space available to the light particles created.

The theory that Fermi developed on this basis, as he himself observed, was immediately confirmed, both with regard to the mean lifetimes of some heavy elements and with regard to the velocity distribution curve of beta rays, in some experiments carried out by Bernice W. Sargent (1906-1993), in 1932 [154] and 1933 [155] respectively.

Fermi concluded his article in "Il Nuovo Cimento" as follows: "To sum up, it seems justified to maintain that the theory, in the form in which we have expounded it here, is in agreement with the experimental data which, by the way, are not always very precise. Even if, in a further comparison of the theory with experience, contradictions should emerge after all it would always be possible to modify the theory without essentially altering the conceptual foundations". Fermi's theory, apart from its flexibility, did not however represent just a remarkable solution to the problem of beta decay. As we shall see, it succeeded in giving a theoretical interpretation to a new phenomenon, which would be discovered soon after, the phenomenon of positive beta decay.

7 The discovery of alpha-particle–induced radioactivity

7.1 Frédéric Joliot and Irène Curie after the Solvay Conference

As has been seen, at the Solvay Conference Lise Meitner had severely criticised the Joliot-Curie's idea according to which, in certain nuclear transmutations produced by alpha particles in aluminium and fluorine nuclei, positrons were also emitted together with neutrons, and so she had aligned herself against the idea that positrons were the product of a nuclear reaction. Meitner's objection was based on the fact that she had never succeeded in seeing the emission at the same time of a neutron and a positron.

Anyway, almost immediately after the Solvay Conference this criticism was withdrawn by Meitner herself, both in a letter sent personally to Irène Curie and in an official announcement sent to the organisers of the Conference which arrived in time to be included as an addition to the Proceedings. On 18 November 1933 Meitner wrote: *Dear Mrs Joliot, after my return a have examined again our measurements on Al and F and I have arrived at the conclusion that the absence of protons [recoil protons, therefore neutrons] in the gas from aluminium can be interpreted very well as statistical errors [...]. I have therefore made some hundreds of measurements with a stronger preparation of polonium and, in the case of aluminium, I have found in the gas about six protons in every hundred measurements and with F two protons in every hundred measurements, so that one of the observations made in the discussion is not correct. I have written to Brussels and I have asked for a small note to be inserted in my observations which I made during the discussion [...]. Now I have arrived at the conclusion that your interesting views on the disintegration process of aluminium are correct and that the positive electrons certainly [wirklich] come from the aluminium nucleus.*

This letter from Meitner was followed a few days later by a thank you letter from Irène Curie. On 22 November Irène Curie wrote as follows: *Dear Miss Meitner, Mr Joliot and I have read your letter with interest and we thank you greatly for informing us of these new results. Could you be so kind as to tell us what is the maximum energy of the neutrons of Po+Al, on the basis of the recoil H rays that you have observed? It would be interesting to know if this energy is in agreement with the energy that we have considered to be probable according to our experiments.*

So already from the end of November 1933 Joliot and Curie knew that their idea of linking the emission of positrons to a nuclear process had also been validated by Meitner's experiments and they were interested in comparing their data on the energy

F. Guerra and N. Robotti, *The Lost Notebook of ENRICO FERMI*,
https://doi.org/10.1007/978-3-319-69254-8_7

of the neutrons emitted with Meitner's.

With regard to Meitner's original criticisms, made during the Solvay Conference and as we have just seen almost immediately withdrawn, Joliot in his preparatory notes written for his course held at the Sorbonne in the year 1948-49 in which he recalled the various stages that had led him, together with Irène Curie, to the discovery of alpha-particle–induced radioactivity observed on page 20: "2 mois de tranquillité à cause de Meitner!!" (2 months of peace because of Meitner!!), as if to say that thanks to Meitner's criticisms research in this field had come to a halt, thus allowing Joliot and Curie to return to work on the question quietly and alone, to finally arrive on 15 January 1934 at the discovery of induced radioactivity.

This claim by Joliot is rather curious (bear in mind however that it was made fifteen years later) if we look at Joliot and Curie's activity at this time. Indeed, from the end of the Solvay Conference until the middle of January 1934 they no longer worked directly on these topics, in particular on the origin of positrons and their link with the neutrons emitted in the nuclear transmutations produced by the α particles in aluminium.

In this period not only did they not publish anything on the subject but the research activity they carried out, in Curie's case, only marginally concerned this question, while in Joliot's case it did not concern it at all. As can be seen in the material kept in the Joliot-Curie Archive, which is the only source of information available, Irène Curie was essentially engaged, starting on 19 October 1933, in the various routine maintenance operations necessary for the conservation of the polonium sources available at the Institut du Radium. Anyway between 18 November and 27 November 1933, she studied with a cloud chamber the "positive electrons of Al emitted by slowed down or not slowed down α particles", while between 2 December and 13 December 1933, returning again to a preceding experiment, she studied the "excitation limit of the neutrons of Al" using a variable pressure ionisation chamber filled with butane and paraffin. In any case, this research of hers, even if it concerned both the production of positrons and the production of neutrons by aluminium bombarded with alpha particles, tackled these two questions completely independently and therefore did not touch the heart of the problem which was the link between the emission of these two types of particles.

For his part, Joliot immediately after the Solvay Conference returned to working experimentally on the dematerialisation properties of positrons obtained by using an "intense positron source" obtained with "80 milliCuries of Po deposited by vaporisation on the Al and covered with Al" and exploiting the previous discovery, his and Curie's, that aluminium emits positrons when bombarded with the α particles of Po. In particular, between the middle of December 1933 and early January 1934, while studying the phenomenon of the annihilation of positrons, Joliot obtained results that fully agreed with Dirac's theory, announced in two notes in "Comptes Rendus" presented on 18 December 1933 and 3 January 1934, only in his name [100, 101].

It is interesting to observe that Joliot used for the first time a relatively recent piece of equipment for this research, a Geiger-Müller counter. Joliot had calibrated it himself, either completely or perhaps only partly, as can be seen in handwritten notes in the Joliot-Curie archive. In any case at that time a young German scholarship researcher was working at the laboratory, Wolfgang Gentner (1906-1980) who had earned his doctorate in Frankfurt under the supervision of Friedrich Dessauer (1881-1963), director of the Institute for the Physical Fundaments of Medicine (Institut für die physikalischen Grundlagen der Medizin). Gentner had considerable experience in constructing Geiger-Müller counters and accessory apparatus such as amplifiers and high value resistors. It can therefore be presumed that Joliot also availed himself of this expertise in order to calibrate his own counters.

The introduction of this counter into the supply of instrumentation available at the Institut du Radium in Paris, as we shall see, would be essential for the discovery a little later of alpha-particle–induced radioactivity.

So in the months immediately following the Solvay Conference, both the research carried out by Irène Curie and that carried out by Frédéric Joliot were not aimed at finding an answer to the various questions raised at the Solvay Conference with regard to the emission of positrons and neutrons in alpha-particle–induced nuclear transmutations in some light elements (aluminium, boron, magnesium).

Suddenly however, after about two months of silence, Joliot and Curie returned to the topic with a note in both names presented on 15 January 1934 by Jean Perrin to the Académie des Sciences in Paris and published in "Comptes Rendus" [59] in which they announced the discovery of alpha-particle–induced radioactivity.

7.2 The discovery of alpha-particle–induced radioactivity

In this note, significantly named "A new type of radioactivity", the Joliot-Curies presented the results of a recent experiment of theirs in which they seemed finally to have adopted the suggestion made by Perrin at the Solvay Conference, according to which the emission of the positrons might have been delayed with respect to the emission of the neutrons, and might represent a new type of radioactive decay involving positrons instead of electrons. In the course of this experiment, after irradiating a sample of aluminium for about ten minutes with a source of alpha particles consisting of 60 mCi of polonium, Joliot and Curie, in order to detect positrons emitted from the sample after it had been moved away from the source, instead of using, as they had done in previous experiments, a Wilson chamber which is an instrument that photographs an event at a given moment, used a Geiger-Müller counter. This was an instrument that, as Joliot had recently been able to ascertain, was able also to provide information on the evolution over time of the event being studied. Moreover it was extremely sensitive, as Fermi, for example, would stress in a Conference in 1949 [68, page 758].

Studying the irradiated sample with this instrument they discovered that the emission of positrons did not cease immediately after it had been removed from the source but it continued over time with an exponential decay, in particular "with a period of 3 minutes and 15 seconds". Moreover, as the energy of the incident alpha particles was reduced the number of positrons was also reduced but the half-life remained unchanged. So the aluminium sample behaved like a natural radioelement.

Similar results were also found with samples of magnesium and boron, with different half-lives which proved to be 2 minutes and 30 seconds for magnesium and even 14 minutes in the case of boron.

Instead, using the elements H, Li, C, Be, N, O, F, Na, Ca, Ni, Ag, Joliot and Curie observed no effect of radioactive type of effect but they justified this as follows: "for some of these elements the phenomenon probably does not occur, for others the period of decay is perhaps too short [to be detected]". To verify that the particles emitted really were positrons (the Geiger-Müller counter is not in fact able to give the sign of the ionising particle that enters inside it) they repeated the experiment, only with boron and aluminium, both with a Wilson chamber to which a magnetic field was applied and with the trochoid method recently developed by Jean Thibaud [160], thus confirming that they really were positrons. Joliot and Curie observed: "Finally for the first time it has been possible with the aid of an external cause to create radioactivity in certain atomic nuclei, which can exist for a measurable time in the absence of the cause of excitation". Once the possibility had been shown of creating by bombardment with alpha particles new radioactive elements never seen in nature that β^+ decayed, Joliot and Curie tackled the problem of explaining their origin. To do this they abandoned their previous hypothesis, according to which a light element such as aluminium, after capturing an alpha particle, could emit a neutron and a positron at the same time in a single nuclear reaction. They assumed instead that the emission of these two particles occurred at two separate times. In a first phase the initial nucleus, after having absorbed an alpha particle, emitted a neutron according to an (α, n) reaction and it was transformed however into an unstable nucleus. This subsequently emitted a positron and was transformed into a stable nucleus. So, according to this new hypothesis, in fact they partly retrieved what they had previously denied, in other words the possibility of dealing with an "unknown" isotope of phosphorus, after the emission of the neutron, with the difference that now the unknown isotope became a radioactive isotope that β^+ decayed. Joliot and Curie explained their results as follows: "These experiments demonstrate the existence of a new type of radioactivity with the emission of positive electrons. We think that the emission process for aluminium will be the following:

$$^{27}_{13}\mathrm{Al} + {}^{4}_{2}\mathrm{He} \rightarrow {}^{30}_{15}\mathrm{P} + {}^{1}_{0}n,$$

the isotope of phosphorus will be radioactive with a period of 3 min 15 s and it will emit positive electrons [ϵ^+] according to the reaction:

$$^{30}_{15}\mathrm{P} \rightarrow {}^{30}_{14}\mathrm{Si} + \epsilon^{+}\text{''}.$$

Similar reactions were put forward for the case of boron and magnesium where the new radioactive isotopes hypothesised were $^{13}_{7}\mathrm{N}$ and $^{27}_{14}\mathrm{Si}$, respectively.

With regard to the fact that these isotopes had never been seen, Joliot and Curie observed: "The isotopes $^{13}_{7}\mathrm{N}$, $^{27}_{14}\mathrm{Si}$, $^{30}_{15}\mathrm{P}$ can only exist for very short times and that is why they are not observed in nature".

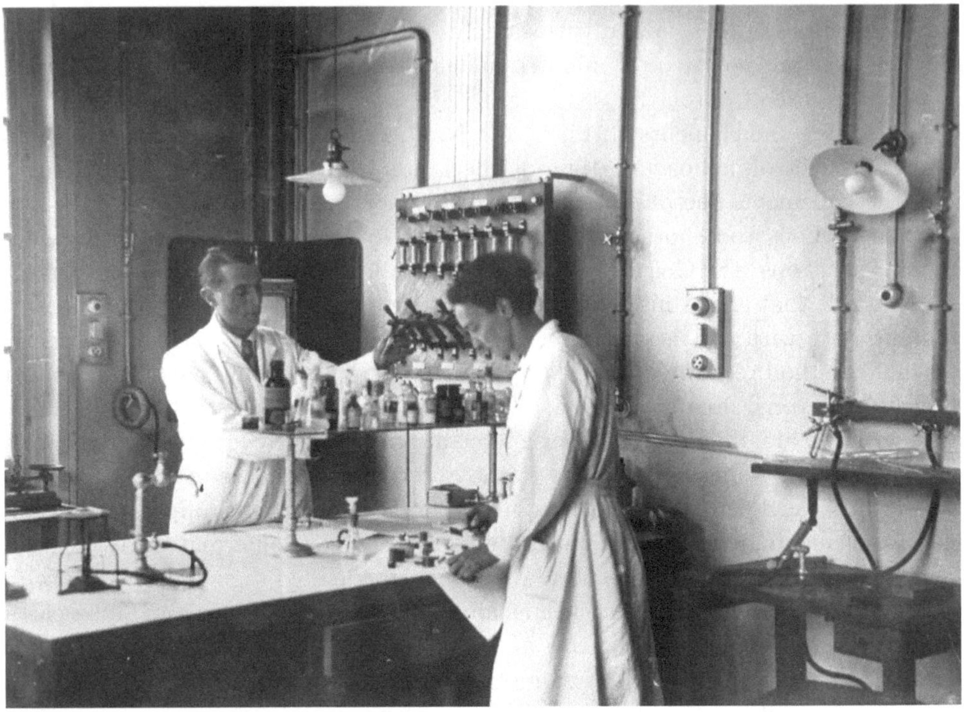

Fig. 41. – Frédéric Joliot and Irène Curie in their laboratory at the Institut du Radium in Paris (1934) - Musée Curie (coll. ACJC) (MC)

It should be noted that the reactions for positive beta decay, where a positron is emitted, were written so as to imply a displacement of the initial element one cell to the left along the periodic table, clearly by analogy with Soddy's "laws on radioactive displacement" from 1911 which for alpha decay, where two positive unitary charges were lost, predicted a displacement of the element two cells to the left and for beta decay a displacement of one cell to the right.

The extraordinary importance of the discovery made by Frédéric Joliot and Irène Joliot-Curie earned them the Nobel prize for Chemistry in the year 1935 "in recognition of their synthesis of new radioactive elements". In Fig. 41 we can see Frédéric Joliot and Irène Curie at work in their laboratory.

7.3 Why alpha-particle–induced radioactivity was discovered only in January

As we saw before, the idea that bombarding some light elements with alpha particles could produce phenomena of positive-beta-type induced radioactivity had already been put forward by Perrin in October 1933, at the Solvay Conference, precisely in reference to Joliot and Curie's experiments on aluminium. So it is natural to ask why this suggestion was only considered by Joliot and Curie in January 1934 and not before, for example at the beginning of December 1933 when Joliot had by then mastered the Geiger-Müller counter technique and was thus able to carry out the decisive test that would be done on 12 January 1934 and that would lead him to the discovery of induced radioactivity. The answer should be sought by looking at what happened on the theoretical front between November 1933 and January 1934.

Immediately after the Solvay Conference, Perrin's suggestion, even if it opened up the possibility of a new physical interpretation for the production of the positron, linking it to a hypothetical new type of beta decay, lacked however any precise theoretical framework. The situation that had emerged at the end of December 1933, just before the Joliot-Curies returned to their research on aluminium, was very different. Indeed in this very month two important facts had occurred on a theoretical level which opened up new possibilities for the interpretation of beta decay (not only negative decay but also positive decay): the publication of an article by F. Perrin himself on the neutrino and the formulation of Fermi's theory on beta decay.

Perrin, who at least until 1949 was in close scientific and personal contact with Joliot, was involved in the middle of December 1933, as we have seen, in an attempt to interpret the phenomena of both positive and negative beta decay [128], even expressing some ideas that anticipated Fermi's theory. At the end of December that same year 1933, as seen before, Fermi published in "La Ricerca Scientifica" his theory on beta decay in which not only was a brilliant solution successfully found for natural beta decay, fully compatible with the experimental data, but the possibility of a positive type of beta decay through the transformation of a proton into a neutron was also predicted.

Clearly, both Perrin's new ideas and Fermi's theory presented a new panorama for the processes of radioactive decay in general and opened up unexpected possibilities of also interpreting the origin of Joliot and Curie's positrons. So, in this new theoretical framework, the suggestion put forward by Perrin at the Solvay Conference assumed a completely different dimension, different enough perhaps to deserve experimental verification, as in fact happened. Becoming aware of these new ideas is certainly the basis of the reason why the Joliot-Curies, after "two or three months of peace", suddenly returned to their old experiment on the emission of positrons. They did so completely modifying it so as to detect any delayed emission and finally discovered induced radioactivity.

That the Joliot-Curies were finally driven to take Perrin's suggestion seriously by the great new developments emerging at a theoretical level, which they were certainly aware of, is made plausible by their very behaviour. Indeed, once positive radioactive beta decay had been discovered, it would have been obvious and natural for Joliot and Curie to tackle the problem of the origin of the positrons emitted in this new type of decay. Instead they did not take a stance. The Joliot-Curies, unlike in the past when, to justify the presence of positrons in their nuclear transformations they had immediately intervened personally putting forward the hypothesis of the "compound proton", here they did not in any way enter into the subject, as if to signal a sort of waiting period. They merely announced their discovery. Obviously, faced with Fermi's recent theory, or simply Perrin's hypothetical theoretical considerations on the reciprocal transformations between protons and neutrons, taking up a position on the origin of the positrons would have been premature and even rash. As we shall see, the solution would only be given at the beginning of March 1934 by Gian Carlo Wick (1909-1992) [165], then a young assistant of Fermi's in Rome, on the basis of Fermi's theory on beta decay.

Even if we do not know for certain if this new theoretical context was the trigger that drove Joliot and Curie in January 1934 to reconsider the suggestion put forward by Perrin at the Solvay Conference, the fact remains that alpha-particle–induced radioactivity was discovered just after Perrin and Fermi's contributions.

As for the timing, it coincided perfectly (see also [84]). Indeed there is no doubt that the Joliot-Curies, because of the friendship that tied them personally to F. Perrin, as well as to the Family, must have heard of the new ideas immediately and in any case through the Proceedings of the Académie de France. Instead, with regard to Fermi's theory, even if it was published for the first time at the end of December in "La Ricerca Scientifica", it immediately spread throughout the scientific world, through the accounts of those who had learned of it, or through Fermi himself, or by indirect communication. For example Pauli heard of Fermi's theory before Christmas 1933 from Felix Bloch, who was in Rome at the time on a study grant. On 24 December 1933 (see [29, page 53]) Bloch wrote to Gregor Wentzel, professor of Theoretical Physics at the University of Zurich: *Fermi has produced a fine theory of beta decay introducing the neutrino, which reproduces the experimental facts so simply that I strongly believe in it. The mass of the neutrino should be exactly zero, or in any case much smaller than that of the electron.* Wentzel in his turn took the news to Pauli, who was at the federal Polytechnic in Zurich (Eidgenssische Technische Hochschule, ETH). Pauli's comment, expressed in a letter to Heisenberg on 7 January 1934 ([29, page 53]) is interesting: *Das wäre also Wasser auf unsere Mühle* (literally: That would be also water to our mill). Finally the physical validity of Pauli's "neutron" was confirmed in the framework of a very precise quantitative theory.

7.4 After the discovery

The discovery of alpha-particle–induced radioactivity, above all that of boron with a half life as long as fourteen minutes, caused such a sensation in the scientific world that the Editor of the journal "Nature", before even having read the article published in "Comptes Rendus" but only having heard of it, on 31 January 1934 explicitly wrote a letter to Joliot, also accompanied by a telegram, to invite him to write an article on the subject for the journal. In this way he attributed all the credit for the discovery to him. The Editor (his name is not given, not even in the journal) wrote as follows: *The Editor of Nature presents his compliments to Monsieur F. Joliot and in confirmation of a telegram just despatched to Monsieur Joliot begs to say that he will be very glad indeed to publish in NATURE an account of Monsieur Joliot's experiments upon the atomic bombardment of boron with the result that a new source of radioactivity is produced. If Monsieur Joliot has already published a paper upon the subject, perhaps he could send a copy of that to the Editor for publication in NATURE, either in the original French language or traslated into English. In any event, the Editor is sure that scientific readers all over the world would welcome any account which Monsieur Joliot may be able to send the Editor on his recent work and the results obtained.*

The requested article reached the Editor, jointly signed by Frédéric Joliot and Irène Curie, presumably around 6 February 1934, the date of the Editor's thank you letter to Joliot for having received the article and it appeared in the pages of "Nature" on 10 February 1934 [102]. The title was "Artificial production of a new kind of radio-element", as if to highlight once again the distinctive feature of the new elements of emitting positrons.

In this article Joliot and Curie summed up the situation, providing an update on their latest results and beginning to align themselves with the hypothesis of the neutrino. In particular they pointed out, in agreement with Perrin's observations at the Solvay Conference with regard to the data reported by Joliot and Curie at the Conference, that "the positrons of aluminium seem to form a continuous spectrum similar to the β-ray spectrum" and that therefore, "as in the case of the continuous spectrum of β rays", perhaps it was necessary to "admit the simultaneous emission of a neutrino [...] in order to satisfy the principle of conservation of spin in the transmutation".

They also announced their recent chemical measurements, aimed at establishing the nature of the new radioactive isotopes produced. These measurements, by the way, had already been partially communicated by Joliot and Curie personally to Rutherford in a letter dated 2 February 1934, and also reported in a note presented by J. Perrin to the Académie des Sciences in the session on 5 February 1934, published immediately afterwards in "Comptes Rendus" [60]. These chemical separations were very important because, through the identification of the chemical elements that had been produced in the nuclear reaction, they allowed the accuracy of the hypotheses made regarding those same reactions to be verified. Given the minute quantity of radioactive material that was produced in the various reactions (of the order of 10^{-16} grammes) and the shortness of the decay times, a chemical analysis was by no means simple. Anyway Joliot and Curie succeeded in "separating and chemically identifying the radioactive elements formed", obtaining "nitrogen and phosphorus in the case of B and Al", in full agreement with the nuclear reactions previously hypothesised. Joliot and Curie observed: "These experiments give the first chemical proof of artificial transmutation, and also the proof of the capture of the α particle in these reactions". Joliot and Curie proposed a specific name for the new radioactive elements produced, composed of their atomic species preceded by the prefix "radio", to stress their diversity or rather, their peculiarity of beta plus decay: for example the name "radionitrogen" for the element derived from the transmutation of boron and which proved to be an isotope of nitrogen, or "radiosilicon" for the element derived from the transmutation of magnesium and which turned out to be an isotope of silicon.

The Joliot-Curies' discovery led immediately to the establishment of a clear distinction between natural radioactivity and artificial radioactivity: according to this distinction, which was tacitly maintained until Fermi's discovery of neutron-induced radioactivity, natural radioactivity manifested itself with the emission of alpha or beta particles while artificial radioactivity was characterised by the emission of positrons. Not by chance, as we have said, was the title of Joliot and Curie's first announcement, published in

"Comptes Rendus", "Un nouveau type de radioactivité" (A new type of radioactivity), as if to stress the fact that it was not only possible to create new radioactive elements artificially but that these behaved in a new way compared to the natural ones.

The announcement of the Joliot-Curies' discovery naturally opened up the idea that these new radioactive elements could be produced in other ways. For example already on 29 January 1934 Rutherford, in a letter typed on notepaper headed "Cavendish Laboratory" sent to Joliot and Curie after their discovery, and in which he congratulated them for *a fine piece of work which I am sure will ultimately prove of much importance*, added at the bottom by hand: *We shall try to see whether similar effects appear in proton-diplon bombardament.* Remember that the isotope of hydrogen with mass 2 had been discovered in December 1931 by H. C. Urey, F. C. Brickwedde and G. M. Murphy [163]. Its nucleus was called in various ways: "deuteron", "deuton" or "diplon".

On 2 February 1934, in a thank you letter sent in reply to Rutherford, Joliot and Curie, fully approving Rutherford's plan to extend research to other particles other than α particles reiterated: *In reality it is very important to try to provoke these types of radioactivity bombarding matter with other projectiles, in addition to alpha particles. In particular it seems to us that the bombardment of carbon (with deuterons) should lead to the production of radionitrogen.* In this way the isotope of nitrogen ${}^{13}_{7}N$ would be produced, the same radioelement that they had produced by bombarding boron with α particles.

And on 10 February 1934, in their article in "Nature" mentioned before, they declared: "These elements and similar ones may possibly be formed in different nuclear reactions with other bombarding particles: protons, deuterons, neutrons".

In any case, results in this regard soon appeared. In fact within just over a month of Joliot and Curie's discovery, in many laboratories equipped to use accelerated protons and deuterons as projectiles this new radioactivity was found almost simultaneously. On 24 February at the Cavendish Laboratory in Cambridge, consistent with the plans anticipated by Rutherford to Joliot and Curie, J. D. Cockroft, C. W. Gilbert and E.T.S. Walton [41] found that by bombarding a graphite (carbon) target with 600 keV protons an activity with a half life time of 10.5 minutes was produced. Three days later two other groups, operating respectively at the Radiation Laboratory, Department of Physics, University of California (Malcolm C. Henderson, M. Stanley Livingston, Ernest O. Lawrence [95]), and at the Kellogg Radiation Laboratory, California Institute of Technology (Horace R. Crane (1907-2007), Charles C. Lauritsen (1892-1968) [45]) simultaneously obtained samples of artificial radioactivity by bombarding with deuterons carbon, boron and, in the case of the first group, also calcium fluoride, calcium chloride, boron oxide, sodium phosphate, lithium carbonate and ammonium nitrate. On 8 March the second group [46] also succeeded in producing this new type of radioactivity in new light elements, both with deuterons and with protons. On 15 March a third group at the Norman Bridge Laboratory of Physics in Pasadena California (S. H. Neddermeyer, C. D. Anderson), using a Wilson chamber equipped with a magnetic field, analysed "several samples of C, B, Be and Al" made available to them by Lauritsen, Crane and W.W. Harper, and "freshly subjected to bombardment". They confirmed [117] that the induced radioactivity was of the beta positive type and they studied its energy spectrum obtaining, as in the case of the radio-elements discovered by Joliot and Curie, "a wide distribution in energy, similar to the β-particles deriving from natural radioactive substances".

So, within less than a month, the possibility of artificially inducing a beta-positive-type radioactivity using not only alpha particles but also protons and deuterons, to which, thanks to new accelerator machines, it was possible to impose large energies, became a confirmed fact.

In any case, as we have seen, protons and deuterons, according to Joliot and Curie, were not the only particles suggested for the artificial production of the new type of radioactivity. According to what they had suggested in the article in "Nature" on 10 February 1934, neutrons were also a possible projectile with which this new type of radioactive elements could be produced. However, as a matter of fact, no one tried to obtain an artificial radioactivity using neutrons. This type of research was hindered mainly by the impossibility of creating neutron sources comparable in intensity to alpha particle sources. Indeed, while alpha particles are the direct product of a radioactive process and are therefore emitted spontaneously and in large quantities by some natural radioactive elements, such as polonium, neutrons instead are the secondary product of a nuclear reaction that occurs with a very low rate, when a beam of alpha particles hits some elements, such as beryllium or boron for example. In his famous Bakerian Lecture in 1932, in paragraph 2 dedicated to the "Production of Neutrons", Chadwick wrote on this subject: "Neutrons have been produced so far only by bombarding certain elements by *alpha*-particles. The process is assumed to be the capture of the α-particle into the atomic nucleus with the formation of a new nucleus and the release of a neutron [...] The yield of neutrons is, of course, very low [...].The greatest effect is given by beryllium, where the yield is probably about 30 neutrons for every million α-particles of polonium which fall on a thick layer of beryllium. For the elements of much higher atomic number the yield is very small, probably of the order of 1or 2 neutrons per million α-particles". For this reason, neutron sources were created using beryllium.

The question of the weakness of the neutron sources was therefore crucial in ensuring that, in the beginning, neutrons were not used as projectiles in experiments on artificial radioactivity. As, for example, Joliot noted in a lecture on 21 February 1934 [103]: "One can believe that neutrons must constitute collision projectiles to transmute heavy nuclei and great efforts should be made to produce intense neutron beams artificially". And again, as Leo Pincherle observed in a review article in 1933 [129]: "Recently, after the discovery of the neutron, attempts were made to use this particle as a projectile for disintegrations. The neutron was suitable for this aim succeeding in splitting even highly stable nuclei like the oxygen nucleus. But the difficulty of obtaining a neutron source intense enough and capable of producing neutrons with a well-known energy does not allow them to be used for systematic studies of nuclear disintegrations". For the disintegration of oxygen the reference is to [113]. On the limitations of neutron sources we can refer to [34] and [124], for example.

As can be seen in Russell's table, already cited and shown in Fig. 28, there were only four elements that, up to the beginning of 1934, had been successfully disintegrated with neutrons: nitrogen, fluorine, neon and oxygen. In any case, despite these difficulties, in March 1934 Fermi chose to bombard elements using neutrons as projectiles seeking exclusively a radioactivity induced by these particles and, as we shall see, he succeeded.

8 The discovery of neutron-induced radioactivity

8.1 A project set aside

In the days following the discovery of alpha-particle–induced radioactivity it was planned to repeat Joliot and Curie's experiments at the Physics Institute in Rome, as was announced in two letters written on the same day by Rasetti to Joliot and by Rasetti and Fermi to D'Agostino respectively.

On 9 February 1934, Rasetti wrote by hand on personal headed notepaper: *Dear Mr Joliot, here in our laboratory nothing else is talked about other than your experiments on radioactivity with positron emission induced in B, Al and Mg. These experiments are really of exceptional interest and we firmly think that the explanation you have given must be correct. We intend to try and reproduce them.*

On the same day, 9 February 1934, Fermi and Rasetti also announced this intention to D'Agostino who, in the meantime as we have said before, in the context of a study grant from the C.N.d.R. had transferred to the Institut du Radium in Paris at the end of January. In this letter the objectives of this new project were specified and information was also given on the alpha particle source, based on polonium, previously created by Rasetti together with D'Agostino and which could now be used. Moreover, from what is written, one can see that in early February 1934 the Physics Institute was not yet equipped with Geiger-Müller counters. Instead, as can be seen in another letter from Rasetti to Joliot, already cited in chapter 4 with regard to D'Agostino's stay at Madame Curie's Laboratory, in Rome interest seemed to be directed to constructing one or more Wilson chambers. In the letter from Fermi and Rasetti to D'Agostino (Fig. 42), typewritten on notepaper headed "Istituto di Fisica della Regia Università di Roma" we read: *Dear D'Agostino,* [...] *The polonium preparations that have been put in the Wilson chamber are very clean and give a fine range of alpha particles all with the same length; the large chamber works well, except for the strong vibrations that rather often end up breaking the seals. The other fine mesh Wilson type chamber is at an advanced stage of construction. We are putting together counters to repeat Joliot's experiments on artificial radioactivity with positron emission, trying if it's not possible in the few minutes of mean lifetime to separate the unstable radioactive product that should form.*

This aim of separating the active product, proposed by Fermi and Rasetti quite independently, had in fact already been achieved before by Joliot and Curie, as reported in the note presented on 5 February 1935, then published in "Comptes Rendus" [60], and in the Letter in the 10 February issue of "Nature" [102].

F. Guerra and N. Robotti, *The Lost Notebook of ENRICO FERMI*,
https://doi.org/10.1007/978-3-319-69254-8_8

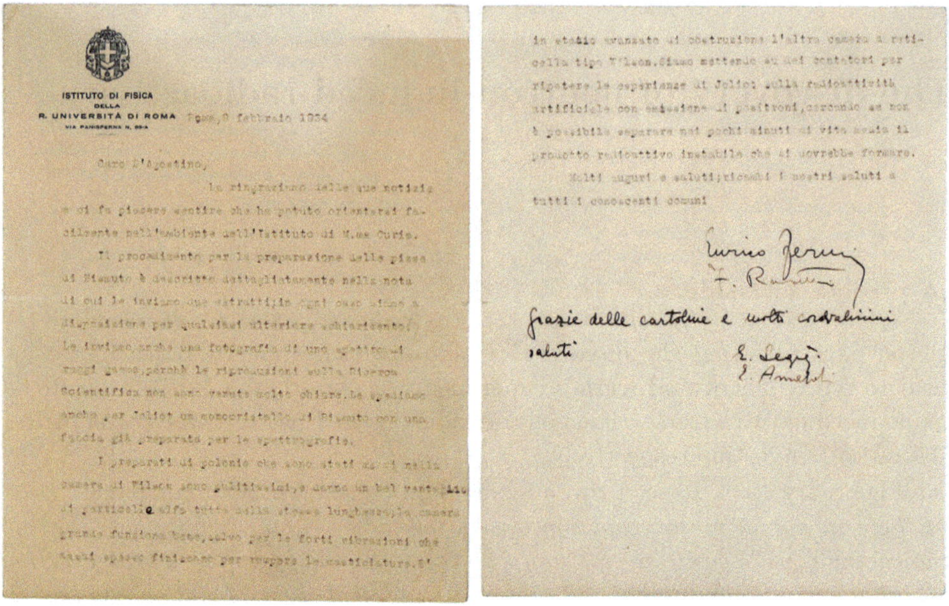

ISTITUTO DI FISICA
DELLA
R. UNIVERSITÀ DI ROMA
VIA PANISPERNA N. 89-A

Roma, 9 febbraio 1934

Caro D'Agostino,

La ringraziamo delle sue notizie e ci fa piacere sentire che ha potuto orientarsi facilmente nell'ambiente dell'Istituto di M.me Curie.

Il procedimento per la preparazione della [illegible] di Bismuto è descritto dettagliatamente nella nota di cui le inviamo due estratti; in ogni caso siamo a disposizione per qualsiasi ulteriore schiarimento. Le inviamo anche una fotografia di uno spettro di raggi gamma, perchè le riproduzioni sulla Ricerca Scientifica non sono venute molto chiare. Le spediamo anche per Joliot un [illegible] di Bismuto con una faccia già preparata per la spettrografia.

I preparati di polonio che sono stati messi nella camera di Wilson sono pulitissimi, e danno un bel ventaglio di particelle alfa tutte della stessa lunghezza; la camera grande funziona bene, salvo per le forti vibrazioni che [illegible] finiscono per rompere la [illegible]. E'

in stadio avanzato di costruzione l'altra camera a reticella tipo Wilson. Stiamo mettendo su dei contatori per ripetere le esperienze di Joliot sulla radioattività artificiale con emissione di positroni, [illegible] se non è possibile separare nei pochi minuti di vita media il prodotto radioattivo instabile che si dovrebbe formare.

Molti auguri e saluti; ricambi i nostri saluti a tutti i conoscenti comuni

Enrico Fermi
F. Rasetti

Grazie delle cartoline e molti cordialissimi saluti

E. Segrè
E. Amaldi

Fig. 42. – Letter sent by Enrico Fermi and F. Rasetti to Oscar D'Agostino, with handwritten greetings from E. Segrè and E. Amaldi (Rome, 9 February 1934) - FODA.

Clearly in an investigation of this type, where specific and highly qualified expertise in the radiochemical field was required, Fermi's participation could only be marginal. The situation would have been different for Rasetti who, as was seen before, had already taken an interest in this field. It should also be said that at that time Fermi was still involved with his theory of beta decay, so much so that in the two papers on the subject, sent just in early January to "Il Nuovo Cimento" [F116] and "Zeitschrift für Physik" [F117], respectively, Fermi concluded by saying: "In a forthcoming communication I hope to be able to specify better the trend of the [velocity] distribution curves for forbidden transitions". Not by chance, as has already been observed, the title of the article on "Zeitschrift" contained the symbol "I", as if to indicate that the work done represented only a first part (see fig. 40). Moreover Enrico Fermi maintained a frequent correspondence with Werner Heisenberg in the early months of 1934 about a brilliant and bold idea by Heisenberg according to which Fermi's theory for beta decay might provide the basis to explain the neutron-proton exchange interactions according to Heisenberg-Majorana, where the "exchanged" object was constituted of the electron-neutrino pair. Even if this idea could not work for solid physical reasons, since it mixed strong interactions with weak interactions, it nevertheless bore witness to the onset of a line of research by Heisenberg that would then culminate in the proposal of the universal four-fermion interaction, as in [93]. Fermi showed interest in Heisenberg's attempt, and offered some comments and suggestions.

In any case, there is no further trace of Fermi and Rasetti's joint project to repeat Joliot and Curie's experiments either in archive sources or in print.

In reality, in the following months, Fermi threw himself alone into a new research programme, aimed at verifying whether neutrons were able to provoke radioactive phenomena in the same way as alpha particles. It should be noted that until then neutrons had never played a part in the experimental projects carried out in Rome and that Fermi, in this research of his, did not use the instrumentation with which the Physics Institute in Rome had been equipped until that moment but, as can be seen from the documentation that has reached us and that we will speak of later, he set up some Geiger-Müller counters by himself, perhaps with the aid of the expertise available at the Institute of Physics of Florence in Arcetri, and used a neutron source based on radon and beryllium, created by him with Trabacchi's support.

8.2 Gian Carlo Wick's contribution

Fermi's choice of applying himself completely independently to a new strictly experimental programme, in the pursuit of a possible neutron induced radioactivity, was heavily influenced by his beta decay theory and, in particular, by a recent extension obtained by Gian Carlo Wick, then a young assistant of Fermi's in Rome. In a note presented by Fermi himself at the session of the Accademia dei Lincei on 4 March 1934 and later published in "Rendiconti" [165], Wick had demonstrated that Fermi's theory of beta decay could be successfully applied to the case of radioactivity β^+ discovered by the Joliot-Curies. Wick's work, which had almost certainly been suggested by Fermi himself, was founded on the theory of beta decay, as expounded by Fermi in his second article, in Italian, which at that time was in the process of being published in the January 1934, volume of "Il Nuovo Cimento". The same numbering of the formulae was used. Indeed, in an introductory note, Wick wrote: "I sincerely thank Prof. Fermi for the chance of seeing the manuscript before publication and for the interest he took in this work".

Wick wrote a letter to Joliot, sent from Rome on 17/2/1934 while he was tackling this problem, in which he asked for clarification of the maximum energy values of the positrons emitted by aluminium and "on the possibility of comparing the order of magnitude given by the formula [which he had calculated] with the experimental results". Wick said: *Monsieur, I have tried to apply a theory of beta disintegration recently developed by M. Fermi to the new phenomenon of radioactivity that you have announced in C.R. and in Nature. According to this theory it is natural to interpret this phenomenon as a kind of inverse process of β disintegration with the destruction of an electron and a neutrino, both with negative kinetic energy. We can then make calculations analogous to Mr Fermi's, with the difference that the positrons are repelled by the nucleus while the electrons are attracted. This implies having to change the sign of the nuclear charge in the formulae. Moreover there are some simplifications due to the fact that they are light elements.* Unfortunately we have not found the letter of reply, probably sent by Joliot.

In his article to the "Rendiconti Lincei" Wick started from the fact that Fermi's theory contained inside it the possibility of two symmetrical and inverse transformation processes — on the one hand the transformation of a neutron into a proton, which gave rise to a beta negative-type decay, and on the other hand the transformation of a proton into a neutron. He therefore set himself the task of verifying whether "the phenomena of induced radioactivity observed by F. Joliot and I. Curie" could be interpreted in the framework of Fermi's theory as a β positive-type decay, which Wick indicated with the symbol $_{-}\beta$, in which a proton was transformed into a neutron with the creation of a positron and an antineutrino, that is to say a "neutrino hole". Wick, referring directly to Fermi's theory, wrote: "The fundamental hypothesis of the theory consists in the introduction into the Hamiltonian system of terms that correspond to transitions in which a neutron is transformed into a proton while an electron and a neutrino are created. A plausible form for these terms can be deduced from considerations of relativistic invariance. Naturally the theory also contains the possibility of the inverse process: transformation of a proton into a neutron and destruction of an electron and a neutrino. For such a process to be able to occur however it is essential that there is a certain density of neutrinos in the vicinity of the nucleus. This density is provided precisely by the neutrinos with negative energy; the destruction of one of these neutrinos is equivalent to the formation of a particle [neutrino hole] perfectly analogous to the neutrino. If the electron that is absorbed is an electron with negative kinetic energy a positron is emitted. It is natural to identify this phenomenon with the phenomenon observed by Joliot and Curie".

Anyway the emission of a positron, or rather the destruction of a negative electron, as Wick pointed out, was not the only way in which "the nuclei considered can disintegrate. Instead of an electron with negative kinetic energy, one of the electrons K, L, M ... that constitute the external structure of the radioactive atom" can be destroyed. Accordingly Wick estimated the probability of both "these transitions", either with the emission of positrons or with the capture of atomic electrons. By means of rigorous theoretical analysis, based entirely on Fermi's theory, Wick deduced that the "$_{-}\beta$ radioactivity" discovered by Joliot and Curie was perfectly explained by this theory and that "for light nuclei the emission of positrons was by a long way more prevalent". Wick concluded as follows: "The same Hamiltonian that produces the transitions in which an electron is created can in fact, if the energy conditions are favourable, produce the "destruction" of an electron that is in a state of negative energy, which according to Dirac's theory is equivalent to the creation of a positive electron. As well as the qualitative explanation of the phenomenon the theory allows us to predict that the positrons are emitted in a continuous spectrum with a shape very similar to the shape of the β spectra. Moreover the estimate of the mean lifetime gives a result that can be reconciled with the experimental result".

These conclusions that Wick arrived at regarding the reliability of Fermi's theory were very important if you think that until that moment the verification of this theory, as even Fermi himself had admitted, was rather qualitative. Wick observed on this subject: "E. Fermi has shown that [...] it is possible to construct a formal framework that can give well defined, quantitative answers to various questions, such as: the shape of the β spectrum, mean lifetimes, etc. For the moment however, the uncertainty that still reigns about eigenfunctions and other nuclear properties does not allow precise data to be extracted from the theory but only estimates of the orders of magnitude. While, following these various uncertainties, it is difficult for now to find sure proofs in favour of this theory, the fact is however interesting that the $_{-}\beta$ radioactivity now discovered by Curie and Joliot perfectly fits [...] in the context of Fermi's theory of β disintegration".

8.3 An excellent point of departure

So Wick's analysis provided Fermi's theory, very soon after its formulation, with a further confirmation actually by means of Joliot and Curie's experiments, and at the same time demonstrated that this theory was able to explain the origin of the "transmutation positrons" seen by Joliot and Curie, linking them to a process within the nucleus of the transformation of a proton into a neutron.

Between 15 January (the Joliot-Curies' discovery) and 4 March 1934 (Wick's analysis), so in a span of less than two months, not only was it established that it was possible to provoke artificially a β positive-type decay but it was shown that this type of decay could also be explained by Fermi's theory.

At this point, if one did not assume, as Joliot and Curie and all those who until then had dealt with the problem experimentally had done, that the characteristic of artificially induced radioactivity was exclusively beta plus decay, that is to say with the emission of positrons, but instead if one fully embraced Fermi's theory and admitted the symmetry between the transformation process of a proton into a neutron and vice versa, a question would naturally arise.

Since it was possible to create artificially elements that had a positive beta decay or, according to Fermi's theory, provoke by external means the transformation of a proton into a neutron, why should it not be possible also to provoke, again by external means, the inverse transformation of a neutron into a proton and so create artificially radioactive substances with a negative beta decay, thus with the emission of electrons?

In order to answer this question, still on the basis of Fermi's theory, the only method was to bombard the nucleus with neutrons. Indeed, once the incident neutron had been absorbed by the target nucleus and, as was known at that time (see Russell's table shown in Fig. 28), an alpha particle had been expelled with a (n, α) reaction, the overall number of neutrons in the nucleus in the balance of the reaction would be increased compared to the number of protons and with it the probability that a neutron might be transformed into a proton, with the subsequent emission of an electron (as well as of a neutrino).

Moreover, in the light of Fermi's theory, a similar mechanism was the basis of Joliot and Curie's experiments and of the experiments that immediately followed by J. D. Cockroft, C. W. Gilbert, E. T. S. Walton [41], M. C. Henderson, M. S. Livingston, E. O. Lawrence [95], H. R. Crane and C. C. Lauritsen [45, 46].

Indeed by bombarding nuclei with α particles, with protons and with deuterons respectively, in the overall balance of the possible reactions provoked, (α, n), (p, n), (d, n), they had increased the number of protons present in the nucleus compared to the number of neutrons and so they had also increased the probability that a proton would be transformed into a neutron with the subsequent emission of a positron, as well as of an antineutrino. So they had favoured the possibility of a beta positive type of decay occurring.

At this point one can understand why it was that Fermi, despite all the difficulties associated with neutron sources, decided to undertake a kind of experiment that until then, although it had already been suggested, had never been attempted even in the most advanced laboratories in the field. Fermi's strength lay all in his theory of beta decay which guaranteed his success, and certainly also in Fermi's determination to obtain a further proof of the validity of his beta decay theory.

So not by chance Fermi, a few days after the presentation of Wick's work, applied himself to the creation of a whole measuring apparatus (which we will amply discuss later) with which to bombard nuclei with neutrons, with the aim of verifying whether it was possible to provoke an artificial beta negative-type decay, and finally, within a few weeks, according to the detailed reconstruction made later, succeeded in discovering radioactivity induced by this type of particles, as announced in his Letter to "La Ricerca Scientifica" dated 25 March 1934 [F120].

In the light of Fermi's theory it was clear that if one wished to induce artificially a beta negative-type decay with a neutron source the presence in the source of any γ rays was irrelevant because they would not provoke any effect in the transformation of a neutron into a proton. This led to Fermi's winning choice of using a neutron source of the radon-beryllium type, which as we shall see better later also emitted strong gamma radiation along with the neutrons but that had the great advantage of producing many more neutrons for the same amount of active product, so for the same number of alpha particles present, compared for example to a polonium-beryllium source, which instead was a practically pure source as far as the emission of alpha particles was concerned, but much less intense.

It was actually Fermi's beta decay theory that drove him to undertake an experiment that no one had attempted until then, giving him the edge, even in the choice of the type of source to use.

8.4 A precedent: fluorine

At this point it should be observed that signals of the possible success of an experiment of this type, and thus the possibility of finding an induced radioactivity, already existed in the literature, in particular with regard to fluorine. In volume 44 of "Physical Review", published on 1 December 1933 and so about a month before the Joliot-Curies' discovery and about four months before Fermi's discovery, a "Letter to the Editor" appeared, sent on 10 November 1933 by three American chemists at the George Herbert Jones Chemical Laboratory of the University of Chicago, William Draper Harkins (1873-1951), David M. Gans, Henry W. Newson (1909-1978), with the very significant title: "Disintegration of fluorine nuclei by neutrons and probable formation of a new isotope of nitrogen (N^{16})" [88]. It can be said that in this Letter there is, in embryo, everything that Fermi would discover in March 1934 regarding fluorine, in particular with regard to the possibility of disintegration with neutrons, to the (n, α) type of reaction, and to the subsequent formation of an unstable nucleus.

This study by Harkins-Gans-Newson was part of a wide ranging project aimed at studying the disintegrations produced by neutrons in some chemical elements. They inserted through a small aperture made in the upper part of a Wilson chamber a neutron source consisting of a blend of beryllium and "mesothorium, thorium X or radiothorium" salts (they had used these salts because "the polonium source available was too weak for this purpose"), and introduced a blend of 30% difluor-dichlor-methane and 70% helium into the cloud chamber. Out of 3200 photographs taken they obtained 10 photographs that showed nuclear disintegrations. These, on the basis of considerations on the momentum and the energy of the traces, were identified for the most part, probably all as "disintegrations of fluorine nuclei". Two of these photographs were included in the article with the caption "Stereoscopic photographs of the disintegration of a fluorine nucleus by a neutron". The authors seemed to have no doubt that they were disintegrations of fluorine. In reality, given the mixture used, they could have been disintegrations of nuclei of chlorine or oxygen. These possibilities were however excluded by the authors: the former on the basis of the energy considerations and the momentum that had led them to opt for fluorine, and the latter because "the disintegrations cannot be due to oxygen, since too little of this element is present". Anyway, to be absolutely certain that they really were disintegrations of fluorine, as the authors announced, new experiments were under way with "carbon tetrafluoride and difluoromethane, neither of which gives the possibility of the disintegration of chlorine".

According to Harkins-Gans-Newson, the photographs previously obtained showed that fluorine disintegrated capturing a neutron and emitting an α particle, according to a (n, α)-type reaction. The final reaction of fluorine, that the authors wrote using the notation invented by Harkins himself in 1915 where the atomic weight M is indicated with superscript and the "isotopic I" number, defined by $I = M - 2Z$ (Z atomic number)(*), with subscript, was thus:

$$\mathrm{F}_1^{19} + n_1^1 \rightarrow \mathrm{F}_2^{20} \rightarrow \mathrm{N}_2^{16} + \mathrm{He}_0^4,$$

which we will write here for clarity using the atomic number Z instead of the "isotopic I" number:

$$\mathrm{F}_9^{19} + n_0^1 \rightarrow \mathrm{F}_9^{20} \rightarrow \mathrm{N}_7^{16} + \mathrm{He}_2^4.$$

This reaction led to the prediction, as they wrote in the very title of the article, of the existence of "a new isotope of nitrogen N^{16}", (that is to say in common notation N_7^{16}), the same isotope that Fermi would predict. Anyway, as the authors noted, the possibility existed that "an electron, the track of which cannot be seen in the photographs, may be emitted at the time of the disintegration [together with the alpha particle], in which

(*) According to the current convention the indices that refer to the isotope and atomic mass are written, respectively, above and below to the left of the symbol of the element. Here and in the following chapters they are kept to the right to follow the convention of the time.

case O_2^{16} would be formed", instead of the new isotope nitrogen N_7^{16}. Obviously in this case the product of the reaction would immediately have been a stable isotope (indeed O^{16} was a known isotope). However for the authors the hypothesis of a (n, α) reaction was "more in accord" with the nuclear reactions provoked by the neutrons that they had obtained until then (they pertained to nitrogen, oxygen and neon) and that they were all (n, α). They ended as follows, aligning themselves conclusively with this type of reaction: "If nitrogen 16 is unstable it may disintegrate subsequently, but this would not affect the calculations given in this paper". Clearly, even if the authors did not state it explicitly, this process of later disintegration would have implied a negative beta decay with the emission of an electron by the unstable isotope, so as to obtain a final stable isotope O^{16} in this way too.

It is interesting to observe what Harkins, Gans and Newson would write in "Physical Review" on 1 January 1935 [89], several months after Fermi's discovery of neutron-induced radioactivity, once having finished the new experiments with carbon tetrafluoride that confirmed the previously hypothesised reaction of fluorine. After having written the reaction of fluorine, using Z and no longer the isotopic number I, they observed: "This reaction is remarkable in that a new isotope of mass 16 and isotopic number 2 appears. In the preliminary announcement of the discovery of this new isotope, the writers considered that it would be likely to be unstable and to emit an electron to form oxygen 16. This has been verified by Fermi who finds nitrogen 16 to be radioactive, and to change to oxygen 16, as was predicted".

The 1933 article by Harkins, Gans and Newson [88] was cited by Russell in the Annual Report of the Chemical Society that same year, which we have already spoken about. Russell reported that fluorine disintegrated due to neutrons and that the process was "possibly $n + F^{19} \rightarrow He^4 + N^{16}$". Anyway, in the table (Fig. 28) in which "the principal disintegration processes which occur with light elements" are summarised this reaction of fluorine was given as "$n + F^{19} \rightarrow He^4 + ?$", with a question mark in the place of the chemical symbol of the final product, as if to stress uncertainty about it or the strangeness of it if it really was a "new isotope" unknown in nature (the isotope N^{16}).

The possibility that fluorine could disintegrate due to neutrons and that it followed a (n, α) reaction was also pointed out by Joliot and Curie in their report at the Solvay Conference in 1933 (where Fermi was also present), without worrying about the implications of such a reaction. Joliot and Curie wrote [99, page 130]: "we have observed in the Wilson chamber a case of transmutation produced by neutrons in sulphur hexafluoride: it is probably the transmutation of a nucleus of fluorine with the capture of the neutron and emission of an α particle. On the other hand, since it is an isolated case, it cannot be asserted that the transmutation observed is not that of a nucleus of oxygen from water vapour".

We do not know if Fermi was aware of these results on fluorine from having followed the Joliot-Curies' report or whether he knew of Harkins-Gans-Newson's article, or if he had read Russell's report. It is plausible however that Fermi, a cautious and responsible person, had amply documented himself before undertaking an experiment on the disintegrations produced by neutrons.

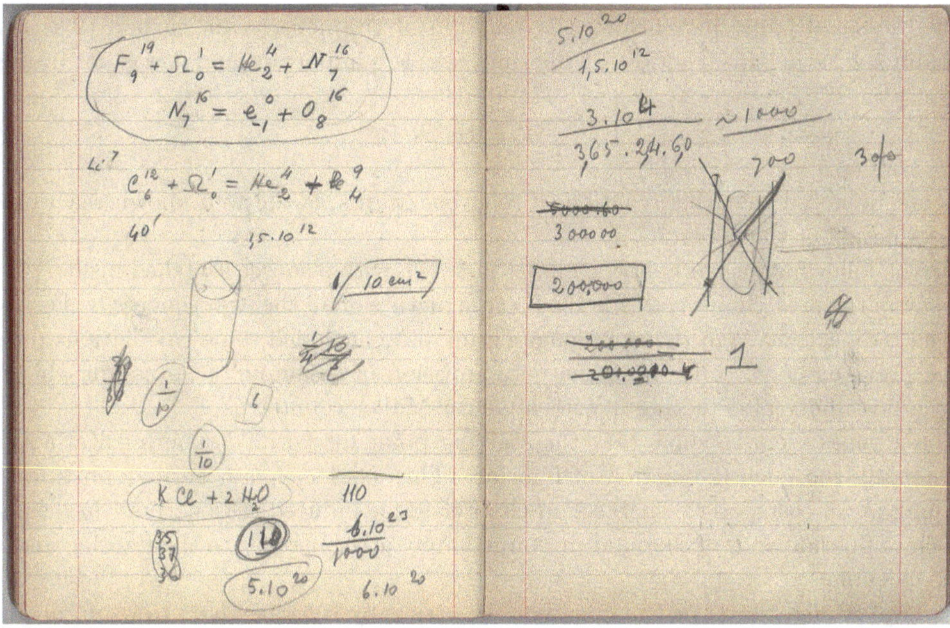

Fig. 43. – Two pages written by Fermi in a laboratory notebook of Amaldi's in 1934 - DFUR.

At the Physics Department of the University of Rome "La Sapienza", a laboratory notebook exists in which Amaldi recorded his atomic spectroscopy results (Archivio Amaldi-Eredi) during the period September 1933 - April 1934. Some pages of this notebook were used by Fermi, in particular for calculations on beta decay. Two pages are particularly significant for our analysis. We show them in Fig. 43. The date is not explicitly given but it can certainly be pinned down as later than 10 February 1934 and prior to 24 March 1934, which are dates recorded a few pages before and after.

On the first page, at the top, Fermi wrote in his own hand the following fluorine reaction, separated into two parts, and framed it with a large oval:

$$F_9^{19} + \Omega_0^1 = He_2^4 + N_7^{16},$$

$$N_7^{16} = e_{-1}^0 + O_8^{16},$$

where the neutron is indicated by the symbol Ω. This expression seems to be the transcription of the usual notations (N, Z) of the nuclear reaction of fluorine suggested by Harkins-Gans-Newson, with the difference that here the later negative beta decay of the active nucleus is specified. With this reaction, Fermi verified that on the basis of his theory it was possible to induce radioactivity in fluorine, that is to say that if a neutron provoked a nuclear reaction in fluorine, as Harkins-Gans-Newson and the Joliot-Curies said they had seen, this led inevitably to the formation of a radioactive nucleus.

On the same page, just below, the reaction of the neutron on carbon-12 is given (probably for Fermi this reaction, at the time, was less important and it was not framed):

$$C_6^{12} + \Omega_0^1 = He_2^4 + Be_4^9.$$

It should be noted that this reaction, if read backwards, is the well known reaction for the production of the neutron.

On the following page a hollow cylinder is drawn with a source sketched inside. As we shall say in the next chapter, this is the basic structure of all the experiments by Fermi on induced radioactivity. On the same page Fermi also concluded some calculations, begun on the previous page, concerning the beta emission of a sample of potassium chloride, and the possibility of detecting it with a Geiger-Müller counter.

These pages seem to show that the starting point for Fermi's experimental activity was actually the nuclear reaction of fluorine. This would explain Fermi's obstinacy in searching at all costs, as we shall see in chapter 9, for induced radioactivity in a sample of calcium fluoride and, once found, in immediately attributing it to the fluorine and not to the calcium.

Perhaps it might even be thought that the article by Harkins-Gans-Newson, in which the possibility of artificially provoking a radioactivity phenomenon can be read between the lines, may have played some role in prompting Joliot and Curie, as we have seen before, finally to take seriously the suggestion put forward by Perrin at the Solvay Conference and to seek in their previous experiments with aluminium and boron possible signs of an artificial radioactivity, in this case induced by α particles. This obviously is only a hypothesis.

In any case it is striking to see the close analogy between the two cases hypothesised by Harkins-Gans-Newson, according to which the neutron could provoke either simultaneously the emission of an α particle and an electron or first the formation of an unstable nucleus followed by a negative beta emission, and the situation created by Joliot-Curie's experiments, in which first it is assumed that the α particle can produce a neutron and a positron simultaneously and then, following Perrin's suggestion, it is instead recognised that in a first phase an unstable nucleus is produced which then undergoes a positive beta decay.

Whatever the role that the article by Harkins-Gans-Newson may have had in directing Fermi's research, or even the Joliot-Curies', it remains the fact that the credit goes to Joliot-Curie and Fermi for having demonstrated experimentally the possibility of provoking an induced radioactivity, with alpha particles and with neutrons, respectively.

8.5 The discovery of neutron induced radioactivity and the sign of the charge

The Letter to "La Ricerca Scientifica" in which Fermi announced his own discovery was very brief. The title is "Radioattività indotta da bombardamento di neutroni" (Radioactivity induced by neutron bombardment) (Fig. 44).

In it Fermi began by directly linking to the Joliot-Curies' discovery. Indeed he wrote that the aim he had set himself was to: "ascertain whether a neutron bombardment dell'ordine di 1 parte su 5000 parti, cioè di una molecola di acqua pesante su circa 5500 molecole di acqua normale.

Concentrazioni dello stesso ordine di grandezza sono state trovate in alcune delle citate celle per elettrolisi.

Dei particolari di questa eseperienza e delle altre tuttora in corso verrà data dal Dr. Lucchi relazione sulla « Gazzetta Chimica Italiana ».

Prof. O. SCARPA

Laboratorio di Elettrochimica e di Chimica-Fisica del R. Politecnico di Milano.

Radioattività indotta da bombardamento di neutroni

Desidero riferire in questa lettera sopra alcune esperienze destinate ad accertare se un bombardamento di neutroni non determini dei fenomeni di radioattività susseguente analoghi a quelli osservati dai coniugi Joliot con bombardamento di particelle α.

Il dispositivo che ho usato è il seguente: La sorgente di neutroni è costituita da un tubetto di vetro contenente polvere di berillio ed emanazione. Usando circa 50 millicurie di emanazione, che mi sono stati forniti dal prof. G. C. Trabacchi che qui desidero ringraziare vivissimamente, si possono così ottenere oltre 100.000 neutroni al secondo, misti naturalmente a una intensissima radiazione γ, che però non dà alcun disturbo per esperienze di questo genere. Dei cilindretti contenenti l'elemento in esame sono sottoposti per un tempo variabile da alcuni minuti ad alcune ore alle radiazioni di questa sorgente.

Essi vengono poi rapidamente disposti attorno ad un contatore a filo, la cui parete esterna è formata da una foglia d'alluminio di spessore di circa 0,2 mm. tale quindi da permettere l'ingresso di eventuali raggi β nel contatore. Fino ad ora l'esperienza ha dato esito positivo per due elementi:

ALLUMINIO. — Un cilindretto di alluminio irradiato dai neutroni per un paio d'ore e posto successivamente attorno al contatore determina nei primi minuti un aumento assai considerevole degli impulsi, che crescono di 30 o 40 al minuto. L'effetto decresce col tempo riducendosi a metà in circa 12 minuti.

FLUORO. — Il fluoruro di calcio, irradiato per pochi minuti e portato poi assai rapidamente accanto al contatore determina nei primi momenti un aumento del numero degli impulsi. L'effetto si smorza rapidamente, riducendosi a metà in circa 10 secondi.

Una possibile interpretazione di questi fenomeni è la seguente. Il fluoro, bombardato coi neutroni, si disintegra emettendo particelle α. La reazione nucleare è probabilmente:

$$F^{19} + n^1 \longrightarrow N^{16} + He^4$$

Si formerebbe così un azoto di peso 16 che, emettendo successivamente una particella β può trasformarsi in O^{16}. Una simile interpretazione potrebbe aversi per l'alluminio, conformemente alla possibile reazione nucleare:

$$Al^{27} + n^1 \longrightarrow Na^{24} + He^4$$

Il Na^{24} così formato sarebbe un nuovo elemento radioattivo e si trasformerebbe in Ca^{24} con emissione di una particella β.

Se queste interpretazioni sono corrette, si avrebbe qui la formazione artificiale di elementi radioattivi che emettono normali particelle β, a differenza di quelli trovati dai Joliot che emettono invece positroni. In particolare nel caso dell'azoto si avrebbero due isotopi radioattivi: N^{13}, trovato dai Joliot, che emettendo un positrone si trasforma in C^{13}; ed N^{16} che, emettendo un elettrone si trasforma in O^{16}.

Sono in corso esperienze per estendere l'esame ad altri elementi e per studiare meglio le particolarità del fenomeno.

Roma, 25 *marzo* 1934-XII.

ENRICO FERMI

Fig. 44. – Letter from Fermi to "La Ricerca Scientifica" in which he announces the discovery of neutron induced radioactivity (25 March 1934) - [F120] (RS).

might cause subsequent radioactivity phenomena analogous to those observed by the Joliot couple with α particle bombardment". He then immediately gave a very brief description of the experimental apparatus used. It consisted of a neutron source, with which the samples being examined were irradiated, and a Geiger-Müller counter with which any beta radiation emitted by the various samples after they had been bombarded with neutrons was measured. Fermi wrote: "The neutron source consists of a small glass tube containing beryllium powder and [radon] emanation. [...] Small cylinders containing the element being examined are submitted to the radiations of this source for a time varying from a few minutes to a few hours. They are then placed around a wire counter, with an outer wall made of an aluminium sheet about 0.2 mm thick, suitable therefore to allow any β rays to enter the counter". He continued reporting the results achieved: "So far the experiment has given a positive result for two elements: aluminium [...] fluorine". In the case of aluminium the sample, after being irradiated for about two hours and placed around the counter, provoked "a rather considerable increase of the impulses which grew by 30 or 40 per minute [compared to background]. The effect decreases over time reducing to half in about 12 minutes". As for fluorine, a sample of calcium fluoride "irradiated for a few minutes and then carried rather rapidly next to the counter" determined "in the first moments an increase in the number of impulses", compared to the background; the effect decreased rapidly "reducing to half in about 10 seconds". For fluorine Fermi suggested as probable the following (n, α)-type nuclear reaction, based on the experience of nuclear disintegrations with neutrons known at that time (remember for example Russell's table in Fig. 28), that is to say with the capture of the neutron and subsequent emission of an alpha particle:

$$\mathrm{F}^{19} + n^1 \rightarrow \mathrm{N}^{16} + \mathrm{He}^4 .$$

And he went on: "So a nitrogen with weight 16 would be formed which, then emitting a β particle, could be transformed into O^{16}". Incidentally, this decay reaction and the previous one are the very reactions written by Fermi in Amaldi's notebook which we talked about earlier. A nuclear reaction of the same (n, α)-type was also proposed for aluminium:

$$\mathrm{Al}^{27} + n^1 \rightarrow \mathrm{Na}^{24} + \mathrm{He}^4 ,$$

where "The Na^{24} thus formed would be a new radioactive element and would be transformed into Ca^{24} with the emission of a β particle" (Fermi made a slip here: the final product of Na^{24} after a β decay is not the nucleus Ca^{24} but the nucleus Mg^{24}).

Fermi continued: "If these interpretations are correct, here we would have the artificial formation of radioactive elements that emit normal β particles, unlike those found by the Joliots which instead emit positrons. In particular in the case of nitrogen there would be two radioactive isotopes: N^{13}, found by the Joliot-Curies, which by emitting a positron is transformed into C^{13}, and N^{16} which, by emitting an electron is transformed into O^{16}".

And he concluded: "Experiments are under way to extend the examination to other elements and to study the peculiarities of the phenomenon better".

At this point a question is natural. Since, as has already been said before, it is not possible to establish the sign of the electrical charge with a Geiger-Müller counter, how could Fermi arrive at the conclusion that in his experiments, unlike those of Joliot-Curie, "normal β particles" were emitted, without making a separate measurement of the sign of the charge.

It should be noted that Joliot and Curie when they had suggested using neutrons too in order to create artificially new radioactive isotopes expected that these would decay with the emission of positrons (β^+), like all the other artificial "radio-isotopes" created up to that time.

It is clear that Fermi's assumption derived directly from his theory of beta decay. Indeed this theory required that, following the bombardment of a nucleus with neutrons, an electron be emitted, so a normal β particle, as a consequence of the transformation of a neutron into a proton. And in fact only two months later, on 10 May 1934 after the discovery, when by now Fermi's research in Rome into neutron-induced radioactivity also made use of the endeavours of his collaborators, an experimental proof of this hypothesis would be sought and found. In a third Letter to "La Ricerca Scientifica" [F123], signed in alphabetical order by da E. Amaldi, O. D'Agostino, E. Fermi, F. Rasetti, E. Segrè, with a title in which reference was finally made to the sign of the charge "Radioattività "beta" provocata da bombardamento di neutroni. III" ("Beta" radioactivity provoked by neutron bombardment. III), we can read: "finally a magnetic device analogous to the one described by J. Thibaud (C.R., 197, 447 (1933)) was constructed to determine the sign of the particles emitted. Because of the intensity of the effect so far it has been possible to study the following elements: Al; Si; P; S; Ga; Cr; Fe; Ag; Br; I. In all cases only negative electrons were observed".

This type of investigation was later pursued by Amaldi and Segrè, in the case of Si, Al, P, Cr, Ag, J, Cr using, as well as the Thibaud trochoid method, as we have seen already used by Joliot and Curie, also a cloud chamber "designed and constructed by Prof. F. Rasetti". Since from the inventory of the Royal Physics Institute it does not emerge that material for a new cloud chamber was ever acquired, apart from the two chambers registered on 2 December 1933, it must be understood that the reference to Rasetti has to be connected to one of these two bought chambers. A magnetic field of "about 180 gauss, ... obtained with two coils in Helmholtz position" was applied to the chamber. With this method too it was confirmed that the new radioisotopes created by neutron bombardment had a negative beta decay [5]. Amaldi and Segrè wrote: "In all cases we have observed only traces of light particles; given the not very great intensity of the magnetic field used the curvature of the trajectories is not always clearly recognisable. For those cases where the sign can be identified it indicates that they are electrons".

Anyway an explicit declaration by Fermi about the link between the predictions of his theory on beta decay and the emission of electrons in the experiments with neutron bombardment, would only be made in his first paper published on the subject in "Il Nuovo Cimento" [F131], which represented the summing up of all the activity carried out until that time. This paper is not dated but presumably it is from mid May since the results given extend up to those announced in the third letter to "La Ricerca Scientifica", dated 10 May 1934, which is also cited. Fermi wrote: "Note that both the absorption of a neutron and the expulsion of a proton or of an alpha particle are processes that tend to raise the number of nuclear neutrons compared to the number of protons. This is most likely the reason why, generally, the emission of negative electrons is observed; indeed the emission of a nuclear electron, which can be interpreted as connected to the transformation of a neutron into a proton, re-establishes the normal ratio between the numbers of neutrons and of protons within the nucleus and determines the passage to a stable isotope".

Regarding the problem of the sign of the particles emitted by the new radioelements produced by neutron bombardment it is significant what Karl K. Darrow of the Bell Telephone Laboratories wrote in a letter to Fermi on 3 May 1934, now in the Domus Galilaeana, referring both to Fermi's first Letter to "La Ricerca Scientifica" [F120], and to the Letter that immediately followed, again sent by Fermi to "La Ricerca Scientifica" [F121]. In this second Letter, which we will speak of more at length later, Fermi, after having pointed out that he had obtained positive results for many chemical elements, as well as aluminium and fluorine, announced that in some cases the "electrons emitted were photographed in the Wilson chamber". It was certainly a normal Wilson chamber without a magnetic field applied, undoubtedly one of the two chambers the Physics Institute was equipped with and which Fermi and Rasetti had spoken of around two months earlier in their letter to D'Agostino.

Darrow wrote: *Dear Dr. Fermi, I thank you for the two notes from La Ricerca Scientifica. I had heard rumors of the earlier of the two by way of Cambridge (England) and Columbia University; in the latter place it produced quite a sensation, as Pegram and Dunning are working there with copious sources of neutrons. The number of elements on which you have already made observations is truly remarkable; you must have worked with terrific industry. One important thing however is not stated: how do you know that the electrons which are given off from the pseudo-radioactive substance are negative? I suppose that you deduce from the aspect of the tracks in the Wilson chamber that the particles are electrons, but how do you know their sign?*

Anyway, as we have just seen, an experimental answer to Darrow's perfectly legitimate question was not long delayed and was actually given with two different methods: with Thibaud's trochoid and with a cloud chamber with a magnetic field applied, thus leading to a further verification of Fermi's theory on beta decay.

8.6 After the discovery

After the first Letter to "La Ricerca Scientifica" on 25 March 1934 [F120] Fermi wrote a later one to the same journal and with a very similar title [F121]. In this second Letter, in which Oscar D'Agostino was cited for having carried out the "chemical separations" and Edoardo Amaldi and Emilio Segrè for collaborating with the "physical part", Fermi announced that "continuing experiments with the method described in the previous letter", he had obtained positive results in the case of "numerous other elements", and he had "studied their radioactivities" (iron, silicon, phosphorus, chlorine, vanadium, copper, arsenic, silver, tellurium, iodine, chromium, barium). He pointed out that there were "indications of effects of varying intensities and period" in the case of the elements sodium, magnesium, titanium, zirconium, zinc, strontium, antimony, selenium, bromine. He added that in the case of phosphorus, silicon and aluminium "the electrons emitted were photographed in the Wilson chamber". Finally he announced that the chemical separations of the radioactive isotopes carried out in the case of iron and phosphorus seemed to indicate a nuclear reaction in which "the neutron was absorbed and a proton emitted", so of the (n, p) type.

The contents of these two Letters were summed up by Fermi in a Letter to "Nature", sent on 10 April, with the title "Radioactivity induced by neutron bombardment" [F122]. The second Letter to "La Ricerca Scientifica" is not dated but since its contents were reported in the Letter to "Nature", we can presume that it too was sent around 10 April. Regarding the communication to "Nature", it should be said that it was not placed as a scientific article but appeared in the section "Letters to the Editor" where it is explicitly stated that: "The Editor does not hold himself responsible for opinions expressed by his correspondents". If one thinks of the great importance that the journal "Nature", as we have seen, reserved to the Joliot-Curies' discovery of α-particle–induced radioactivity, so much so that the Editor was driven to write personally to Joliot to ask him for an article to publish on the subject, it is clear that Fermi's discovery was not immediately fully understood by the journal and considerably underestimated. The error is incredible. Indeed, despite its enormous scientific importance, from the point of view of possible applications the Joliot-Curies' discovery is reduced basically to a minor curiosity while instead Fermi's discovery opened up the door for developments of applications of staggering significance.

On 10 May 1934, as we have already mentioned, a third letter was sent to "La Ricerca Scientifica" [F123]. This Letter officially introduced the formation of a real working group in Rome around the figure of Fermi made up of those who signed the Letter. Later, after the summer of 1934, Bruno Pontecorvo, a young man who graduated in Rome in 1933 and who had been working until then in spectroscopy research (see for example [86]) was included. The aim of this working group, which would remain operative until about the middle of 1935, was essentially the carpet bombardment of the entire periodic table to pinpoint the elements that were activated and to identify any new radioactive isotopes that were thus formed.

To organise the group research better special cards were printed, to be compiled front and back for each of the measurements taken. On the front were indicated the substance being examined, the date of the measurement, the length and conditions of the irradiation, the chemical operations carried out, the equipment used, any observations. Instead on the back the counts obtained were recorded as a function of time. Every card was generally accompanied by one or more graphs on squared paper where the counts were recorded. In this way the results obtained, instead of being scattered in multiple laboratory notebooks, were gathered together centrally and easy to consult. Fermi could thus follow the whole research activity directly. Figures 45-47 show an example of a card. In this case unfortunately there were impurities present in the irradiated sample so the measurements were unreliable. Fermi signalled his disapproval by writing “Balle!” (Balls!) on the graph complete with an exclamation mark. It is one of the few explicit comments by Fermi on the recordings of the results obtained.

The work carried out was immense and the results arrived at were sensational. Within a few months activation was successfully demonstrated for a large number of elements of the periodic table, including uranium, and an enormous quantity of new radioactive isotopes (erroneously it was believed that elements outside the periodic table, the so called “transuranium elements” had also been produced). In the summary article [F135], sent in July 1934, it was announced that “more than forty elements out of about sixty investigated could be activated by this method”. In fact 63 elements were subjected to irradiation, of which 37 proved to be certainly active while doubts remained for 9 elements. Instead 17 elements proved to be certainly inactive. These results, if compared to those obtained with alpha particle induced radioactivity, where the number of elements successfully activated was just under ten, were striking. Later, with the discovery of the effects of slow neutrons it would be possible to add new elements to the list of those that could be activated.

Naturally in this group activity the undisputed boss was Enrico Fermi. This is confirmed, for example, by the policy adopted with regard to publications, in particular the way that the scientific papers that step by step were produced were signed (in total, between May 1934 and May 1935, 12 Letters were published in “La Ricerca Scientifica”, 2 articles in “Il Nuovo Cimento”, 3 articles in “La Gazzetta Chimica”, 3 publications in “Nature”, 2 articles in the “Proceedings of the Royal Society” in London on the problem of neutron-induced radioactivity).

Routine articles, in which the list of new elements that had been successfully activated were announced, generally furnished with a brief description of their properties, were always signed by all the members of the group listed strictly in alphabetical order, Fermi included. Instead, the most important and significant articles which summed up the situation, also providing a theoretical interpretation, or papers in which a new discovery was announced, such as for example the “discovery of the transuranium elements” or the effect of slowing down the neutrons, were only signed by Fermi, or by Fermi followed by the names of the other members of the group, given in alphabetical order with no priority.

With regard to this period of group activity what Fermi declared in the speech given in December 1938 when he was awarded the Nobel prize for Physics is significant [F173]: “A systematic analysis of the behaviour of the elements through the Periodic Table was carried out by myself with the help of several collaborators, namely Amaldi, D’Agostino, Pontecorvo, Rasetti, and Segrè”. Note the lack of first names and the failure to respect academic rank. And he concluded: “I must thank in particular all my collaborators that have already been mentioned; the Istituto di Sanità Pubblica in Rome and especially Prof. G. C. Trabacchi, for the supply of all the many radon sources that have been used; the Consiglio Nazionale delle Ricerche for several grants”.

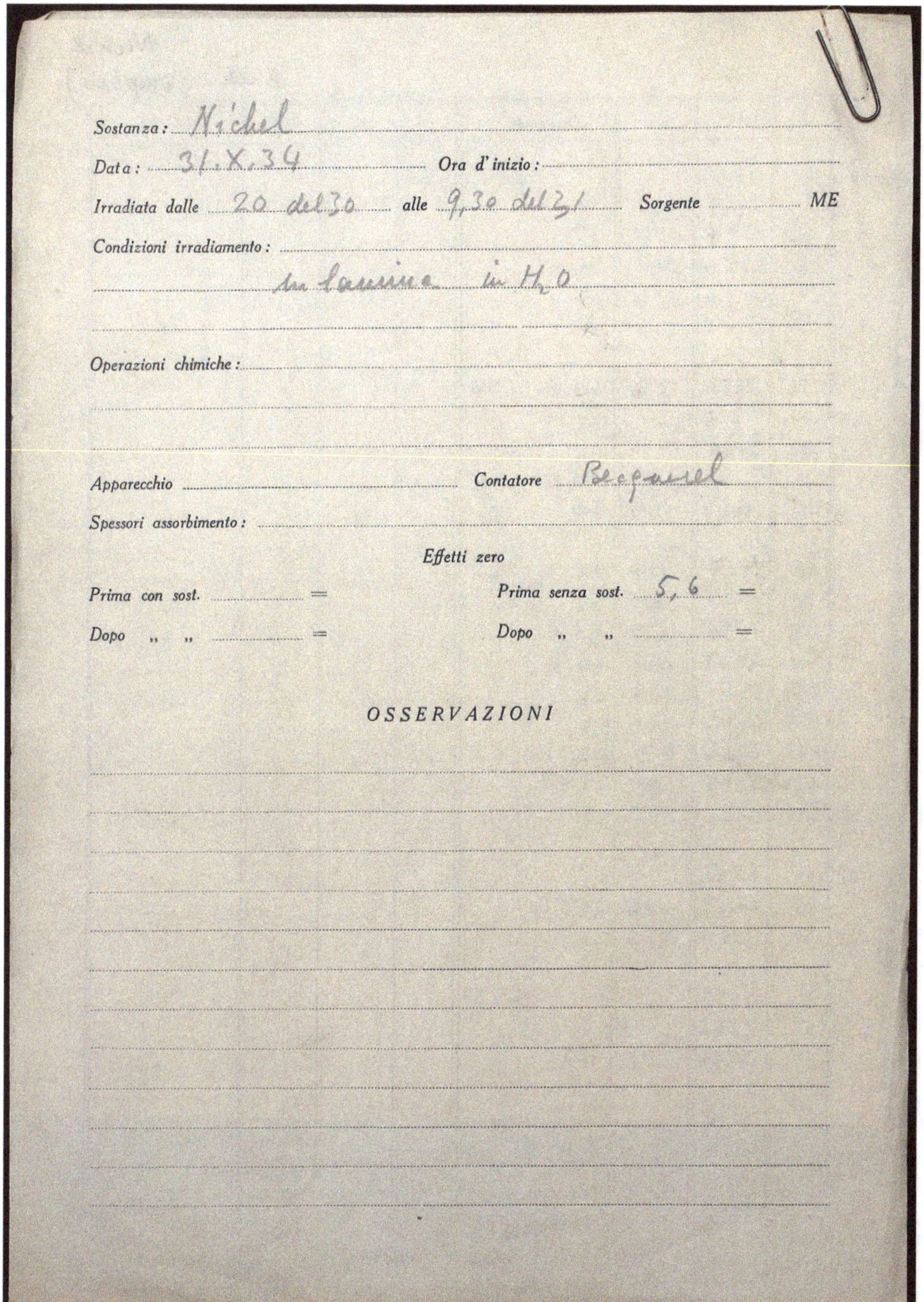

Sostanza: Nichel

Data: 31.X.34 *Ora d'inizio:*

Irradiata dalle 20 del 30 *alle* 9,30 del 31 *Sorgente* ME

Condizioni irradiamento:

in lamina in H_2O

Operazioni chimiche:

Apparecchio *Contatore* Becquerel

Spessori assorbimento:

Effetti zero

Prima con sost.	=	*Prima senza sost.* 5,6	=
Dopo ,, ,,	=	*Dopo ,, ,,*	=

OSSERVAZIONI

Fig. 45. – Example of a pre-printed card for the study of neutron-induced radioactivity (front) (1934) - FDG.

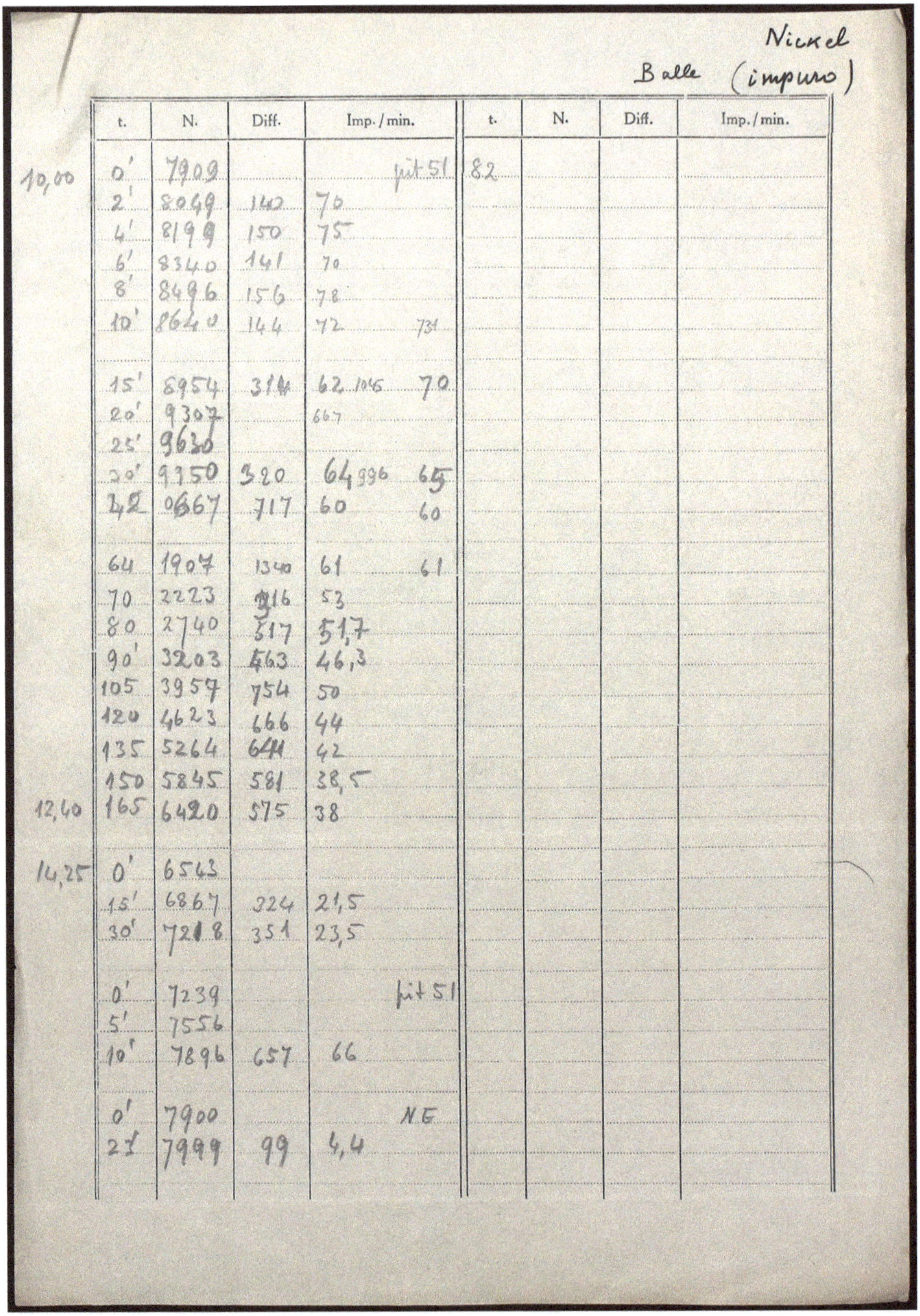

Nickel

Balle (impuro)

	t.	N.	Diff.	Imp. / min.		t.	N.	Diff.	Imp. / min.
10,00	0'	7909			pit 51	82			
	2'	8049	140	70					
	4'	8199	150	75					
	6'	8340	141	70					
	8'	8496	156	78					
	10'	8640	144	72	73				
	15'	8954	314	62 1045	70				
	20'	9307		667					
	25'	9630							
	30'	9950	320	64 996	65				
	42	0667	717	60	60				
	64	1907	1340	61	61				
	70	2223	316	53					
	80	2740	517	51,7					
	90'	3203	463	46,3					
	105	3957	754	50					
	120	4623	666	44					
	135	5264	641	42					
	150	5845	581	38,5					
12,40	165	6420	575	38					
14,25	0'	6543							
	15'	6867	324	21,5					
	30'	7218	351	23,5					
	0'	7239			pit 51				
	5'	7556							
	10'	7896	657	66					
	0'	7900			N.E.				
	21'	7999	99	4,4					

Fig. 46. – Example of a pre-printed card for the study of neutron-induced radioactivity (back) (1934) - FDG.

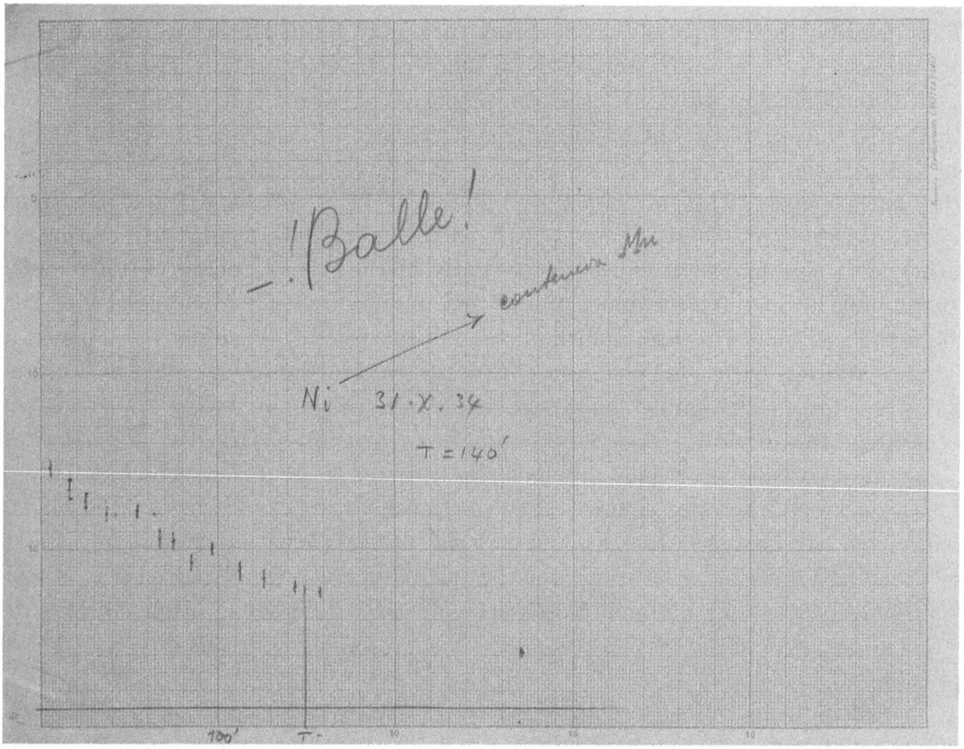

Fig. 47. – Graph of the measurements of neutron-induced radioactivity in Nickel (1934) - FDG.

From the very beginning Fermi's research into neutron induced radioactivity was widely acclaimed, even in the media. The importance of Fermi's discoveries was highlighted solemnly on 3 June 1934 in the Royal Session of the Regia Accademia dei Lincei, which took place in the presence of the King and Queen at the conclusion of the academic year. Orso Mario Corbino gave a lecture with the title "Prospects and results of Modern Physics" in which, after recalling the various discoveries made until then at an international level in nuclear Physics, he emphasised the enormous contribution made by Enrico Fermi in this new field, stressing how he had succeeded, using neutrons as projectiles, in activating a great number of elements ("45 of the sixty two tested so far"), even those elements that had seemed invulnerable to any attack, "even uranium which represents the extreme of the series of known elements", while in the rest of the world, using other projectiles such as alpha particles or protons, "only three or four elements" had been successfully activated.

Furthermore Corbino, in particular, placed great significance on the alleged discovery of the new element 93, derived from the activation of uranium, considering it to be "already certainly verified" despite "Fermi's prudent circumspection in continuing his investigations before announcing the discovery as final".

The scientific importance of Fermi's results of nuclear transmutation through neutron bombardment, as Corbino stressed, was obvious and it was enormous. Corbino then also took into consideration the "practical consequences" to which these results could "open up the path, that is to say the possibility of transforming matter in practically efficient quantities".

He also recognised the fact that "the power to carry out such transformations in sufficient measure would give man, as well as the immediate availability of the rarest elements, also dominion over a practically unlimited source of energy". With great lucidity, perhaps also as a consequence of long and intense scientific discussions with Fermi, Corbino realised that "nuclear transmutation of matter [...] is subject to quantitative limitations which, at least for now, must be considered insuperable". And he continued: "Other procedures that rapidly engage the whole mass of the body to be transformed will be necessary. In a certain sense the intranuclear reactions of this new Superchemistry can be compared to combustions in which the neutron is the natural combustive agent, that is to say it is the only element that can reach the nucleus without the obstacles deriving from electrical repulsion. [...] It follows that currently we can procure this precious agent only by chasing it out from the nuclei where it is lodged, this is laborious and can only be done for minimal quantities. [...] So a different solution presents itself, or rather is required; for example, the solution of producing neutrons directly. Will that be possible?".

These words from Corbino really seem to anticipate the chain reaction that would only be possible after the discovery of fission. Anyway it seems clear that the prospects opened up by Fermi's research were extremely wide ranging, well beyond "the systematic bombardment of the periodic table".

All the major newspapers in Italy and abroad, and even some periodicals and magazines, reported parts of Corbino's talk with great emphasis, so Fermi's discoveries and their importance became known everywhere. The headlines, in giant letters, had great impact. They went from "Fascist victories in the field of culture. A great Italian scientific discovery that opens up the path to the transformation of matter, announced and illustrated in its theoretical and practical importance at the Lincei in the presence of the King" ("Il Giornale d'Italia", 5 June 1934), to "A new record for Italian Science. The experiments of academician Fermi have ripped away a secret from the mysteries of the atom" ("Il Secolo XIX", 5 June 1934), to "The King attends the conclusion of the Academic Year of the Lincei. An interesting communication on Prof. Fermi's research into the disintegration of the nucleus" ("La Tribuna", 5 June 1934). Even the "Corriere dell'Irpinia", taking into account the fact that Corbino had mentioned Fermi's "valiant collaborators" (Rasetti, Segrè, Amaldi, D'Agostino, in academic order), published on 7 July 1934 a long article with the title "The great successes of Italian Science and the useful contribution of a valiant young man from Irpinia".

Amongst foreign newspapers, we cite the case of "Candide" which in an article on 5 July 1934, with the headline "Au coeur de l'atome - La création de l'élément 93", reported the discovery by the "grand savant de Rome, Fermi", whose "victory is certain" since the Joliots had repeated and confirmed his results. Instead on 16 June 1934 "The

Times" used the headline "20th-CENTURY ALCHEMY - A NEW RADIOACTIVE ELEMENT - THE ITALIAN DISCOVERY', and reported the "fascinating" results of the young Italian physicist, "Signor Enrico Fermi", pointing out that science knows no frontiers, in a tone critical with the Italian press that had talked of a "Fascist victory in the field of culture".

Corbino's talk was then published in full, not only in "La Ricerca Scientifica" [44] but also for example in "Nuova Antologia", a quarterly journal of Arts, Sciences and Fine Arts, and in the magazine "L'Energia Elettrica", thus obtaining a broad spectrum national distribution.

9 Fermi's strategic choices

9.1 Confidence in the neutron

Already in 1933 before he had developed his own theory of beta decay Fermi had taken an interest in the properties of the neutron, showing great confidence in the use of this particle as a projectile with which to bombard nuclei and produce artificial disintegrations. In his talk "The last fundamental particles of matter" presented at the annual Conference of the Italian Society for the Progress of Science, held in Bari from 12 to 18 October 1933 [F112], just before leaving for the Solvay Conference that same year, he spoke as follows about neutrons: "They are electrically neutral and have a mass fairly close to the mass of a proton, that is to say about one unit of atomic weight. The strong penetrating power of neutrons, which are able to pass through a thickness of lead of several centimetres, is easily explained by the fact that they are electrically neutral; indeed, for this reason when neutrons pass through matter they are in no way affected by the effect of the electrical fields due to the electrons of the body passed through, which would have a braking action on them, but can only interact with a nucleus when by chance they pass at an extraordinarily small distance from it. The fact that these corpuscles are electrically neutral also justifies their effectiveness as agents for the production of artificial disintegrations of nuclei. Indeed, if one wishes to produce a nuclear disintegration, launching against the nucleus a corpuscle with a positive electrical charge such as a proton or an alpha particle, the electrical field of the hit nucleus repels the hitting particle, obstructing its approach; while a neutron, since it has no electrical charge, is not repelled and can arrive undisturbed at the nucleus".

When in 1934 Fermi started to seek a possible radioactivity induced by these particles, despite his great confidence in neutrons, he was well aware that the neutron sources available to him, precisely because they depended on whether a certain nuclear reaction did or did not occur, were much weaker than alpha particle sources (according to Fermi, "in terms of order of magnitude 100000 times less intense" [F143]), even if one could always hope for a certain "compensation" due to the fact that the neutrons, not having any charge, "are much more effective" than charged particles in producing artificial disintegrations.

To obviate this limit inherent in neutron sources, Fermi prepared a whole brilliant measurement strategy, only barely touched upon in the Letter to "La Ricerca Scientifica", which drove him to adopt very precise choices regarding the type of neutron source to

F. Guerra and N. Robotti, *The Lost Notebook of ENRICO FERMI*,
https://doi.org/10.1007/978-3-319-69254-8_9

use and regarding the geometry of the apparatus, and which finally led him to success. What Otto Robert Frisch recounted on this score a few years later is significant: "I remember that my reaction and probably that of many others was that Fermi's was a silly experiment because neutrons were much fewer than alpha particles" [72].

9.2 The choice of source

One of the most important factors in Fermi's success was the choice of source [81]. Instead of using a polonium-beryllium source, commonly used in nuclear disintegration experiments with neutrons, he used a radon-beryllium source. This type of source, as Chadwick had already revealed, is one of the most intense: the rate of neutron production, which in the case of a Po-Be source is 30 neutrons for every 100000 alpha particles, in the case of a Rn-Be source almost doubles. This is due to the fact that radon ($^{222}_{86}$Ra or $^{222}_{86}$Em) (which belongs to the uranium-radium decay chain, Fig. 23) from which polonium in the strictest sense or RaF ($^{210}_{84}$Po) derives, is in equilibrium, not only with with polonium itself but also with two other strong alpha particle emitters, both isotopes of polonium, RaA ($^{218}_{84}$Po) and RaC' ($^{214}_{84}$Po). These by the way emit alpha particles with much greater energy than those emitted by polonium, that is to say 6.11 MeV and 7.83 MeV respectively, compared to 5.40 MeV for those from polonium. As a result, with an equal initial active deposit, and thus with an equal number of alpha particles present, the Rn-Be sources, thanks to the presence of RaA and RaC' and the greater energy of the alpha particles, produce many more neutrons compared to Po-Be sources.

Fermi, when he chose to use an Rd-Be source, was well aware of the special effectiveness of this kind of source. In his Letter on his discovery to "La Ricerca Scientifica" on 25 March he wrote concerning the rate of his source's neutron production that with 50 mCi of emanation "over 100000 neutrons per seconds could be obtained". Bearing in mind that 1 mCi are $3.7 \cdot 10^7$ disintegrations per second and that therefore 50 mCi of emanation correspond to $185 \cdot 10^7$ alpha particles per second, this meant, on the basis of Fermi's numbers, the origin of which is not known, that to produce a neutron $1.85 \cdot 10^4$ alpha particles were necessary. Instead in the case of a polonium-beryllium source, on the basis of Chadwick's famous estimate according to which 1000000 alpha particles produced 30 neutrons, about $3 \cdot 10^4$ alpha particles were necessary to produce a neutron, therefore about twice as many.

Anyway, despite this advantage, Rn-Be sources had a great snag which Fermi knew about: radon, unlike polonium which emitted only alpha particles, also emitted a strong gamma radiation, and as a result the neutrons produced were accompanied by this strong gamma radiation. In many applications, to obtain a "clean" neutron source, a platinum screen was used to absorb the gamma rays present.

This characteristic of Rn-Be sources was not however a problem for Fermi thanks to his theory on beta decay. On the basis of this theory, indeed, the gamma rays would not produce any effect in activating or not activating a process of beta-type induced radioactivity, that is to say that their presence would not interfere with a possible transformation of a neutron into a proton.

Fermi was so convinced of this prediction that he would only seek verification, as we shall see better in section 10.8, after he had already discovered neutron-induced radioactivity in aluminium and fluorine and all the material was ready for the first Letter to "La Ricerca Scientifica", or even when this Letter had already been sent. The verification is contained on page 24 of Fermi's laboratory notebook (Irpinia notebook) which we will discuss later. It is not possible to date this page exactly: the only information is that two pages later the date 27 March 1934 is indicated.

Another disadvantage of Rn-Be sources compared to Po-Be sources was due to the fact that radon ($^{222}_{86}$Rn) emits alpha particles with a mean lifetime of 3.82 days, much shorter than that of polonium ($^{210}_{84}$Po) which is 138.4 days. Therefore these sources had to be replaced fairly often which required the very frequent and repeated extraction of radon (from radium salts) because the quantity of radon that could be extracted every time was very small.

Anyway for Fermi this too was not a problem. Indeed for the supply of radon and the replacement of sources he could count, as indeed he did, on the cooperation of the Physical Laboratory of the Istituto di Sanità Pubblica and its Director (G. C. Trabacchi), in particular on the radon extraction system of the Laboratory. As can be seen from a laboratory notebook now kept at the Domus Galilaeana which we will discuss later (the "Thesaurus Elementorum Radioactivorum"), the sources were renewed weekly, or, if they were very intense, every two or three weeks, in any case always on Tuesdays.

Already earlier Fermi had availed himself of the precious cooperation of the Istituto di Sanità Pubblica for the supply of small radon sources. Indeed, as we have seen, at the end of 1933, when Fermi and Rasetti [F143] undertook the construction of "a bismuth crystal spectrograph for γ rays", in order to test this instrument they had used as gamma ray sources "small glass capillary tubes, 15 mm long and with an internal diameter of 0.2–0.3 mm, filled with radium emanation, expressly prepared by Prof. G. C. Trabacchi".

Certainly this past experience must also have influenced Fermi in the choice of a radon-based neutron source, that is to say of the Rd-Be type, and must have guided him in his planning and construction. Indeed it was enough to place the beryllium powder in the glass capsule before filling it with radon. This in fact was what Fermi did in collaboration with Trabacchi. On this subject Fermi in the first letter to "La Ricerca Scientifica" [F120] wrote: "the neutron source consists of a small glass tube containing beryllium powder and emanation. Using about 50 mC of emanation, which were provided to me by Prof. G. C. Trabacchi whom I would like to thank here most profoundly, it is thus possible to obtain 100000 neutrons per second, mixed of course with an extremely intense γ radiation, which however does not cause any disturbance for this type of experiments".

It should be noted that the intensity of the radon sources used previously by Fermi and Rasetti [F143] to test the bismuth crystal spectrograph varied between 100 and 150 mCi, and was therefore double or triple compared to the quantity of radon used here by Fermi (50 mCi) to construct his first Rn-Be source and that would lead him to the discovery of neutron induced radioactivity.

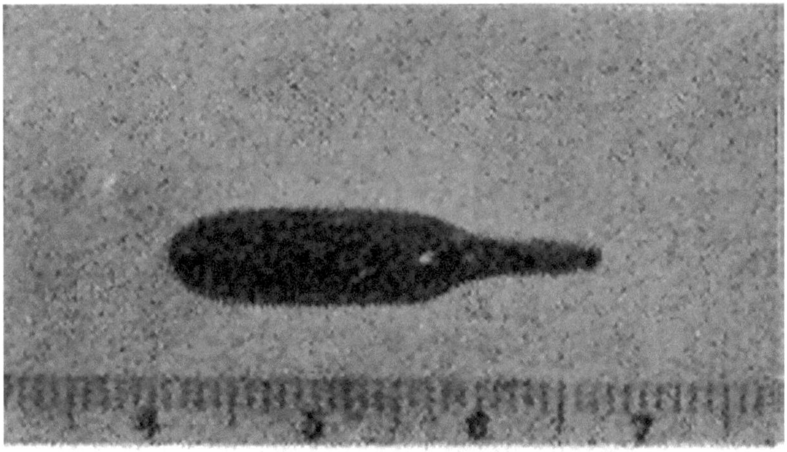

FIG. 2. - Sorgente di neutroni.

Fig. 48. – Neutron source used by Fermi - in [F171] (ISP).

In the first article sent to "Il Nuovo Cimento", with only his signature, Fermi would specify, something that he had never done before, that "the small sealed glass tube" had "an external diameter of about 6 mm and a length of about 15 mm", so the same dimensions as the "small glass tube" that, filled only with radon, had earlier been used by Fermi and Rasetti [F143]. Not only that. This same type of "small glass tube" would be used to create all the neutron sources used by Fermi in his later experiments on neutron induced radioactivity. As Fermi observed in 1938 in a lecture at the Istituto di Sanità Pubblica [F171, page 426]: "The neutron sources that I used consisted of small glass tubes filled with beryllium powder (Fig. 2) and radium emanation that were generously prepared by the Physics Laboratory of this Istituto di Sanità". In Fig. 48 we show the photograph of an Rd-Be source, included by Fermi in his report as Fig. 2, from which it can be seen that the dimensions of the capsule are always the same. The diminutive size of the source caught the attention of an anonymous interviewer of Fermi who in "Il Messaggero" on 3 September 1934 wrote: "The terrible battery capable of shooting as many as a million neutrons per second is all in small glass test tube that could be hidden in a fist".

While cataloguing Fermi's material at the Domus Galilaeana we found [106], abandoned on top of a cupboard, a wooden box measuring 93.0 cm × 32.5 cm × 5.0 cm high, lined inside with red velvet, wrapped in a lead sheet 1.5 mm thick, "acting as a screen", containing 12 sources used by Fermi and his collaborators (Fig. 49, top).

These sources were originally part of the collection of instruments "used by Fermi in the period 1932-1938" and consigned to the Domus Galilaeana by Amaldi, together with Fermi's documents, in the spring of 1956, and which then, except for the sources, were

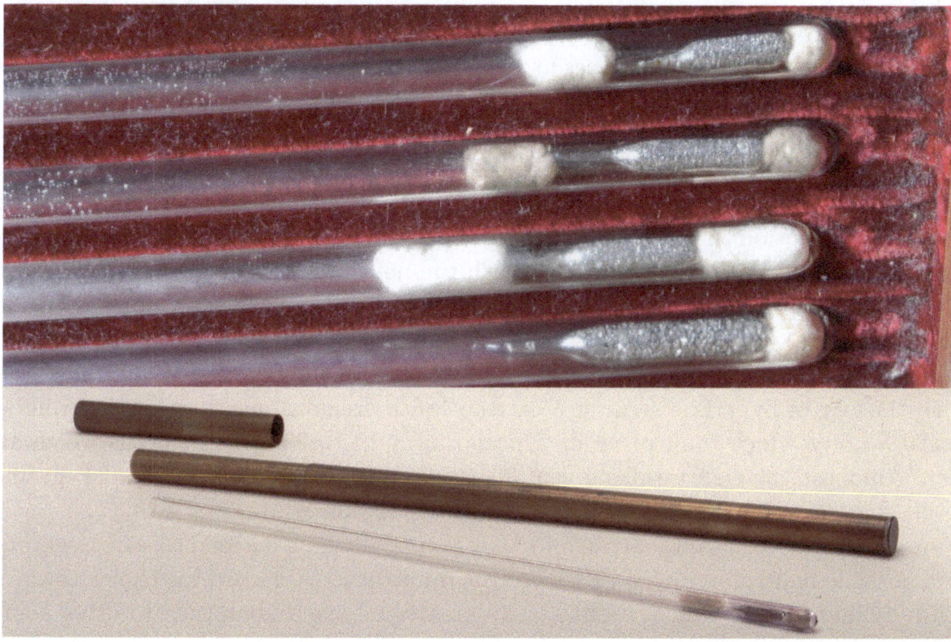

Fig. 49. – Top: Partial view of the collection of glass 12 tubes of varying lengths (from 30 to 80 cm) containing the neutron sources used by Fermi - FDG. Bottom: The additional source at the Smithsonian Institute - SI.

taken back to Rome in the summer-autumn of 1983, as is borne witness to by archive documents (Amaldi's first official request is dated 13/3/1980; the declaration that the items had been collected is in the minutes of the Administrative Board of the Domus Galilaeana dated 25/10/1983).

The sources were inserted at the end of small glass tubes of varying lengths, even some tens of centimetres, so that they could be easily moved by hand in safety. After the discovery of these sources an assessment was ordered, carried out on 9-10-11 February 2009 and consigned to the Domus Galilaeana, by qualified experts of the Department of Fire Fighters. The sizes recorded coincide with those communicated by Fermi on several occasions, thus confirming that they are the same sources. They are still very weakly active. An additional source of the same kind is kept at the Smithsonian Institute in Washington, donated by Emilio Segrè in 1963 (Fig. 49, bottom).

Returning to Fermi's idea of creating an Rd-Be source by mixing beryllium powder and radon in a "small sealed glass tube" it must be said that it was not new. Already earlier Rn-Be sources with a structure similar to Fermi's had been made and used immediately after the discovery of the neutron in at least two laboratories in Paris, in Maurice de Broglie's and Jean Perrin's, and in the laboratory of John Dunning (1907-1975) at Columbia University in New York.

In fact, articles regarding this type of sources had been published in very broadly distributed journals, such as "Comptes Rendus" and "Physical Review". Moreover Fermi himself had taken part in Conferences at which Rd-Be sources had been widely discussed. For example, as we have seen, in July 1932 he had attended in Paris the International Conference on Electricity, where he had presented a talk on the latest developments in nuclear physics [F99]. He had also chaired the discussion, which followed on after the talk by Marie Curie, and in which de Broglie and Leprince-Ringuet reported their experiments on neutron absorption in matter comparing the results obtained with the neutrons of a radon-boron source with those obtained with the neutrons of a radon-beryllium source, and which Fermi praised as "very interesting facts" (page 981 in the "Comptes Rendus" of the Conference).

Moreover, a year later in June 1933, Fermi was present and also gave a public lecture on the "Theory of hyperfine structure" at the 186th "regular meeting "of the American Physical Society which took place in Chicago from 19 to 24 June 1933, as we recalled earlier. Amongst the eighty talks given [36], there was one by Dunning and Pegram on this very subject: "On Neutrons from a Beryllium-Radon Source".

From what has been said it is obvious that Fermi, when he decided to seek experimentally a possible neutron induced radioactivity, immediately directed his choice towards a radon-beryllium source which, thanks to its greater intensity compared to other sources that he might have had access to, such as for example a Po-Be type source that could be created using polonium made available in Rome by Rasetti [F143] and D'Agostino, guaranteed him *a priori* a greater margin of success. It would only be after the discovery of neutron induced radioactivity, at a date corresponding to 14 April, as we shall see, that a test would be made irradiating silicon with a polonium-beryllium source and obtaining, as was predictable, negative results.

9.3 The use of "geometry"

If the choice of the type di source favoured Fermi in his discovery a no less important role was played by the use that Fermi made of geometry in designing the experiment, both with regard to the irradiation phase and with regard to the phase of measuring any induced radioactivity.

An Rd-Be source of the type used by Fermi produces, as for example de Broglie and Leprince-Ringuet had already stressed in 1932, "an isotropic [neutron] radiation in all directions" [62]. To exploit this property of the source as best he could Fermi arranged the substances to be examined "in the form of small cylinders" and he irradiated them placed "around the source". When the substances were in powder form they were placed inside "paper containers of appropriate shape" and were irradiated with the same system. In this way, thanks to the isotropy of the source, the various substances were subjected to neutron bombardment almost over a whole solid angle, and so the action of the source was considerably increased compared, for example, to the typical layout in which samples were placed in front of the source.

This experimental layout would be maintained by Fermi throughout his research activity in this field. Regarding the paper containers Fermi, in his article published in "Il

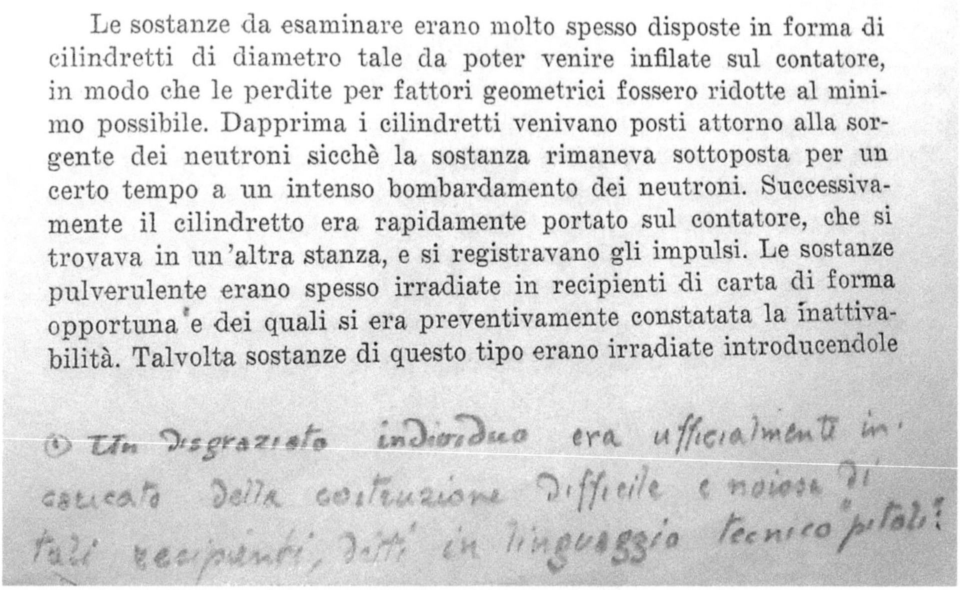
Le sostanze da esaminare erano molto spesso disposte in forma di cilindretti di diametro tale da poter venire infilate sul contatore, in modo che le perdite per fattori geometrici fossero ridotte al minimo possibile. Dapprima i cilindretti venivano posti attorno alla sorgente dei neutroni sicchè la sostanza rimaneva sottoposta per un certo tempo a un intenso bombardamento dei neutroni. Successivamente il cilindretto era rapidamente portato sul contatore, che si trovava in un'altra stanza, e si registravano gli impulsi. Le sostanze pulverulente erano spesso irradiate in recipienti di carta di forma opportuna (1) e dei quali si era preventivamente constatata la inattivabilità. Talvolta sostanze di questo tipo erano irradiate introducendole

(1) Un disgraziato individuo era ufficialmente incaricato della costruzione difficile e noiosa di tali recipienti, detti in linguaggio tecnico "pitali"!

Fig. 50. – A page from "Il Nuovo Cimento" 1934 [F131] in Rome with a comment by Fermi - DFUR.

Nuovo Cimento" in 1934 [F131], specified that before filling them with the "powdery substances" "he had previously ascertained [that they] could not be activated". As we shall see this practice was not adopted in the initial phase of discovery, for example in the case of calcium fluoride which, as is well known, occurs in the form of powder: here Fermi aimed directly at the search for any radioactivity induced in the sample without worrying particularly about the receptacle used.

We also report a curious fact. In the copy of the volume of "Il Nuovo Cimento" that contains this article by Fermi at the Physics Department in Rome, which dates back to Fermi's time, at the bottom of the page of the article where these paper containers are spoken about there is a note written in pencil (Fig. 50): "An unfortunate individual was officially assigned the difficult and boring task of constructing such containers, called in technical terms 'pitali' (chamber pots)".

The conspicuously common word pitale is never used in the papers published by Fermi and his collaborators. Instead we do find it, always accompanied by a serial number, in the laboratory notebooks or measurement cards belonging to Fermi and his collaborators to indicate some samples being examined. On the basis of analysis of the handwriting, even if the author tried to modify it by using lower case block letters, it was possible to identify who wrote the phrase: it was demonstrably Enrico Fermi. We do not know who the "unfortunate individual" was.

At the Museum at the Physics Department of the University of Rome "La Sapienza" eight of the original cylinders used by Fermi and his collaborators are kept. Two of them

Fig. 51. – A cylindrical sample used by Fermi in his neutron irradiation experiments - DFUR.

are of the "chamber pot" type. One of the cylinders is shown in Fig. 51.

Geometric factors were also crucial in the measurement phase. To determine any induced radioactivity Fermi used a Geiger-Müller counter that he himself had designed, adopting a remarkable experimental innovation. The various substances to be studied, which as we have seen were shaped as hollow cylinders, after irradiation in which the source was placed inside them, were "slipped over" the counter which was also cylindrical, naturally after their respective diameters had been appropriately chosen. In this way the losses in intensity, due to geometric factors, were "reduced to the minimum possible" since about half the solid angle could be relied on with the result that even weak activity could be easily detected. In Fig. 52, top, we show the photograph of one of the Geiger-Müller counters used at the Royal Physics Institute in Rome in the Thirties to study neutron induced radioactivity.

The experimental layout used by Fermi to measure the neutron induced radioactivity, with the counter placed inside the sample did however have some precedents. For example in 1931 Bruno Rossi, then assistant to Antonio Garbasso at the Physics Institute of the University of Florence, had suggested to Giuseppe Occhialini, a young fellow worker at the same Institute, that he should study the weak natural beta emission of some elements such as rubidium and potassium by wrapping a hollow cylinder internally with these elements and placing it at the centre of a Geiger-Müller counter so as to increase the irradiating surface, as Occhialini himself recalled in a letter to Rossi on 18 September

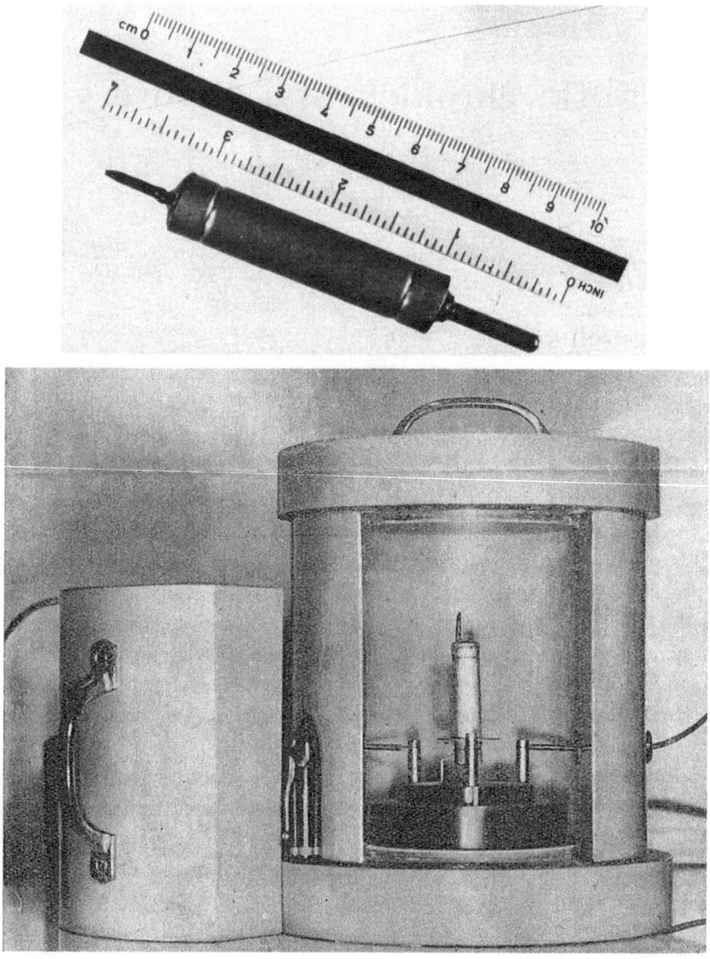

Fig. 52. – Top: A Geiger-Müller counter used by Fermi in his neutron irradiation experiments - DFUR. Bottom: The counter with the lead protection shield - in [F171] (ISP).

1970 (Occhialini-Dilworth Archive). This layout, which is very similar to the layout used by Fermi, would be made by Occhialini in the same year 1931 [121]. So very probably in this case too regarding the design of the geometry of the experiment, as in the case of the construction of the counters, there was an input, indirect or even direct (we do not know), by Bruno Rossi on Fermi's way of operating.

In any case even if the type of sources had already been used by others and even if the geometric layout was similar to Occhialini's, nevertheless Fermi's great originality shows through in the harmonious use of all the ideas available, tending towards the success of the research.

Figure 52, bottom, shows how Fermi protected the counter agaist natural radioactivity and cosmic rays, through lead shielding.

10 Fermi at work: chronicle of a discovery

10.1 The Irpinia notebook

As we indicated in the first chapter, a few years ago we found in the "Oscar D'Agostino" Archive in Avellino, to be precise in the summer of 2002, a notebook of Fermi's that on the basis of the dates and above all the contents we could also define as the "notebook of the discovery of neutron-induced radioactivity" since it was used to record the results of the measurements.

The Irpinia notebook represents the only direct testimony of Fermi's discovery and, as we shall try to show in this chapter, it affords us a complete reconstruction of its initial phase.

10.2 The "upright" side of the notebook

As we have said, the Irpinia notebook is written from both ends. The front of the notebook, compiled over 15 numbered pages with numbers in circles, without dates, is devoted to problems relative to β decay. In particular it begins (page Qf2) with formulae concerning atomic eigenfunctions that are then used to calculate, in the framework of Fermi's theory, the probability that a beta particle (an electron), emitted by a nucleus, is captured in an atomic orbital K. This formula, of the "Fermi's Golden Rule" type, is reported on the last page but one on this side of the notebook (page Qf15), under the horizontal line. Here g represents Fermi's coupling constant, h Planck's constant, c the speed of light, $\mathrm{d}\tau$ the volume element for the nuclear eigenfunctions and p_σ the momentum of the neutrino. The eigenfunctions of the neutron, the proton and the electron are, respectively, u_n, ν_m, ψ_s. This formula was then simplified by treating the electron in the non-relativistic approximation and assuming that the neutrino emitted had an energy almost equal to that of the electron. As a result of this calculation, Fermi came to the conclusion that the probability of capture for the β electron was of the order of $10^{-20}Z^7$ and was thus negligible for any element for any atomic number Z.

The significance of this calculation is clear. Fermi wanted to ensure that, once a beta decay had occurred inside the nucleus, the electron created would have a very small probability of being captured on an orbital and could anyway exit the atom and be seen in an experimental setting. After the first fifteen pages, in which these theoretical calculations are given and this important result was found, which by the way does not

F. Guerra and N. Robotti, *The Lost Notebook of ENRICO FERMI*,
https://doi.org/10.1007/978-3-319-69254-8_10

seem ever to have been published by Fermi, the notebook was turned upside down to the other side and was used as a laboratory notebook for Fermi's following experiments on neutron-induced radioactivity. This time sequence between arriving at a result that theoretically guaranteed the possibility of seeing an artificial decay and the beginning of laboratory activity, aimed at experimentally searching for this type of decay, cannot plausibly be considered accidental. This further reinforces our idea that in undertaking this new experimental research Fermi was driven by, and then guided by, his theory of beta decay without which it is unlikely that he would have embarked on an experiment that seemed "lost" from the very beginning. Not only that. This theory, as we shall try to show, was a constant point of reference for Fermi, his safe guide, both in the choice of the first elements to be irradiated and in the interpretation of the results arrived at and, finally, in predicting new nuclear reactions.

10.3 The counter

The back of the notebook, as has already been said, is written over 141 pages, numbered by Fermi and with recorded dates from 27 March 1934 (the first date given on page Q44) to 24 April 1934 (the last date given on page Q140) and it contains both the preparatory procedures for the experiments and the records of the measurements.

It begins with notes from which we see that Fermi, before starting to irradiate the various samples with a neutron source, proceeded to build some Geiger-Müller counters in order to be able to measure any induced radioactivity.

That means that in Rome, when Fermi began his experiments, these counters had not yet been built or at least they were not available, even though in a letter on 9 February (section 8.1) Fermi and Rasetti had written to D'Agostino that they were "putting together some counters".

In any case the fact that at that time a Physics Institute like the one in Rome was not officially equipped with counters is not unusual. The Geiger-Müller counter was a fairly recent instrument, of complex construction, handcrafted and specific for particular measurements in which it was necessary to fine-tune every single elementary ionising event (photon or charged particle) with great sensitivity. As we have seen Rasetti, during his first stay at Berlin-Dahlem in 1931-32, had used this instrument to study the "penetrating" radiation emitted by Po-Be. It is one thing however to learn how to use an instrument, such as the Geiger-Müller counter, already built and adjusted in the host laboratory, and it is another to learn how to build one. In any case we have not found any trace of measurements made by Rasetti with the Geiger-Müller counter after his return to Rome and not even any attempt to build one.

Instead we know with certainty, through the Irpinia notebook, that it was Fermi who did this, equipping himself with the counters he would then use. Indeed in the first 17 pages of the Irpinia notebook we find the electrical diagrams of some amplification circuits for Geiger-Müller counters with the reconstruction of the characteristics of the valves employed: plate current in accordance with the grid voltage for Telefunken RE 604 and Philips A 425 (page Q7) and background recordings made by counters in various screening conditions and with various voltages applied.

On page Q4 we find a three-valve amplifying circuit drawn (repeated on page Q6). It is a very sophisticated circuit with direct couplings between the various stages made with batteries, V_1 and V_2, without capacitors. This type of coupling had been promoted, for example, by Gustav Ortner and Georg Stetter in [126] for the amplification of transient signals in a form called "Gleichstromverstärkung" (direct current amplification), which is identical to the type then adopted by Fermi. At the time direct coupling amplification circuits had also been used in electrophysiology experiments where it was necessary to amplify small voltages of physiological origin without distortion. These circuits are also very much in use today, sought after by hi-fi enthusiasts for their characteristic of being very effective and faithful in the amplification of low frequencies.

From the Institute's accounting documents it emerges that in the previous years some "low frequency, three stage amplifiers for photoelectric cells" had been acquired. Fermi could have adapted one of these as a signal amplifier for his counters.

The tuning of these devices is very delicate. After a few attempts Fermi obtained "good operating conditions" for direct coupling voltages on the grids equal to $V_1 = 72$ V and $V_2 = 99$ V, respectively. In these conditions, a variation in the entry signal on the grid G of the first valve (on the right), from $V_3 = -9$ V to $V_3 = -4.5$ V made the plate current of the last valve (on the left) vary from $I = 0$ mA to $I = 48$ mA, sufficient to activate a telephone numbering device.

To have an idea of the documentary force of the notebook consider that from the calculations given on the following page (page Q5), which give the operating conditions of the amplifier, measured and calculated, one can infer that the value of the resistance r_1 through which the plate of the first valve (on the right) is powered is 500000 Ω.

In the pages immediately following, Fermi proceeded to connect the Geiger-Müller counter, powered by high voltage, to the amplifier. From a natural interpretation of the circuit on page Q15 it seems that this connection was also made directly, connecting the central wire of the counter directly to the grid of the first valve without a capacitor in between. A battery provided for the correct grid polarisation. If the counter was subjected to a discharge the grid underwent a sharp variation in potential which was then amplified, finally activating a telephone numbering device. Fermi made other attempts to find the optimal operating conditions, also in accordance with the potential applied to the counter. These conditions finally seem to have been reached on page Q17. At this point, starting from the following page, Fermi began measuring.

Certainly Fermi while preparing his counters must have made use of the expertise of Bruno Rossi, his long-time friend [145].

In Italy Bruno Rossi was the pioneer in the use of the Geiger-Müller counter (remember "Rossi's coincidence circuit" which we have already spoken about) and also one of the greatest experts on its construction. In 1930, with a study grant from the C.N.d.R., Rossi had gone to Walther Bothe's laboratory at the Physikalisch-Technische Reichsanstalt in Charlottenburg (Berlin) precisely in order to acquaint himself with Geiger-Müller counter techniques. On this occasion, as he himself told, he was informed directly by Bothe of the "secret" that ensured the prompt and reliable operation of the Berlin counters: the central wire was not made of steel, as was the custom officially established even in publications, but was made of appropriately treated aluminium.

It seems that Rossi had let Fermi into this secret. Indeed, as Fermi would declare in the concluding article in "Il Nuovo Cimento" in May 1934 only under his name [F131], the counters had been constructed with the central wire in aluminium and thus exactly following the procedure that Bothe had confessed to Rossi. Fermi wrote on page 431: "The device used to detect any activations of the bombarded substances consisted of a Geiger-Müller wire counter. Its tube was made of a sheet of aluminium 0.1 or 0.2 mm thick, such as to allow even electrons with not very great energy to enter the counter. The dimensions of the counters were generally about 5 cm long and 1.4 cm in diameter. The aluminium wire with a diameter of 0.1 or 0.2 mm was connected in the usual way to a system that amplified the impulses that put into operation a numbering device on which, from time to time, the number of impulses was read. The pressure of the air inside the counters was at 5 to 10 cm of mercury so as to have working voltages from 1000 to 1500 volts".

It really seems that "Bothe's secret" had arrived in Rome from Berlin, passing via Arcetri.

The numbering device used by Fermi was a telephone counter, as would later be specified. Since the values of the number of the recorded counts in the Irpinia notebook are all with four digits, we can easily deduce that the numbering device was also a four digit one. Some examples are currently kept in the Physics Department of the University of Rome "La Sapienza".

Generally, during the periods of measurement, as the notebook shows, the Geiger-Müller counter was always left switched on, even in pauses during work (indeed the counts continued to grow because of the presence of background radiation). As a result, as we shall see, its numbering can be used by us as a sort of clock to interpret the moment when the various measurements recorded in the notebook were taken.

Regarding the Geiger-Müller counters, it is interesting what Fermi would observe, again in 1949 [68, page 758]: "Now radioactivity can be measured with experimental methods of such refined sensitivity, Geiger and Müller's counters in particular, that even when a few radioactive atoms disintegrate every second or every minute, the counter manages to discover them; which is why the phenomenon, although quantitatively rather small, can actually be easily observed".

We also call your attention to some information provided by E. Amaldi about the counters available to the laboratories in Rome. Amaldi wrote in [24]:

"I would like to add a final small recollection that dates back to the spring of 1934. In Rome Fermi, at the end of March that year, had discovered artificial radioactivity provoked by neutrons and our group had begun to carry out a systematic study of this phenomenon in most of the elements. For this purpose we made use of very small Geiger counters with thin walls that we had learned to build ourselves, also after consulting Bruno Rossi who by then was in Padua, while Gilberto [Bernardini] and Daria [Bocciarelli] had stayed in Florence. They had much longer and more established experience than us since they had been using that same technique for years for the study of cosmic rays.

One weekend in April or May 1934 Bernardini, Occhialini, Daria Bocciarelli and Emo Capodilista came to Rome; they brought boxes full of Geiger counters and proportional counters: they were a gift to help us in our work. They had built them for us, devoting considerable time to making them and testing them, at least ten or fifteen days. However I have to say that we never used them. They were beautiful and they worked really well but unfortunately the geometry was not suitable because we needed much smaller ones. [...] Even after years I have to say that I have never heard of any other episode of such generosity and real friendship as that shown by our friends in Florence on that occasion".

Naturally if the visit reported by Amaldi had actually happened in March 1934 than the whole reconstruction relating to the counters used in Rome would have to be revised and the role of Arcetri would be amplified.

On the character and reliability of the reconstructions based on the memory of the protagonists it is important to remember what Edoardo Amaldi lucidly observed in a letter to Emilio Segrè on 5 July 1965, in a context in which the two of them, during the drafting of the biography in [9], found themselves in disagreement about the reconstruction of the well-known alleged episode when Majorana checked whether the table of atomic functions constructed by Fermi was correct. Amaldi wrote: *But for me too they are good memories but not reliable ones in the sense that my mind (like yours) may have unconsciously reworked them later.*

In the Irpinia notebook, in the initial pages Q1-Q15, Fermi recorded operational tests carried out on a series of counters which he called *Al 3/10 counter, Large brass counter, Al 2/10 counter, 4 Al P 301 counter, 5 Al P 201 counter, Al 1 small counter.* It seems an excessively long list, rather incongruent with the idea that simple counters had been built independently. In particular it seems strange that Fermi would also have built a "large brass counter", which it would have been more logical to use in research into cosmic rays. If the hypothesis that the counters had been supplied by Arcetri were true then pages Q1-Q15 have a natural interpretation: Fermi carried out tests to establish which were the best counters, the most suitable for his research, and in what operating conditions. Naturally other evidence would be necessary in order to reach certain conclusions on this important question.

10.4 The backwards periodic table

After the first 15 pages in which he concerned himself exclusively with adjusting the counter, on page Q16 Fermi, when he was about to arrive at conditions where his devices worked well and could therefore begin to bombard the various substances with his neutron source, drew up a periodic table that was "anomalous" or at least strange. This table is very illuminating with regard to Fermi's method of proceeding when choosing the elements to be irradiated first and it further strengthens our belief that for Fermi the theory of beta decay was his constant point of reference.

This periodic table, precisely because of its strategic position in the notebook, in other words "put there" in the moment when Fermi had to choose which elements to irradiate first, certainly represented Fermi's starting point. This table presents two remarkable aspects: in many points it is incomplete and it is backwards, that is to say that the succession of the elements is given with the atomic number decreasing (it begins with bismuth ($Z = 83$) and ends with hydrogen ($Z = 1$)).

Taking as a reference a periodic table of the time, as provided for example in [120], we can see that in Fermi's table all the highly radioactive elements (from $Z = 92$ to $Z = 84$) are missing, as are the rare earths (from $Z = 71$ to $Z = 59$) , the elements from $Z = 47$ to $Z = 44$, Nb ($Z = 41$), the other two radioactive elements, Rb ($Z = 37$) and K ($Z = 19$), Ge ($Z = 32$) and finally all the noble gases. Moreover Rb ($Z = 37$) was included but then cancelled. So Fermi excluded from his "backwards" periodic table the elements for which it would have been practically impossible for him to see any radioactivity induced by neutrons, such as indeed the elements that are themselves already radioactive (which in any case apart from potassium were difficult to find), and the elements that were extremely expensive or rarely found (apart from niobium and germanium), and all the rare gases which were unsuitable for an experiment of the type planned by Fermi since in normal conditions they only existed in a gaseous state.

The characteristic of this table of beginning with the elements with a high atomic number shows us how Fermi was interested in the heavy elements as a first step, to then move on to the lighter ones. This supposition of ours agrees perfectly with what Fermi would then do. Indeed, two pages after this table, Fermi, at the moment when he had everything ready and had to decide which element to irradiate first, chose platinum ($Z = 78$) which is one of the elements with the highest Z given in his "backwards" table and fairly easy to obtain.

Fermi's interest in the heavy elements and his subsequent choice of beginning with one of them was clearly dictated by his theory on beta decay. Indeed, it was well known in Fermi's time that as the atomic number increased the nuclei of the various elements not only contained, obviously, a greater number of neutrons but they also had an ever greater excess of them compared to the number of protons. This is because the presence of an adequate number of neutrons was essential to keep the nucleus bound together against the growing Coulomb repulsion of the protons present, as was also explained by the Heisenberg-Majorana theory.

The tendency of this neutron enrichment emerged clearly from the list of chemical elements officially adopted at the Solvay Conference in 1933, or from the various tables reproduced by Fermi in his articles, in which alongside each element he stated, as well as the atomic number Z which represents the number of protons, also the mass number A corresponding to the number of the nucleons (protons and neutrons) that composed its various isotopes then known. In the case of elements up to sulphur ($Z = 16$), the excess of neutrons was very small (from zero to 2) and then increased significantly as Z increased for any isotope considered, from the most abundant neutrons to the rarest. For example already in the case of yttrium ($Z = 39$), which had only one isotope ($A = 89$) the excess of neutrons was 11, in the case of barium ($Z = 56$), considering the most frequent isotope ($A = 138$), it was 26, while in the case of lead ($Z = 82$), with regard to its most frequent isotopes ($A = 206, 207, 208$) the excess of neutrons actually increased to a value between 42 and 44.

So, since Fermi's theory predicted that in order to have a beta negative type decay it was necessary that a neutron be transformed into a proton, it was clear that by using elements with a high atomic number, and thus with a much greater "excess" of neutrons than light elements, the probability of such a process would be greater. The heavy elements, in the light of Fermi's theory, appeared much more suitable to undergo a negative beta decay compared to the light elements. Hence the choice to start with platinum, one of the heavy elements most accessible to Fermi.

The interest shown by Fermi in bombarding the heavy elements was also justified by the fact, already recalled, that the neutrons, since they have no charge, would not be affected by any Coulomb repulsion from the nucleus as they passed through matter. They would therefore succeed in easily penetrating inside any type of nucleus, even the heaviest ones, therefore presenting themselves as an excellent projectile and guaranteeing certain success, theoretically, unlike the other charged particles (alphas, protons or deuterons).

10.5 The moment of the discovery

After page Q16 devoted to the backwards periodic table, on page Q17 we find the records of various tests necessary to adjust the counter and for the measurements of background radiation. Fermi began his actual experiments on the following page Q18, which opens with the note of a time which is apparently very strange ($5^{\mathrm{h}}57'$) but which is perfectly comprehensible in the setting of the precise chronological reconstruction that we will give in the later section 10.10. After testing the counter on a sample of potassium chloride (KCl), which was known to have low beta radioactivity, Fermi, contrary to what is commonly told in historical reconstructions based on memories, and in step with his "backwards" table, chose platinum as the first substance to irradiate. Remember that platinum was also used as a screen for radon-beryllium neutron sources against γ rays, and so it was important to ascertain that no spurious effects could arise due to any possible β activation.

To start with he took a background measurement over 35 minutes with non irradiated platinum. In the first column the time is recorded, in the second column the numbers given by the counter. As we can see, the counts were taken every $6'$ three times, then after $7'$ and finally after $5'$ twice. The counts of the counter are recorded alongside the times, starting at the beginning (8400) up to (8757) at the end. He then took the differences between the later readings and divided them by the time elapsed in the third and fourth columns. He established that in the first $30'$ there were 291 counts in total, corresponding to an average of about 9.7 counts per minute. The counts per minute calculated in the time intervals of the intermediate recordings agree with this value, bearing in mind the inevitable statistical fluctuations.

Then he irradiated the platinum for $15'$. The difference between the counts from the end of the background measurement (8757) and the beginning of the measurements on the sample of irradiated platinum (8923) corresponded approximately to $17'$, just the time necessary to go from the room where the neutron source was, irradiate the sample for $15'$ and return to the room where the counter was. Indeed, the intense γ activity from the radon-beryllium source would certainly have disturbed the counter. As a result the source and the counter had to be kept well separated. Moreover the irradiated substance had to be placed "around" the counter as quickly as possible since the mean lifetime of the new radioactive isotopes was not known a priori and presumably could also be very short, as would in fact be discovered for example in the case of fluorine where it turned out to be about 10 s. Note that in interpreting these measurements, knowing the background level, we were able to use the counter as the equivalent of a clock, although subject to statistical fluctuations.

The counts on irradiated platinum were made at intervals of 3 minutes, 5 minutes, 10 minutes and showed no effect. Indeed the values observed were statistically equivalent to the background, as Fermi could already establish from the first 3 minutes when only 32 counts were recorded.

Anyway, after trying with platinum and finding a negative result, on the next page (Q19) after just over 17 minutes (in fact the last count of the counter in the case of platinum was 9029, while the first count recorded on page Q19 from which he started for a new background measurement was 9200) he moved on to analyse aluminium. As a first operation the background was measured in the presence of non irradiated aluminium, and it proved to be about 10.5 counts per minute over a measurement of $30'$. Later he proceeded to irradiate the aluminium. The irradiation time is not explicitly recorded in the notebook, unlike the previous case of platinum. Nevertheless, on the basis of the counts recorded in the meantime it can be estimated at around $20'$.

The effect of the irradiated aluminium on the counter was immediately visible. Indeed in the first 5 minutes 82 counts were detected which reduced in the following 5 minute intervals to 74, 59, 57 counts, respectively. The values in the first minutes were significantly higher than background, which in the same $5'$ time interval gave about 50 counts. The results were fully compatible with a β activity, provoked by irradiation, which decays with a mean lifetime of about 10 minutes. A new β radioactive element had been artificially produced! This page can rightfully be indicated as the page of the discovery of neutron induced radioactivity. On this page however no sign, no comment, no exclamation was given. Only a red mark indicates that this measurement would be taken into account in the final elaboration before publication. It is as if nothing unexpected had occurred. Everything seemed to be predicted and everything seemed to be under control.

And yet in other situations Fermi did not hesitate to comment on the results of measurements. We have one example, already cited, on a card with records of the measurements made by other members of the group, where in his handwriting an explicit "Balle!" (Fig. 47) appears. In that case we know that the measurements were unreliable because of the presence of impurities in the sample examined.

The recording of the measurements for aluminium show a procedure typical for these experiments, perfectly analogous, even from the point of view of layout, to the procedure followed by Joliot for the discovery of alpha-particle–induced radioactivity. You must consider that the source was kept far away from the counter. The sample was subjected to irradiation by the source and then swiftly taken close to the counter for the measurements, which had to take place regularly at various time intervals to observe any activation and then the decay of the induced activity. This is a very different logic to that applied instead in "on line" experiments where source, irradiated sample and detectors occupy fixed positions, and the intensities uniform in time of the effects of the source on the sample are recorded. This conceptual leap had been made in a few months by the Joliot-Curies as they progressed from their experiments on the production of neutrons and positrons on samples irradiated by the alpha particles of polonium to the induced radioactivity experiment in which the sample was first irradiated, then moved away from the source and underwent measurements with the Geiger-Müller counter spread out over time.

It is surprising how Fermi, who in reality was a novice as far as nuclear measurements are concerned, immediately seized upon this logic of "off line" experiments and adopted straight away the most effective and simplest procedures. Looking at things in detail, one can note a residual trace of the old logic in the Joliot-Curies' method of proceeding. In order to verify the alpha-particle–induced radioactivity they irradiated the aluminium sample with their precious and sophisticated polonium source, while it would have been much simpler to put the aluminium sheet close to one of their vials containing the radium salt preparations to obtain the same effects and with greater intensity. Fermi's choice to use radon-beryllium sources, and not polonium-beryllium, is a clue to his full understanding of activation and measurement experiments.

In the following days the measurements of aluminium were repeated four times, irradiating aluminium for 12 hours, for 3 hours and 30 minutes, for 90 minutes and finally for 15 hours, respectively, and filtering the neutrons through a 1mm thickness of lead. The results are summed up on page Q36. Here the activation effect is fully evident, with a decay time that can be estimated at around 10 minutes, as was communicated by Fermi in the first Letter to "La Ricerca Scientifica". So at this point neutron induced radioactivity in the case of aluminium was a well established fact.

10.6 From aluminium to fluorine and more

So Fermi, after having tried with platinum without success, at once went on to irradiate aluminium ($Z = 13$), immediately finding a positive result. Fermi's choice of aluminium seems obvious: indeed it was known that this element became radioactive using alpha particles and protons and deuterons and that it was therefore an extremely "wobbly" element, particularly sensitive to bombardment. Actually, if he had not been influenced by the theory of beta decay the natural choice of first element for Fermi should have been aluminium.

In fact, straight after the success obtained with aluminium, therefore with a light element, Fermi moved on again to a heavy element, in particular lead ($Z = 82$). In this case too, as for platinum, he obtained a negative result. After this new failure Fermi turned again to a light compound, calcium fluoride. He persisted with this compound until he found a positive effect which he attributed to fluorine. The measurements for calcium fluoride are extremely interesting. We know that in this case the induced effect corresponds to a mean lifetime of β activity of the order of 10 seconds and that it is therefore very difficult to discover.

The first experiment appears on page Q21 and was carried out with a sample of CaF_2 after it had been irradiated for 37 minutes. The counts were taken at one minute intervals. Initially the results were negative.

Success was finally achieved in the series of experiments recorded on pages Q27, Q28 and Q29 of the notebook. After having irradiated the sample for some minutes each time, Fermi recorded the counts every minute, subsequently carrying out numerous tests. At first sight the results seemed inconclusive. Fermi however noticed that in the first minute there were systematically more counts than in the following minutes. At this point he split the first minute into intervals of 20 seconds each and began to record the counts every twenty seconds in the first minute. This was done for the first time in the fourth experiment on page Q28, in the one minute interval between counts 063 and 077. The results every 20 seconds are recorded as "$5+6+3[=]14[=077-063]$". Repeating this procedure he found evidence that the number of counts every twenty seconds decreases exponentially during the first minute. Indeed, for example, in the sixth, eighth, tenth experiments, on the right hand side of page 28, we find the series "$9+4+3[=]16$", "$5+4+2[=]11$", "$7+3+3[=]13$", respectively. Adding up the counts in every twenty second interval for these four experiments he found the following sensational result: a total of 26 counts for the first twenty second interval ($5+9+5+7=26$), 17 counts for the second twenty second interval ($6+4+4+3=17$), and 11 counts for the third 20 second interval ($3+3+2+3=11$), as noted in the middle of page Q29. Fermi finally estimated the background radiation by taking the average of all the five minute experiments, as summed up at the top of page Q29, finding about 64 counts in nine minutes, equivalent to 7.2 counts per minute. So the background radiation for the four consecutive twenty second intervals is 9.6 counts ($7.2 \cdot (4/3) = 9.6$). Subtracting these background counts Fermi thus found that the counts of beta particles in the samples activated in these successive twenty second intervals were $16.4[=26-9.6], 7.4[=17-9.6]$, respectively, fully compatible with an exponential decay with a mean lifetime of around 10 s for neutron induced radioactivity on fluorine.

This result is confirmed on page Q35 by ten series of systematic measurements, with counts taken every twenty seconds in the first minute and then after two minutes. At this point neutron-induced radioactivity became a scientifically well established fact for calcium fluoride too.

Fermi, even if he found induced radioactivity in a sample of calcium fluoride, as can be seen in the Letter to "La Ricerca Scientifica", attributed it to fluorine without bothering to check with a separate experiment whether perhaps it was the calcium that became radioactive. A test of this type would be carried out by Fermi (D'Agostino was entrusted with taking the data and it would be his first written contribution, see page Q48 of the Irpinia notebook) only on 27 March, so two days after the discovery of neutron-induced radioactivity in aluminium and fluorine had been communicated to "La Ricerca Scientifica". D'Agostino would use a sample of CaO and obtain a negative result. A reason for this position of Fermi's, distinguished by his obstinacy in seeking radioactivity in calcium fluoride at all costs and, once having found it, his choice to attribute it to fluorine without troubling further even in the absence of a separate test on calcium, may have been due to the fact that Fermi was aware of the probable successes concerning the activation of fluorine previously achieved by Harkins-Gans-Newson and the Joliot-Curie, which we spoke of in section 8.4.

In any case the choice of fluorine and not calcium as the element that had become radioactive also received significant support from his theory of beta decay. Starting from the premise, then taken for granted, that the neutron provoked an (n, α) reaction, in the case of fluorine which only possesses the isotope F_9^{19} there would have been the following reaction, for that matter already written by Fermi in Amaldi's notebook:

$$F_9^{19} + n_0^1 \rightarrow N_7^{16} + He_2^4,$$

while in the case of calcium, considering its most frequent isotope, Ca_{20}^{40} at 96.9%, this would have been the reaction:

$$Ca_{20}^{40} + n_0^1 \rightarrow Ar_{18}^{37} + He_2^4.$$

Both Na_7^{16} and Ar_{18}^{37}, produced in the two reactions respectively, represented two new unknown isotopes with the difference that the former had an excess of neutrons compared to the most frequent isotope of nitrogen (N^{14} at 99.7%), and thus could have a negative beta decay, while the latter had fewer neutrons compared to the most frequent isotope of argon (Ar^{40} 99.6%) and so, according to Fermi's theory, could not be radioactive. Hence the inevitable choice of fluorine and not calcium as the element that had been activated.

The measurements on calcium fluoride which we have just seen were interspersed with various attempts on a heavy element compound (HgCl), on a light element compound (KCl), again on lead, and finally on a light element (Cu) respectively. These attempts proved fruitless however.

Anyway, at a certain point corresponding to page Q36, where he concluded the measurements on aluminium, Fermi suspended this exploratory investigation and, having obtained a positive result in the case of two elements (aluminium and fluorine), on 25 March 1934 he wrote the Letter to "La Ricerca Scientifica" to communicate the discovery of neutron induced radioactivity.

The manner of the discovery, as we have reconstructed it through an analysis of the Irpinia notebook, excludes categorically that "as a methodical person" Fermi, as too often has been reported, tried to irradiate the elements in their order in the periodic table, starting with hydrogen then moving on to helium and the others, always obtaining negative results until he came to fluorine, in ninth place, with positive results, then confirmed in the case of aluminium in thirteenth place. We have seen that the order of irradiation was completely different and dictated by profound theoretical reasons. In particular, without first having discovered radioactivity induced in aluminium it would have been impossible to discover it in fluorine, where the active product has a mean lifetime of only about 10 seconds, and would therefore have been completely overlooked in a first analysis. Naturally when publishing the summary of the results of a series of experiments, as for example in [F135], the order of the periodic table, which is not necessarily the order in which the experiments were carried out, was followed.

10.7 The study of the periodic table and the beginning of work as a group

Once induced radioactivity in aluminium and fluorine had been discovered and the results arrived at had been published Fermi launched a wide ranging project concerning systematic research on the whole periodic table. Pages Q37-Q38-Q39 of the Irpinia notebook testify to this.

Here Fermi gives a complete periodic table, put in the usual order according to increasing atomic number. This periodic table represents a sort of general index to the notebook, even if very incomplete. As we can see, some elements are marked by one or more numbers and by some notes. The marked elements are the ones that Fermi tried to activate, and the numbers are the numbers of the pages of the notebook where the corresponding measurements are given, taken both before page Q37 (where the periodic table begins) and after page Q39 (where the periodic table ends). Note the words "yes" or "nothing" next to the symbol of some elements to indicate whether a positive result was obtained or not. This list however is incomplete.

Certainly one reason is that at a certain point, starting from about 24 April, the measurements taken are not only recorded in this notebook but also, as far as we are aware so far, on the 16 loose pieces of paper in Avellino and in the Amaldi notebook, and so a list referring only to the Irpinia notebook would no longer make sense. Working using at least three different laboratory notebooks at the same time was clearly due to the fact that Fermi was now no longer alone but had four collaborators to whom he had also entrusted the recording of the measurements. From the handwriting they were, in order, D'Agostino, Amaldi, Segrè and Rasetti, who had returned from an official visit to Morocco around 8 April. We can see the collaborators in photographs of the time Fig. 53-56. The last meeting between Bruno Pontecorvo and Enrico Fermi is recalled in Fig. 57.

In this research across the board on the periodic table, begun immediately after the communication to "La Ricerca Scientifica", Fermi's aim is clearly to succeed in activating the greatest possible number of chemical elements. In this second research phase too Fermi, in agreement with his theory of beta decay, tended to favour the heavy elements for which however there would be full success only after the discovery of the effects of slow neutrons. Fermi tried to activate about thirty elements ranging from sodium ($Z = 11$) to bismuth ($Z = 83$). For these elements positive results were given for 22 elements, of which no less than four with an atomic number greater than 50 (antimony ($Z = 51$), iodine ($Z = 53$), barium ($Z = 56$) and lanthanum ($Z = 57$)), all with an impressive excess of neutrons (of the order of about twenty). In this case too the theory of beta decay seems to have been a reliable guide for Fermi. The results of this new phase of research would be published, again only with Fermi's name, in a second Letter to "La Ricerca Scientifica" (in which, as we said before, D'Agostino, Amaldi and Segrè are

Fig. 53. – Oscar D'Agostino (~1934) - DFUR.

Fig. 54. – Edoardo Amaldi (~1933) - DFUR.

mentioned) with a slightly different title from the first Letter, "Radioactivity provoked by neutron bombardment" (the adjective "provoked" is used instead of "induced").

10.8 The gamma rays of the source

As we have said, one of the decisive factors in Fermi's success was that he used a radon-beryllium–type neutron source, despite the fact that it was known that radon along with alpha particles also emitted strong gamma radiation that would certainly compromise the purity of the neutron source. In the Letter to "La Ricerca Scientifica" on 25 March in which he communicated the discovery of neutron induced radioactivity Fermi observed on this point, when talking of his Rd-Be source in a very general way, that the "very intense γ radiation [...] causes no disturbance for experiments of this kind".

Fermi did not specify, as instead he would do later, whether this observation was based on some experimental data. Indeed, as we can see from the Irpinia notebook, verification of this in the laboratory came very late, probably even after the letter to "La Ricerca Scientifica" had been sent. Clearly what drove Fermi to make such a bold and decided statement was yet again his theory of beta decay, on the basis of which gamma rays would have no effect in inducing beta decay.

As proved by the Irpinia notebook, from the very beginning Fermi used a Rd-Be source without worrying about the problem. Only on page Q24, after having discovered radioactivity in aluminium, did Fermi seem interested in studying, even in a very ap-

Fig. 55. – Emilio Segrè (∼1934) - DFUR.

Fig. 56. – Franco Rasetti (∼1934) - DFUR.

proximate way, any action produced by gamma rays emitted by his source. In fact he did an experiment on calcium fluoride (in which he had not yet discovered induced radioactivity), irradiating the sample first with a source of only radon and then with the Rd-Be and taking the counts every minute, for a period of about ten minutes. The results he arrived at however were indistinguishable one from another and therefore inconclusive.

Indeed, as he would discover a few pages later, in the case of fluorine the radioactive element produced had a very short mean lifetime, less than a minute (of the order of 10 s), and so could not be detected taking the counts every minute. Certainly the choice of calcium fluoride, precisely because it was not yet known whether it became radioactive or not, is at least odd for a test of this type.

Fermi was so sure of the lack of influence of gamma rays that only on page Q40, when by then he had already discovered radioactivity in aluminium and fluorine, did he do a test irradiating a sample of aluminium with gamma rays and finding, as he expected, that it did not become radioactive. It is not possible to date this page of the notebook: the only information we have is that two pages later the date 27 March 1934 is reported.

Anyway, it would be much later, when the group had already started working on the whole periodic table, that Fermi would return to the problem and specify the experimental reasons that led him to consider gamma rays ineffective in provoking induced radioactivity.

As he observed in the first paper sent to the "Proceedings of the Royal Society" in July 1934, also in the name of Amaldi, D'Agostino, Rasetti and Segrè: "The neutrons are accompanied by a very intense γ radiation. This fact, however, should not produce any difficulty since the radioactivity induced is measured after irradiation and it has been shown that radon without beryllium produces no effect".

Fig. 57. – Bruno Pontecorvo with Fermi (1949) - in [86] (AFP).

Regarding the use of a source other than Rd-Be to produce induced radioactivity, that is to say a Po-Be source, one of the 16 loose sheets of paper bears witness that an attempt of this type was only made on 14 April. Indeed on the back of sheet 15 (S15a) the counts on the silicon in "chamber pot 51", irradiated with Po+Be, are recorded with uncertain results. On the other hand silicon had already been activated by Fermi with a radon-beryllium source and it had shown a very intense effect, with a period of about 3 minutes, as communicated in the second Letter to "La Ricerca Scientifica" [F121].

10.9 First chemical analyses and new nuclear reactions

As we saw in section 8.5, when interpreting the results regarding aluminium and fluorine Fermi assumed that the target nucleus, after capturing the neutron, emitted an alpha particle, undergoing an (n, α)-type reaction, thus being transformed into an unstable nucleus which, in turn, would undergo negative beta decay and would be transformed into a stable nucleus. The hypothesis of an initial (n, α)-type reaction, even if it was in agreement with the transmutations produced by neutrons known up to that moment (see Russell's table, Fig. 28), still had to be verified. To do that, the only possible way was to succeed in identifying chemically the active element that had been formed and then, once having identified it, to establish what reaction bound it to the starting element, thus recognising if it really was an (n, α)-type reaction that had occurred or another type. In reality one might have thought in principle of identifying, again chemically, the final stable element that had been formed. But this path was unfeasible because the quantities involved were extremely small, for example of the order of $10^{-18} - 10^{-15}$ grammes, and in consequence were such as "to challenge any chemical reagent one might want to use for direct research on them".

So the only way was to aim at identifying the active isotopes produced since, in that case, a whole technique derived from radiochemistry, which allowed "imponderable quantities" of matter to be analysed, would have been available. Clearly this was a very complex and delicate operation, also bearing in mind the generally very short decay times of the new radioactive substances produced.

This task, as we see in the Irpinia notebook, was entrusted by Fermi to the "chemist Oscar D'Agostino" just a very few days after the discovery. From the end of January 1934 D'Agostino, as we have already said, had gone to the Institut du Radium in Paris to familiarise himself with "the chemistry of radioactive substances" and just at the end of March he was in Rome. Somewhat worried about the fate of the study grant that he should have used in Paris, D'Agostino wrote a letter headed "Rome, 25 June 1934" to the Chairman of the Physics Committee of the C.N.d.R. (Fig. 24): *having returned to Rome for the Easter holidays I was kept, as I still am, at the physics Institute of the Royal University by Prof. Fermi for the research being carried out there on artificial radioactivity. I inform Your Excellency of my activities and I ask you to kindly arrange that the second instalment of the grant can be paid to me.* Fermi confirmed in a letter on 26 June (Fig. 25): *Dr O. D'Agostino, who has a C.N.R. study grant for Paris, as I mentioned to you has been kept by me to work here on the new artificial radioactivities. I know he has written to you asking that the second instalment of the study grant be paid to him. See if it is possible to help him.*

As the Irpinia notebook bears witness, D'Agostino began to work with Fermi around 27 March and was his first collaborator. Indeed, as we have already said, the first handwriting other than Fermi's that appears in the notebook is D'Agostino's. On page Q48, which corresponds to the date 27 March which was the Tuesday of the week before Easter (the date 27 March is given on page Q44 and is followed on page Q50 by the date 28 March), so only two days after the communication of the discovery, the counts are recorded regarding a sample of "CaO irradiated 1h" (written by Fermi).

D'Agostino, as is obvious, was immediately involved not only in taking the data but also in the preparation of samples to irradiate and in all other questions concerning chemistry. For example the next day (page Q60) D'Agostino wrote about Al_2O_3: "It is put back to cook at 10^h and 40′", "it is removed at $12^h2'$"; or again, regarding this aspect, on sheet S7 we find written: "Porcheriola di De Agostino" (De Agostino's muck), where the handwriting seems to be Rasetti's, including the mispelling of D'Agostino's name.

From this moment on D'Agostino took part in the research full time: his only days off were the Saturday before Easter (31 March), Easter Sunday (when Fermi worked however) and Easter Monday (when not even Fermi worked).

On the Tuesday after Easter (3 April, page Q93) D'Agostino began to apply himself to the chemical identification of the new active elements produced, beginning with iron. The results of this research would be published by Fermi in the second Letter to "La Ricerca Scientifica" (only in his name) around 10 April, making explicit reference to the fact that "The chemical separations were carried out by Dr. O. D'Agostino".

The criterion that D'Agostino adopted for his chemical identifications, and that he would then continue to use, was "equal" to that used in natural radioactivity when from a mixture it was necessary to separate out radioactive substances "present in imponderable quantities", as D'Agostino would explain, in his first publication on the subject [61]. It was based on the principle that if in a solution an "imponderable quantity" of a radioactive element was present together with a "ponderable quantity" of one of its stable isotopes, and if this stable isotope through a precipitation reaction was separated from the solution, it would drag along the radioactive element with it. As D'Agostino wrote: "[...] the mixture is brought into solution and to this is added a ponderable quantity of the inactive isotope of the radioelement (that one wishes to separate). When this isotope is separated from the solution with a precipitation reaction or another, the active element follows its fate".

When identifying new produced radioelements the design of the operations that D'Agostino carried out was as follows: the substance being examined, after being activated, was brought into solution adding a "ponderable quantity" of the elements suspected of possibly being isotopes of the active element that had been formed; then, with a precipitation reaction, the various elements, including the original one, were separated and examined at the counter. The element that proved to be active was the one to which the new radioactive isotope belonged and so it could immediately be identified.

The crucial point of the whole operation was, at this point, the choice of the elements to add to the active product. The criterion of the choice to follow, obviously, had to be established on the basis of the expectations that one had of the possible nuclear reactions produced by the neutrons on the bombarded element; this choice, which was beyond D'Agostino's chemical expertise, was certainly established by Fermi. If we analyse this criterion of choice through the Irpinia notebook and the 16 loose sheets of paper, we can see that Fermi, in this case too, was guided from the very beginning by his theory on beta decay. Two elements were added that preceded the irradiated element by one or two cells in the periodic table (on the basis of an analogy between all the nuclear transmutations known at that time and due to protons, neutrons, alpha particles and deuterons), in order to take into account respectively a possible (n, p) reaction (a shift one cell to the left) or an (n, α) reaction (a shift two cells to the left). Instead an element that followed the irradiated element by one cell was also added, as required by Soddy's displacement law in the case of natural beta decay. So this choice indicated that Fermi had predicted that the starting nucleus could not only undergo the two known reactions (n, p) and (n, α) but also a reaction, predicted by his theory on beta decay, in which the neutron was absorbed and the new nucleus, which had an excess of neutrons, at once underwent a negative beta decay (without emitting alphas or protons), so shifting one cell to the right. If Fermi had not been able to rely on his theory of beta decay certainly the choice of a verification of this type would not have made any sense. So the theory of beta decay represented a constant guide for Fermi throughout his path towards the search for neutron induced radioactivity, even when it was a question of identifying the chemical nature of the new radioactive isotopes produced.

On page Q93 of the notebook (corresponding to 3 April) the chemical analysis of the active product of iron began. Mn ($Z = 25$), Cr ($Z = 24$) and Co ($Z = 27$) were added to the iron ($Z = 26$) "irradiated 14 hours"; they then proceeded (page Q94) with the "separation of Mn Cr Co with the Rolla method" and the "Mn Cr Co aqueous extract" was examined at the counter, finding that this extract had become radioactive. At this point it was a question of choosing, from these three elements, which had dragged the active element away. Through further precipitations, and also repeating the operation with $FeCl_3$, D'Agostino showed that the active element followed Mn ($Z = 25$), and that therefore, as Fermi communicated in his second Letter to "Ricerca Scientifica", the active product was an isotope of this element, and that the reaction was of the (n, p) type.

A similar procedure was applied in the case of phosphorus on 9 April, using the "Chamber pot 37". The measurements are reported partly in the Irpinia notebook and partly on the loose sheets of paper. The handwriting is both D'Agostino's and Fermi's.

On page Q119 of the notebook, the radioactivity of the "phosphorus irradiated overnight, taken out at $9^h43'$" was measured. At $11^h40'$ (page S3a), as can be established by examining the next page S4, silicon ($Z = 14$), aluminium ($Z = 13$) and sulphur ($Z = 16$) were added to the phosphorus ($Z = 15$). The silicon (in the form of silicon dioxide) was extracted at $12^h37'$ and proved to be radioactive; the residue (phosphorus + aluminium + S) was analysed by the counter at $16^h14'$ (page S4) and showed no sign of radioactivity. As a result the active element, as Fermi again communicated in his second Letter, was identified as silicon and the reaction was also recognised as of the (n, p) type. Regarding these two reactions, which until that moment had never been seen for neutrons, Fermi wrote as follows: "The chemical separations carried out in the case of iron and phosphorus seem to indicate that at least in these two cases the neutron is absorbed and a proton emitted. The unstable product, manganese and silicon, respectively, with the emission of a beta particle reverts to being equal to the starting element". Note the term "beta" to indicate the particle emitted.

As far as the (n, α) reaction hypothesised by Fermi at the beginning is concerned, this was found for aluminium starting from 15 April (page S12).

Moreover a first reaction, in which the active product proved to be an isotope of the starting element (a reaction later indicated as (n, γ), was found on 12 April (page S8) for the case of bromine.

So we can see that D'Agostino's contribution showed itself to be of a high scientific level. It is not by chance that the Fondo D'Agostino houses the scientific correspondence that Fermi had in the years 1934-35 on chemical questions with important personalities such as the nuclear chemist Aristid von Grosse (1905-1985), as if to indicate that on these themes D'Agostino was the point of reference, almost a sort of official delegate.

10.10 The date and time of the discovery

Rd-Be type neutron sources, because of the brief mean lifetime of radon, had to be replaced frequently, on average every week. Thanks to this very characteristic and thanks to the fact that Fermi recorded scrupulously the sources used one by one in a notebook that we found at the Domus Galilaeana in Pisa, we are now able to give a precise indication of the date when Fermi began his measurements and discovered neutron-induced radioactivity for the first time.

In this notebook, which Fermi called "Thesaurus Elementrum Radioactivorum" (Fig. 3), is the systematic summary of some results achieved regarding the study of artificial radioactivity induced in the elements of the periodic table, from hydrogen all the way to uranium (pages 1-92: an element on each page). It was compiled immediately after the first letter to "La Ricerca Scientifica" on 25 March, presumably starting on 27 March (the first date given, the same date on which Fermi, as can be seen in the Irpinia notebook, planned his research on the whole periodic table), until 2 December 1934 (the last date given).

This notebook at the end contains ten unnumbered pages with the title "Register of sources", where the activities of the various sources used and the days corresponding to their use are given. The first page, shown in Fig. 58, is particularly significant for our purposes.

In the first column the days of the month are shown starting on 20 March (20 III) and in the following columns the activities of the sources used, with their progressive number. Note that the exact values (calculated on the basis of the rule of radioactive decay, written by Fermi at the top right: "every day divide by 1.7") are only given starting on 27 March referring to source 2 which makes us think that this Register was begun, along with the Thesaurus, on 27 March. It is not by chance that Fermi attributed to source 1, which corresponded to the week before the week beginning on 27 March, an approximate value, written slanted, equal to about 30 mCi, without specifying the daily performance of the source. This page of the Thesaurus, starting with the date of 20 March (remember that the first of Fermi's Letters to "La Ricerca Scientifica" was on Sunday 25 March) and with a source indicated with the number "1", covers all of the initial phase of Fermi's research. So it follows that Fermi began his measurements on 20 March, the day when the source was made available to him, using indeed source "1".

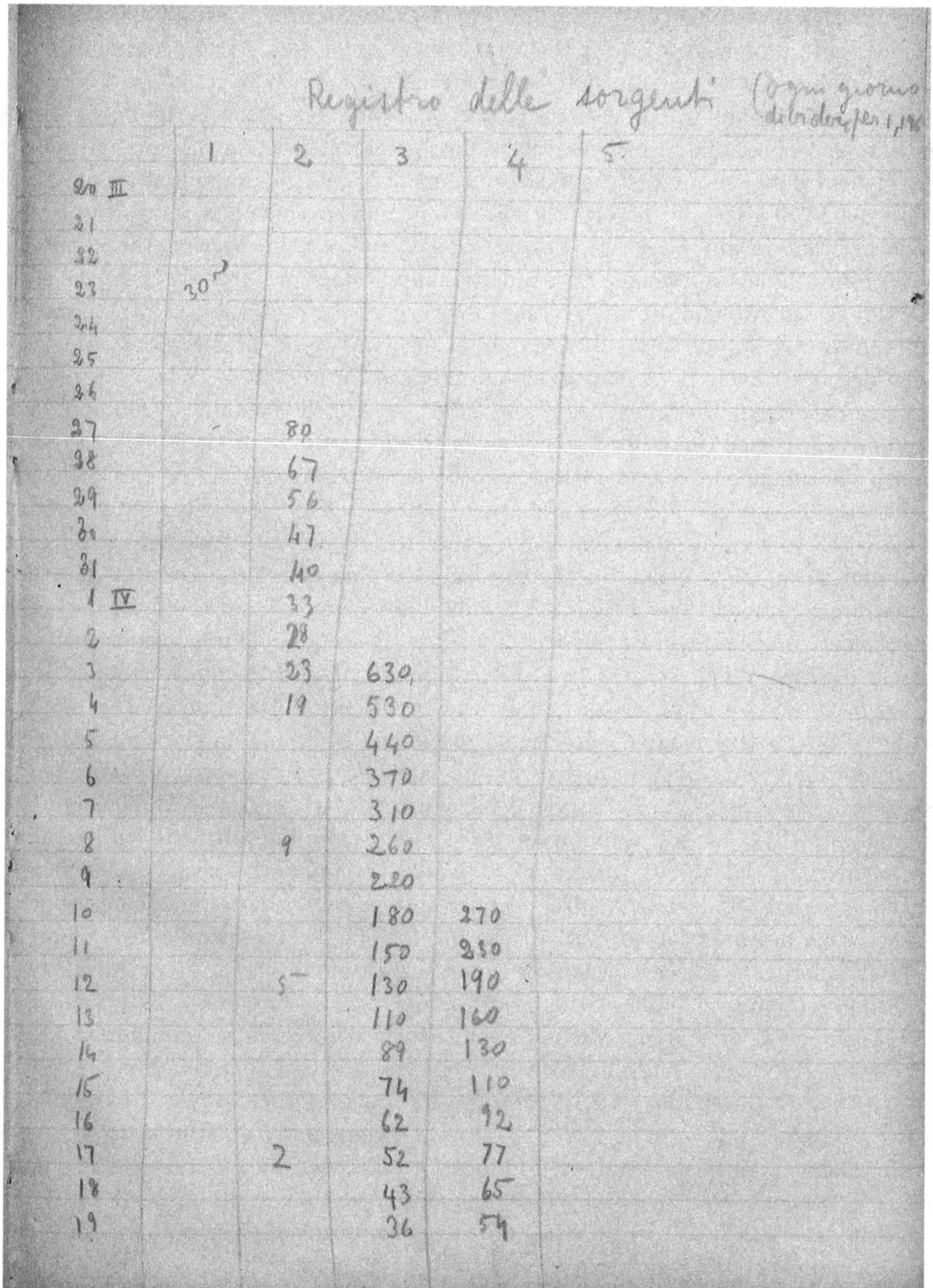

Registro delle sorgenti ([illegible])

	1	2	3	4	5
20 III					
21					
22					
23	30'				
24					
25					
26					
27		80			
28		67			
29		56			
30		47			
31		40			
1 IV		33			
2		28			
3		23	630		
4		19	530		
5			440		
6			370		
7			310		
8		9	260		
9			220		
10			180	270	
11			150	230	
12		5	130	190	
13			110	160	
14			89	130	
15			74	110	
16			62	92	
17		2	52	77	
18			43	65	
19			36	54	

Fig. 58. – First page of the "Register of sources" in the "Thesaurus Elementrum Radioactivorum" (1934) - FDG.

Besides, the approximate value given by Fermi to this source, "30 $\sim$", is in very good agreement with an initial value for the source equal to 50 mCi, as was communicated by Fermi in his first Letter to "La Ricerca Scientifica".

From pages Q18 and Q19 of the Irpinia notebook we can see that Fermi began his measurements on nonirradiated platinum $9^{h}56'$ in the morning, when the counter indicated 7600. Instead the beginning of the measurements on irradiated aluminium corresponded to the value 9750 given by the counter. So about four hours passed, as we can estimate considering that in any case the background gives about ten hits on the counter per minute. Since the measurements on irradiated aluminium immediately show the success of the activation, we can state that the discovery of neutron-induced radioactivity on aluminium occurred precisely on 20 March, at around two in the afternoon.

The day of Tuesday 20 March was characterised by frenetic activity. On page Q18 the notebook records the times $9^{h}35'$ and $9^{h}56'$, to which the values 7200 and 7600 of the numbering device correspond. They were then cancelled but they are still useful to establish the timeline of the operations. Moreover on page Q22 the time $18^{h}55'$, late in the afternoon, is given when the numbering device showed 2450, after having returned automatically to zero after passing 9999, as we see at minute $19'$ of the measurements on aluminium on page Q19. So there were 4850 "clicks" of the counter in 9 hours, corresponding to an average of 9 clicks a minute, fully compatible with all the background measurements taken. This also confirms that it really is possible to consider the counter as a sort of clock, albeit subject to stochastic background fluctuations, to the possible variations in the operating conditions of the whole apparatus, and to the deviations introduced by possible manual approximations at the beginning of the counts.

In these nine hours Fermi carried out in particular the following measurements in succession. With potassium chloride, with platinum, with irradiated platinum, with aluminium, with irradiated aluminium (success), on the background, with irradiated lead, with irradiated CaF_2, with HgCl. Bearing in mind the necessity of recording the counts for all of these manually, only five possible pauses of significant length are identified, one beginning at around 10, lasting 800 clicks and therefore about an hour and a half, another of only about twenty minutes starting around 12.30, a third for about half an hour starting at 14.30, then one of perhaps half an hour from 15.30, and a final one of about an hour starting at 18.00. With regard to Fermi's legendary traditional lunch break, respected according to witnesses even during the activation phase of the nuclear pile on 2 December 1942, it appears that on this frenetic day it really was not possible. The need to exploit as far as possible the rapidly volatile source of radon-beryllium neutrons will have made Fermi opt for quick "sandwich" breaks. The nuclear chain reaction could wait, a radon-beryllium source does not wait.

But there is more. A disturbing continuity in the counts of the numbering device well before the number 7600, which as we have said corresponded to the time $9^{h}56'$, shows that Fermi spent much of the night between 19 and 20 March in the laboratory, busying himself with final adjustments to the apparatus (counter and amplifier), and with background measurements, so as to be ready for the actual measurements on irradiated samples when he received the neutron source on the morning of 20 March. The first figure recorded by the numbering device on page Q17 is 2600, and that would correspond to about one in the morning of 20 March. A background measurement for 23 minutes followed (9.04 clicks per minute), then again a background measurement for 48 minutes (9.89 clicks per minute), perhaps to test the apparatus, then a measurement for three minutes perhaps with the KCl (30 clicks per minute), followed by a very long background measurement, perhaps with altered operating conditions, for which the results were recorded at the 24th minute, at the 30th minute (only 7.97 clicks per minute overall), and after 2 hours and 54 minutes from the beginning (1390 hits over 174 minutes corresponding to about 8 per minute). Page Q17 also contains indications of the values used to power the counter and for the values of the coupling batteries of the amplifier. A later background measurement was made at around 8 in the morning with the counter that went from 6500 to 7080 in 67 minutes (8.65 hits per minute). During the night there was only a possible pause of about 2 hours and twenty minutes, when the apparatus could work alone during the final part of the long data collection, when the counter went from 4239 to 5390, around 3.30 a.m. There was another pause for a couple of hours between six and eight in the morning before the data was taken over 67 minutes. So it was certainly a very exacting night for Fermi but bearing in mind what we have seen happened during the day it was certainly worth the trouble: neutron-induced radioactivity on aluminium was discovered. Even after $18^h55'$, written on page Q22, the measurements continued through the following night, with only short breaks. The radon-beryllium source rapidly lost its intensity and it was therefore best to exploit it as far as possible straight after it had been prepared.

As can be seen from the Registers kept in the Archive of the University of Rome "La Sapienza", on the same day that he discovered neutron induced radioactivity Fermi had actually taught the final lessons of the courses on "Mathematical Physics" and "Theoretical Physics" (devoted, respectively, to "Developments in series of spherical harmonics" and "Discussion of contemporary measurements") before the Easter holiday break (the courses would then start again on 8 May), see Figs. 59-62. The times that day of the lessons on Theoretical Physics and Mathematical Physics may correspond to the pauses that started at 10 a.m. and 6 p.m..

It is interesting to observe also that the day of Fermi's discovery coincides precisely with the last day in which Franco Rasetti taught his last lesson in Rome (on "Moseley Diagrams for terms K, L, M", Figs. 63, 64) of his course on "Spectroscopy", before departing for Rabat in Morocco where, as the representative of the Società Italiana per il Progresso delle Scienze (SIPS), he would attend the Conference of the "Société Française pour l'Avancement des Sciences" planned for 24 March to 4 April 1934.

The fact that Rasetti, although he was in Rome not only on the day of the discovery but also on the previous days, had not been involved in this discovery, despite the plans expressed to Joliot and D'Agostino for possible experiments in common with Fermi on alpha-particle–induced radioactivity, and despite his experimental expertise, shows us yet again how much the search for possible neutron-induced radioactivity was a line of research that Fermi had thought of alone and that he wanted to conduct by himself, which is what he did.

Mod. n. 10

R. Università degli Studi di Roma

Facoltà di FACOLTA' DI SCIENZE matematiche, fisiche e naturali

LIBRETTO

delle lezioni di Fisica teorica

dettate dal sig. prof. Fermi Enrico

nell'anno scolastico 1933 1934

Fig. 59. – Register of the lessons of the course in "Theoretical Physics" given by Fermi in the academic year 1933-34 -URS.

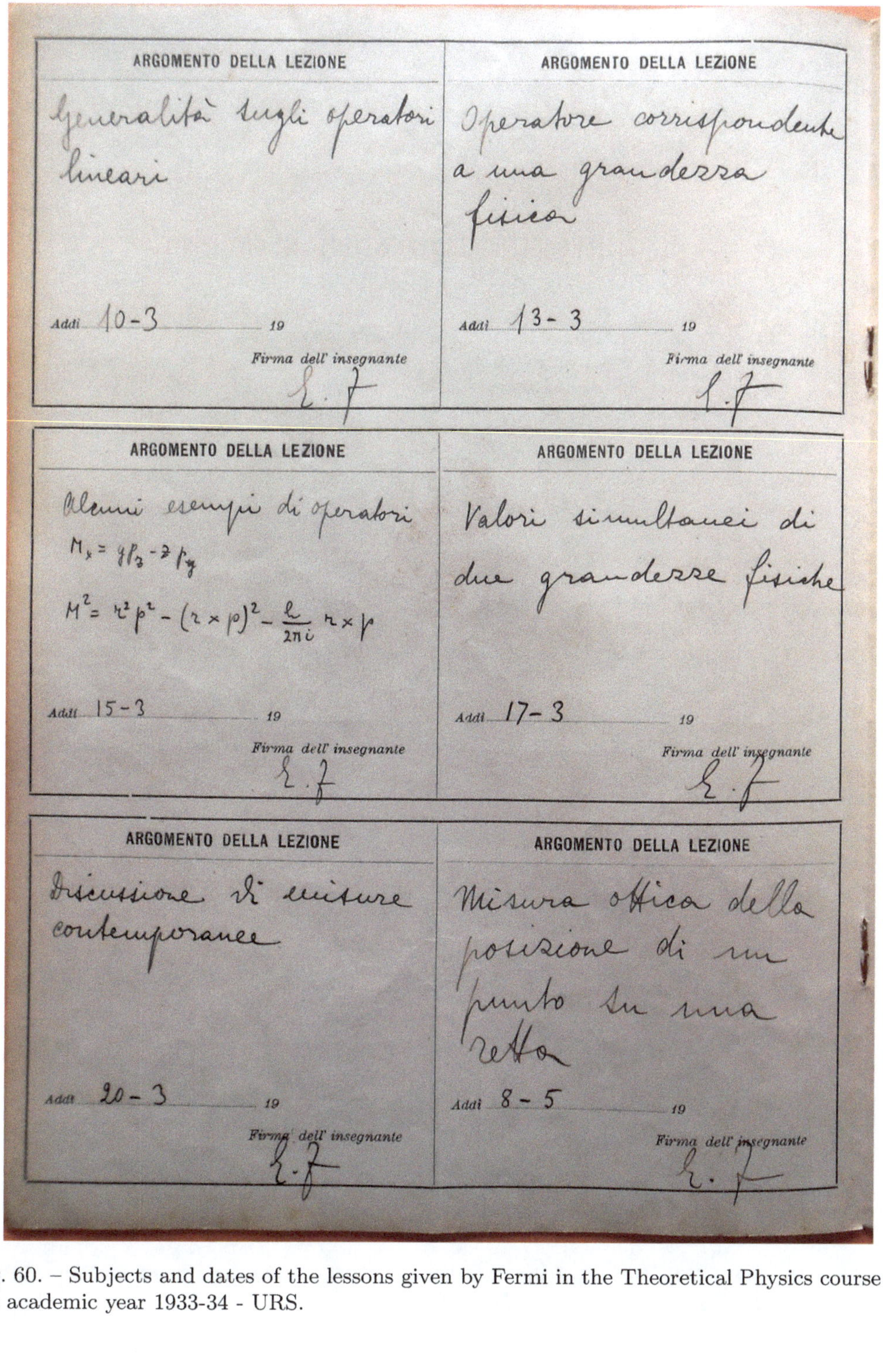

ARGOMENTO DELLA LEZIONE	ARGOMENTO DELLA LEZIONE
Generalità sugli operatori lineari	Operatore corrispondente a una grandezza fisica
Addì 10-3 19	Addì 13-3 19
Firma dell' insegnante E. F.	Firma dell' insegnante E. F.

ARGOMENTO DELLA LEZIONE	ARGOMENTO DELLA LEZIONE
Alcuni esempi di operatori $M_x = yp_z - zp_y$ $M^2 = r^2p^2 - (r \times p)^2 - \frac{h}{2\pi i} r \times p$	Valori simultanei di due grandezze fisiche
Addì 15-3 19	Addì 17-3 19
Firma dell' insegnante E. F.	Firma dell' insegnante E. F.

ARGOMENTO DELLA LEZIONE	ARGOMENTO DELLA LEZIONE
Discussione di misure contemporanee	Misura ottica della posizione di un punto su una retta
Addì 20-3 19	Addì 8-5 19
Firma dell' insegnante E. F.	Firma dell' insegnante E. F.

Fig. 60. – Subjects and dates of the lessons given by Fermi in the Theoretical Physics course in the academic year 1933-34 - URS.

Mod. n. 10

R. Università degli Studi di Roma

Facoltà di FACOLTA' DI SCIENZE
matematiche, fisiche e naturali

LIBRETTO

delle lezioni di Fisica matematica

dettate dal sig. prof. Fermi Enrico

nell'anno scolastico 1933 1934

5

Fig. 61. – Register of the lessons of the "Mathematical Physics" course given by Fermi in the academic year 1933-34 - URS.

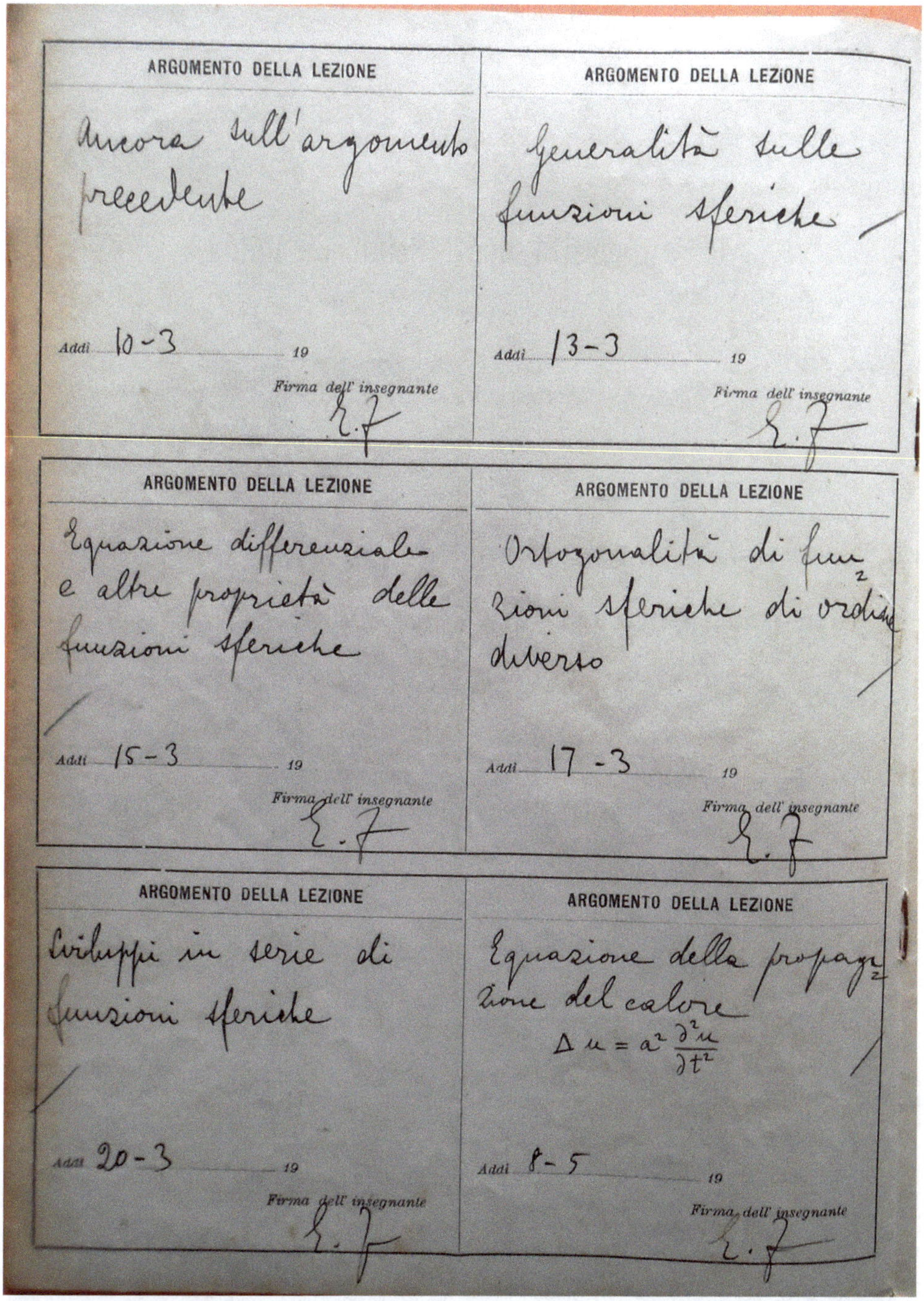

ARGOMENTO DELLA LEZIONE	ARGOMENTO DELLA LEZIONE
Ancora sull'argomento precedente Addì 10-3 19 Firma dell'insegnante E.F.	Generalità sulle funzioni sferiche Addì 13-3 19 Firma dell'insegnante E.F.
Equazione differenziale e altre proprietà delle funzioni sferiche Addì 15-3 19 Firma dell'insegnante E.F.	Ortogonalità di funzioni sferiche di ordine diverso Addì 17-3 19 Firma dell'insegnante E.F.
Sviluppi in serie di funzioni sferiche Addì 20-3 19 Firma dell'insegnante E.F.	Equazione della propagazione del calore $\Delta u = a^2 \frac{\partial^2 u}{\partial t^2}$ Addì 8-5 19 Firma dell'insegnante E.F.

Fig. 62. – Subjects and dates of the lessons given by Fermi in the course of "Mathematical Physics" - URS.

Mod. n. 10

R. Università degli Studi di Roma

Facoltà di FACOLTA' DI SCIENZE matematiche, fisiche e naturali

LIBRETTO

delle lezioni di Spettroscopia

dettate dal sig. prof. Rasetti Franco

nell'anno scolastico 1933 1934

20

Roma - Tip. Ditta F.lli Pallotta

Fig. 63. – Register of lessons of the "Spectroscopy" course given by Franco Rasetti in the academic year 1933-34 - URS.

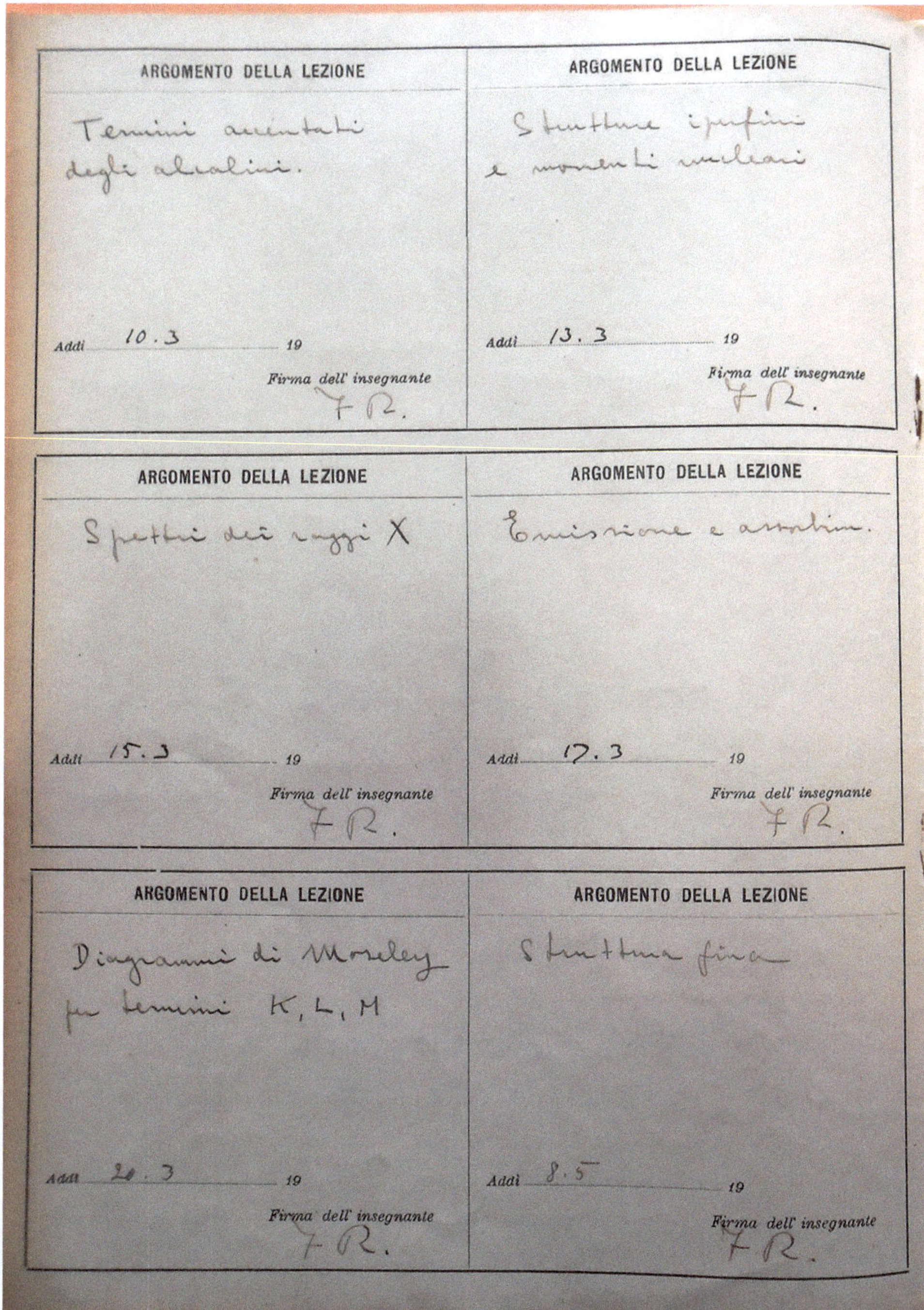

ARGOMENTO DELLA LEZIONE

Termini accentati degli alcalini.

Addì 10.3 19

Firma dell' insegnante
F R.

ARGOMENTO DELLA LEZIONE

Struttura iperfina e momenti nucleari

Addì 13.3 19

Firma dell' insegnante
F R.

ARGOMENTO DELLA LEZIONE

Spettri dei raggi X

Addì 15.3 19

Firma dell' insegnante
F R.

ARGOMENTO DELLA LEZIONE

Emissione e assorbim.

Addì 17.3 19

Firma dell' insegnante
F R.

ARGOMENTO DELLA LEZIONE

Diagrammi di Moseley per termini K, L, M

Addì 20.3 19

Firma dell' insegnante
F R.

ARGOMENTO DELLA LEZIONE

Struttura fina

Addì 8.5 19

Firma dell' insegnante
F R.

Fig. 64. – Subjects and lessons given by Franco Rasetti in his "Spectroscopy" course - URS.

11 Pages from the Irpinia notebook and the loose sheets of paper

Here we reproduce some pages of the Irpinia notebook and of the loose pages of paper thanks to the kind permission of the Fondazione Oscar D'Agostino, Avellino.

From the upright side of the notebook we include pages: Qf0 (cover), Qf2, Qf15, Qf16.

From the upside down side the pages: Q1-Q7, Q15-Q24, Q27-Q29, Q35-Q40, Q44, Q46, Q48, Q50, Q60, Q93, Q94, Q119, Q140.

From the loose sheets of paper pages: S3a, S4, S7, S8, S12, S15, S15a, S16a.

F. Guerra and N. Robotti, *The Lost Notebook of ENRICO FERMI*,
https://doi.org/10.1007/978-3-319-69254-8_11

Qf0

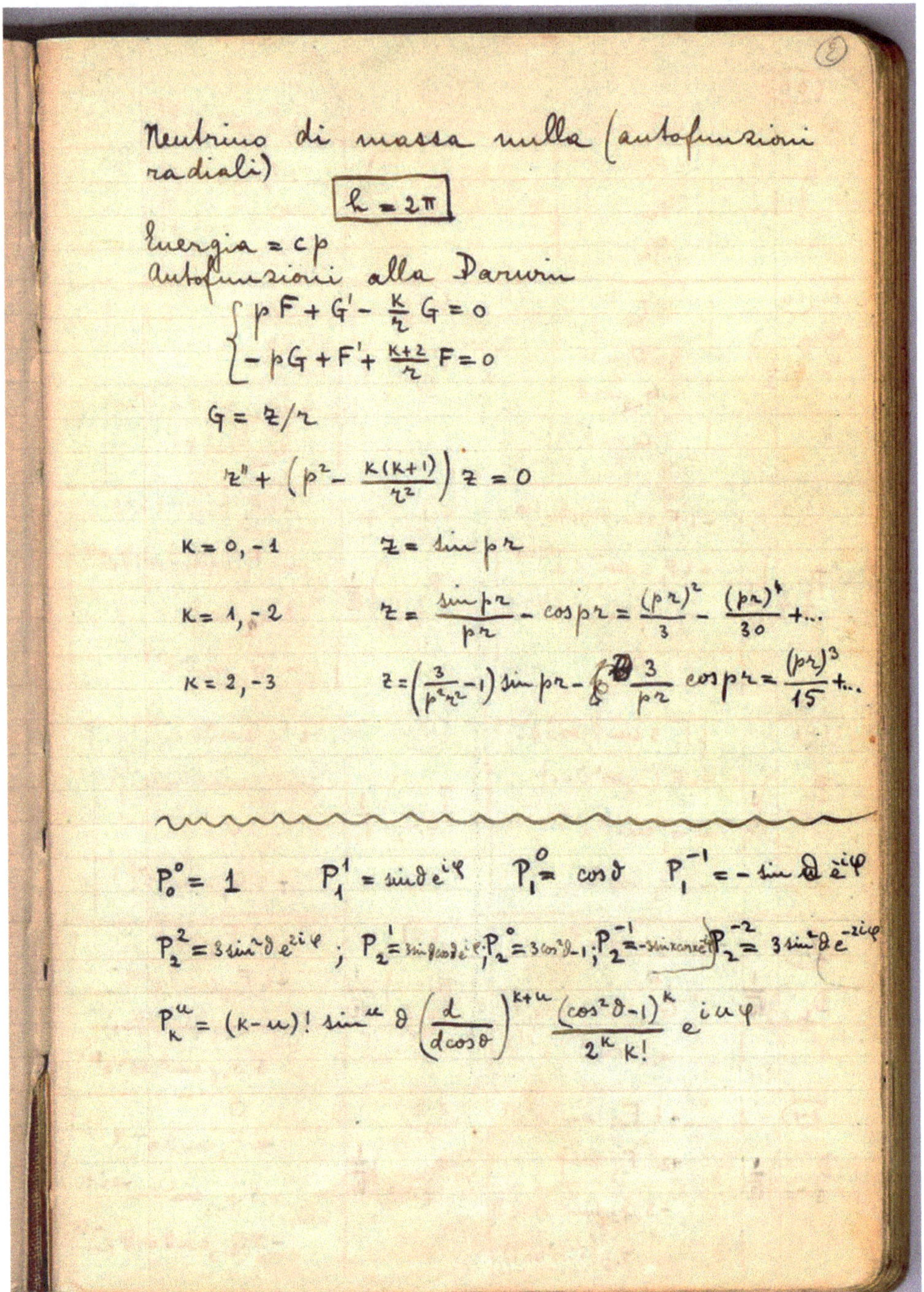

Neutrino di massa nulla (autofunzioni radiali)

$$\boxed{h = 2\pi}$$

Energia = cp

Autofunzioni alla Darwin

$$\begin{cases} pF + G' - \frac{k}{r} G = 0 \\ -pG + F' + \frac{k+2}{r} F = 0 \end{cases}$$

$$G = z/r$$

$$z'' + \left(p^2 - \frac{k(k+1)}{r^2}\right) z = 0$$

$k = 0, -1$ $\quad z = \sin pr$

$k = 1, -2$ $\quad z = \frac{\sin pr}{pr} - \cos pr = \frac{(pr)^2}{3} - \frac{(pr)^4}{30} + \dots$

$k = 2, -3$ $\quad z = \left(\frac{3}{p^2 r^2} - 1\right) \sin pr - \frac{3}{pr} \cos pr = \frac{(pr)^3}{15} + \dots$

$P_0^0 = 1 \quad P_1^1 = \sin\vartheta\, e^{i\varphi} \quad P_1^0 = \cos\vartheta \quad P_1^{-1} = -\sin\vartheta\, e^{-i\varphi}$

$P_2^2 = 3\sin^2\vartheta\, e^{2i\varphi}$; $P_2^1 = \sin\vartheta\cos\vartheta\, e^{i\varphi}$; $P_2^0 = 3\cos^2\vartheta - 1$; $P_2^{-1} = -\sin\vartheta\cos\vartheta\, e^{-i\varphi}$; $P_2^{-2} = 3\sin^2\vartheta\, e^{-2i\varphi}$

$$P_k^u = (k-u)!\, \sin^u \vartheta \left(\frac{d}{d\cos\vartheta}\right)^{k+u} \frac{(\cos^2\vartheta - 1)^k}{2^k\, k!}\, e^{iu\varphi}$$

Qf2

(15)

Termine 1s relativistico

$$\begin{cases} F = f\, r^s e^{-\lambda r} \\ G = r^s e^{-\lambda r} \end{cases} \qquad s = \sqrt{1-\gamma^2} - 1 \;;\quad \gamma = Z/137$$

$$f = -\frac{s}{\gamma}$$

$$\lambda = \frac{mc^2\gamma^2}{Ze^2}$$

$$\tilde{\psi}\psi = \frac{(2\lambda)^{3+2s}}{4\pi\,\Gamma(3+2s)}\, r^{2s} e^{-2\lambda r}$$

Probabilità di transizione in cui la particella β viene legata in una orbita K

$$P = \frac{8\pi^3 g^2}{h^4} \left|\int v_m^* u_n \, d\tau\right|^2 \frac{p_\sigma^2}{c}\, \tilde{\psi}_s \psi_s$$

Trattando l'elettrone K non relativisticamente e ammettendo che il neutrino emesso abbia energia quasi eguale risulta

$$P = \frac{2048\,\pi^{12} g^2 m^5 e^{14}}{c^3 h^{14}}\, Z^7 \left|\int v_m^* u_n \, d\tau\right|^2$$

$$P \sim 10^{-20} Z^7 \left|\quad\right|^2$$

Qf15

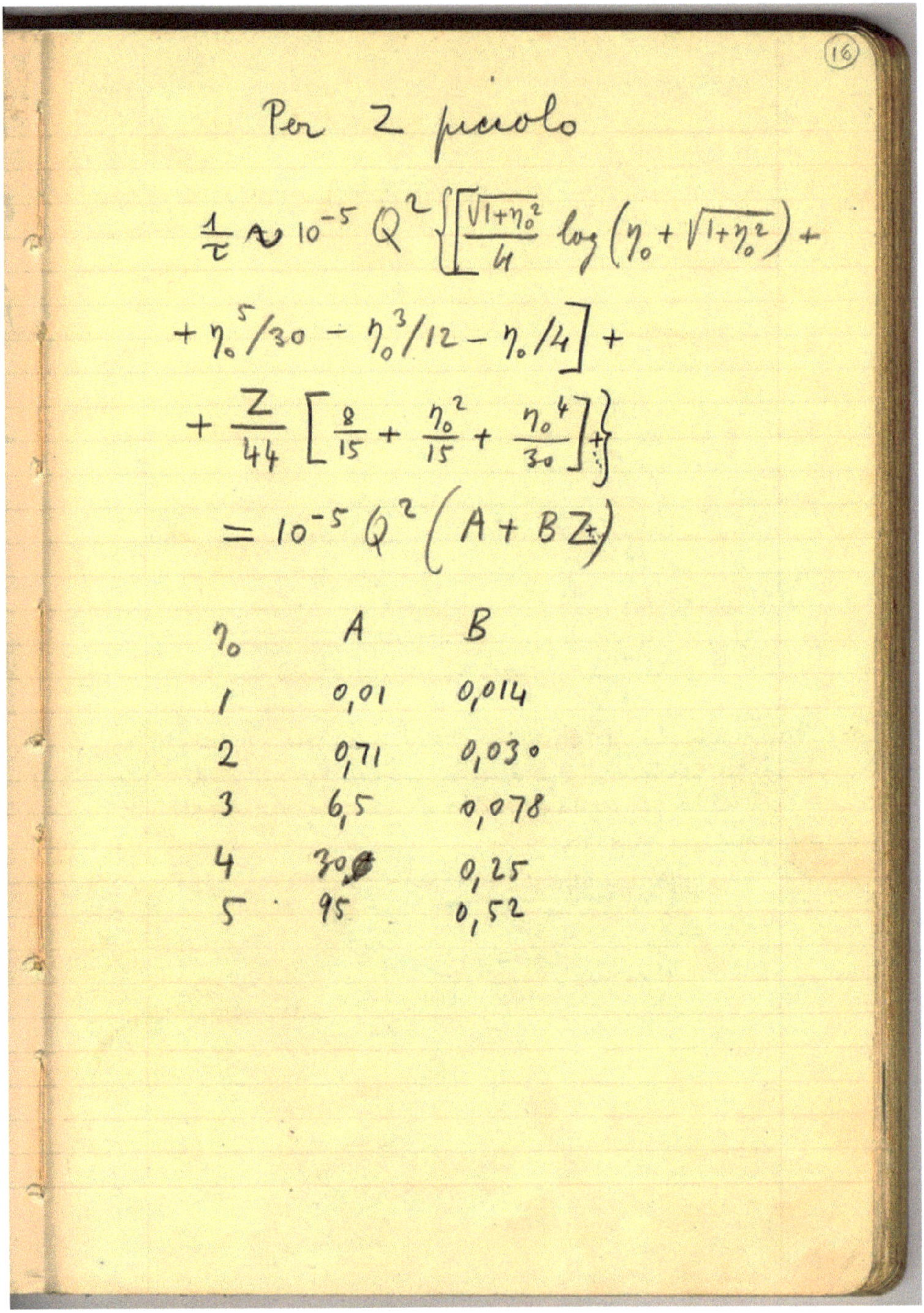

(16)

Per Z piccolo

$$\frac{1}{\tau} \approx 10^{-5} Q^2 \left\{ \left[\frac{\sqrt{1+\eta_0^2}}{4} \log\left(\eta_0+\sqrt{1+\eta_0^2}\right) + \eta_0^5/30 - \eta_0^3/12 - \eta_0/4 \right] + \frac{Z}{44}\left[\frac{8}{15}+\frac{\eta_0^2}{15}+\frac{\eta_0^4}{30}\right] + \dots \right\}$$

$$= 10^{-5} Q^2 (A + BZ + \dots)$$

η_0	A	B
1	0,01	0,014
2	0,71	0,030
3	6,5	0,078
4	30	0,25
5	95	0,52

Qf16

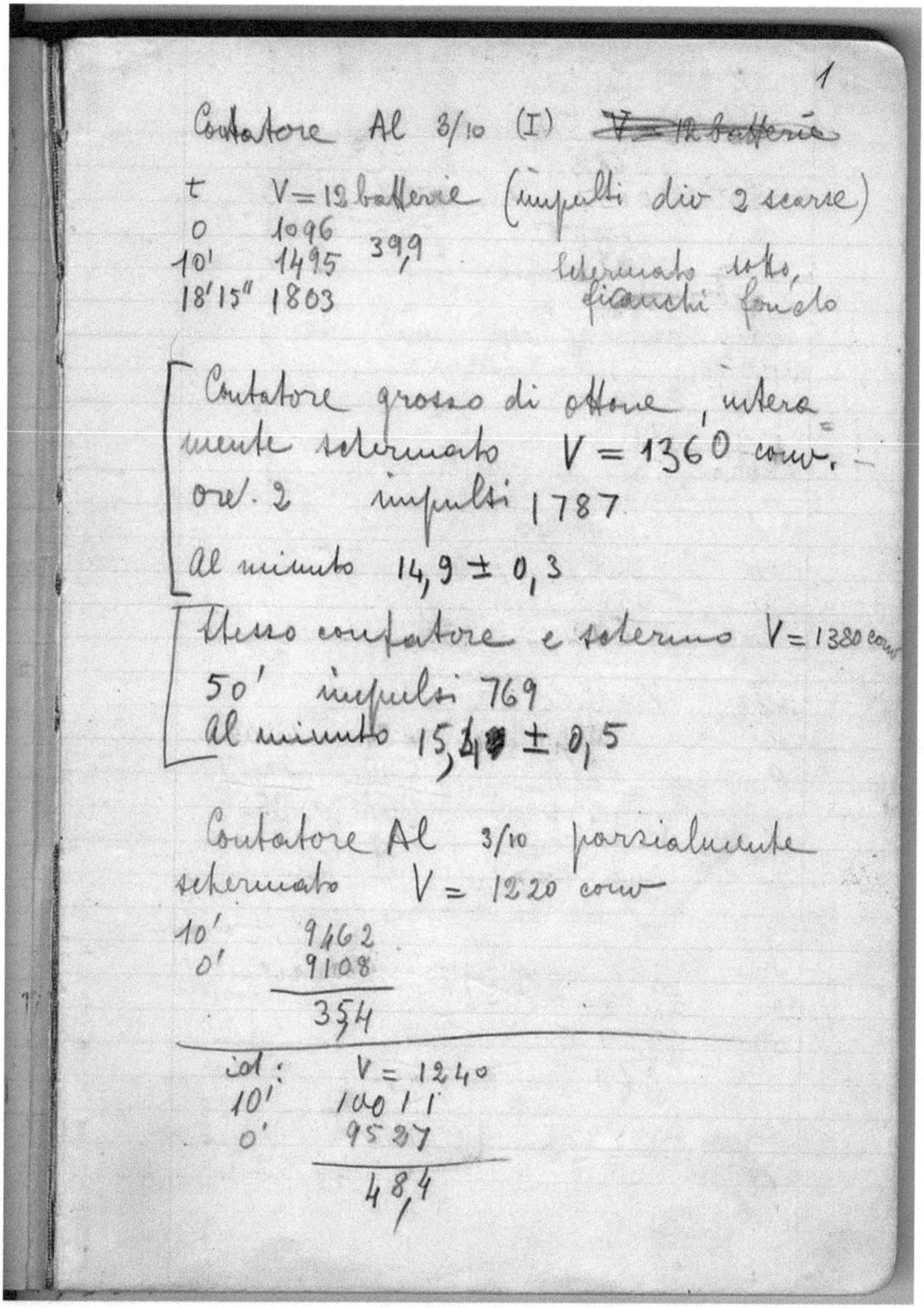

1

Contatore Al 3/10 (I) ~~V = 12 batterie~~

t V=12 batterie (impulsi div 2 scarse)

0 1096

10' 1495 39,9 schermato sotto, fianchi fondo

18'15" 1803

Contatore grosso di ottone, interamente schermato V = 1360 conv.

ore 2 impulsi 1787

al minuto 14,9 ± 0,3

Stesso contatore e schermo V = 1380 conv

50' impulsi 769

al minuto 15,4 ± 0,5

Contatore Al 3/10 parzialmente schermato V = 1220 conv

10' 9462

0' 9108

35,4

id: V = 1240

10' 10011

0' 9527

48,4

Q1

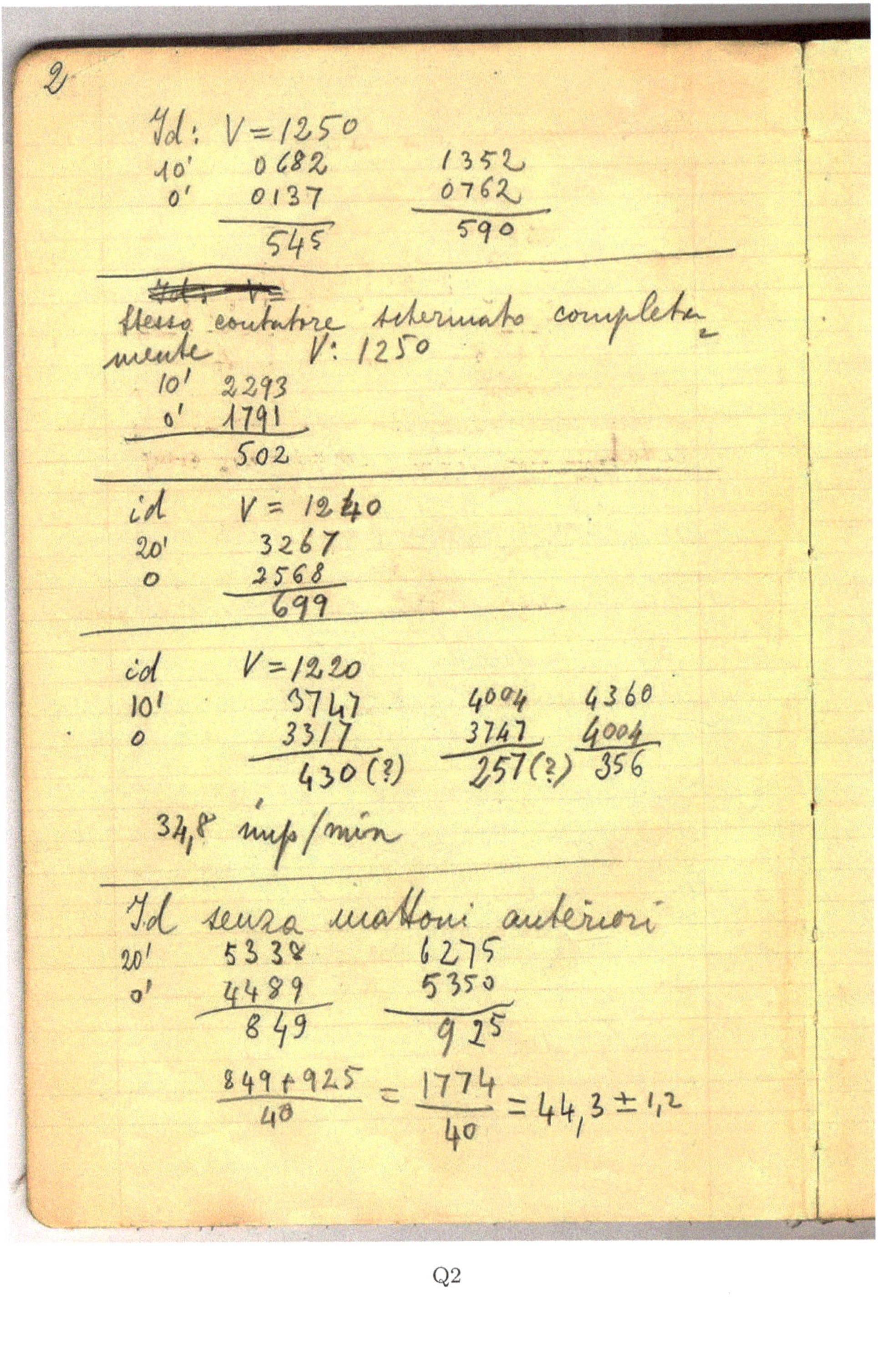
2

Id: V=1250

10'	0682	1352
0'	0137	0762
	545	590

~~Id: V=~~

stesso contatore schermato completamente V: 1250

10'	2293
0'	1791
	502

id V = 1240

20'	3267
0	2568
	699

id V=1220

10'	3747	4004	4360
0	3317	3747	4004
	430 (?)	257 (?)	356

34,8 imp/min

Id senza mattoni anteriori

20'	5338	6275
0'	4489	5350
	849	925

$$\frac{849+925}{40} = \frac{1774}{40} = 44{,}3 \pm 1{,}2$$

Q2

3

Id con vaschetta soluzione

0′	6500			
5′	6733	233		
10′	6995	262	959	$\frac{1908}{40} = 47{,}7$
15′	7240	245		
20′	7459	219		
40′	8408			

Contatore Al 2/10 schermato completamente
V = 1120 conv

0′	8870	
10′	9020	150
20′	9369	349

Q3

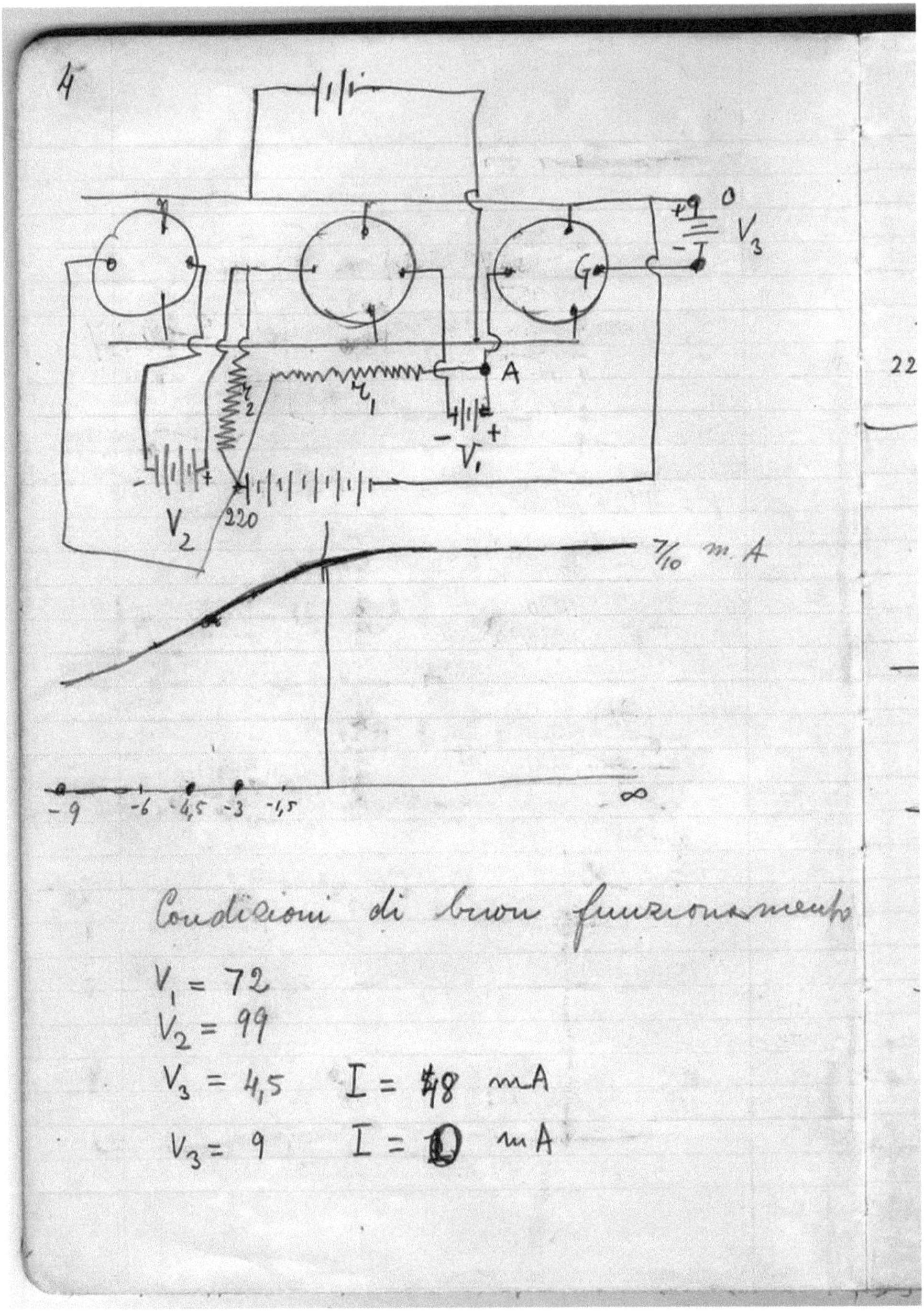
4
V_3
A
V_1
V_2
220
G
7/10 m. A
-9 -6 -4,5 -3 -1,5
∞
Condizioni di buon funzionamento
V_1 = 72
V_2 = 99
V_3 = 4,5 I = 8 mA
V_3 = 9 I = 0 mA

Q4

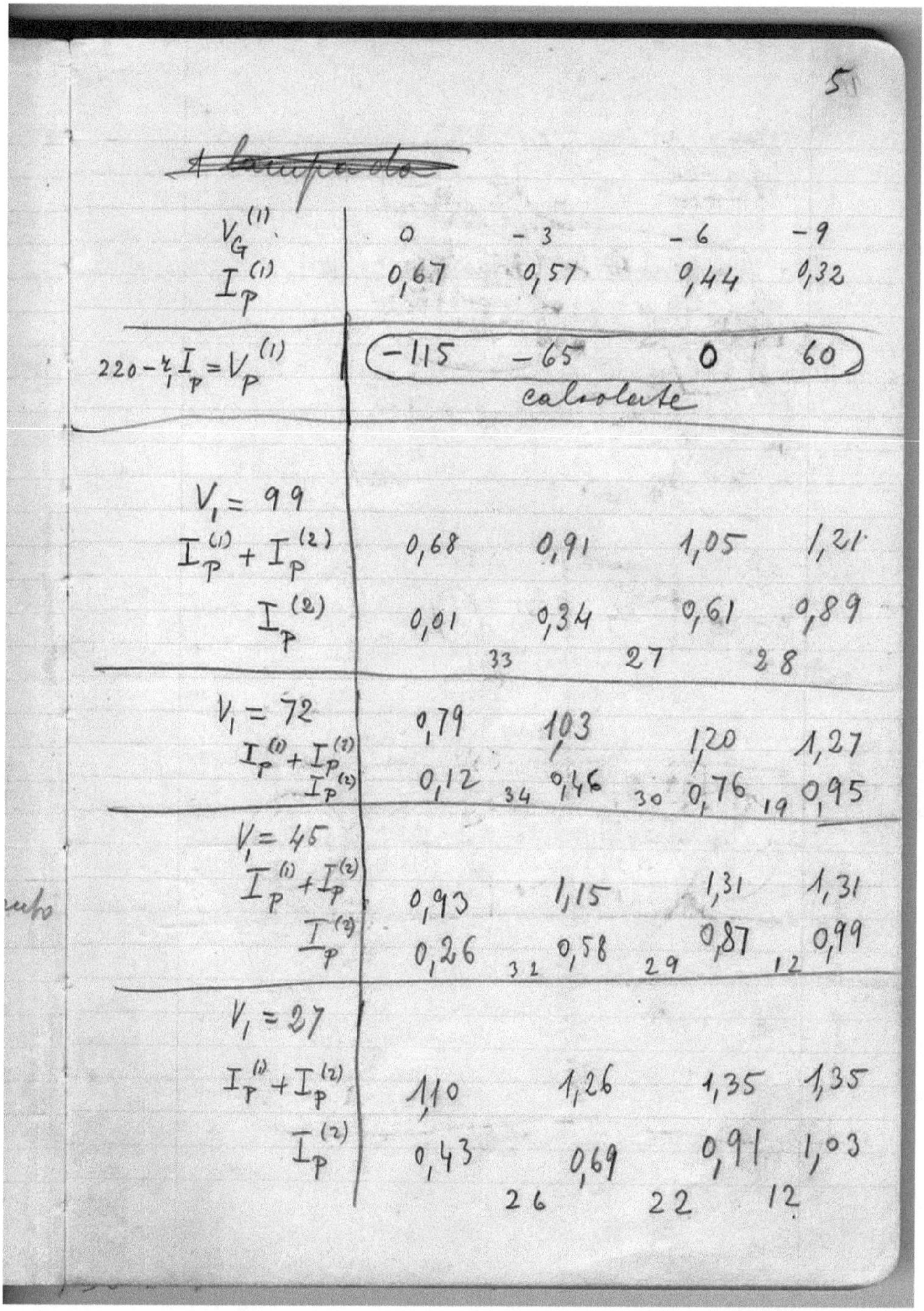

5

~~A lampada~~

$V_G^{(1)}$	0	−3	−6	−9
$I_p^{(1)}$	0,67	0,57	0,44	0,32
$220 - r_1 I_p = V_p^{(1)}$	−115	−65	0	60
		calcolate		
$V_1 = 99$				
$I_p^{(1)} + I_p^{(2)}$	0,68	0,91	1,05	1,21
$I_p^{(2)}$	0,01	0,34	0,61	0,89
	33	27	28	
$V_1 = 72$				
$I_p^{(1)} + I_p^{(2)}$	0,79	1,03	1,20	1,27
$I_p^{(2)}$	0,12	34 0,46	30 0,76	19 0,95
$V_1 = 45$				
$I_p^{(1)} + I_p^{(2)}$	0,93	1,15	1,31	1,31
$I_p^{(2)}$	0,26	32 0,58	29 0,87	12 0,99
$V_1 = 27$				
$I_p^{(1)} + I_p^{(2)}$	1,10	1,26	1,35	1,35
$I_p^{(2)}$	0,43	0,69	0,91	1,03
	26	22	12	

Q5

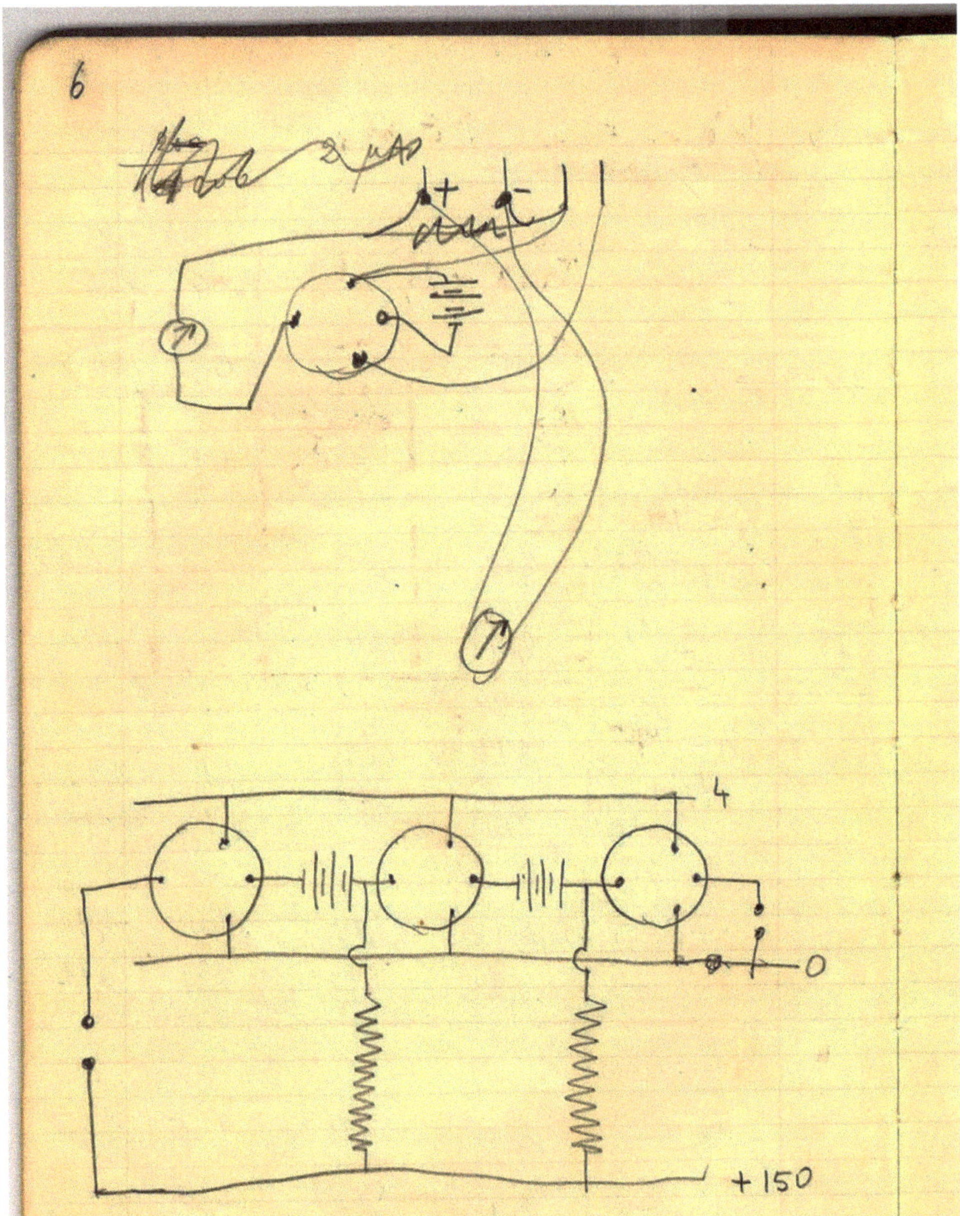

Q6

7

Telefunken RE 604

$V_p = 150$

V_G	I_p (mA)	V_G	I_p
−45	0	−42	0
−36	1	−32	2
−30	5,5	−18	3
−27	9,9	−13	15
−22,5	17,6		
−18	27,3		
−13,5	39		
−9	51		
−4,5	65		
−0	80		

Philips A 425

$V_p = 150$

V_g	I_p (mA)
−27	0
−18	0,02
−9	0,2
−7,5	0,42
−6	0,94
−4,5	

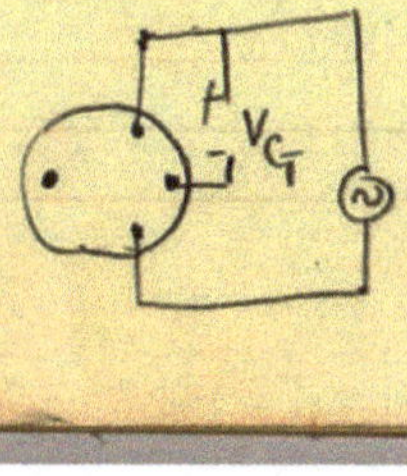

Q7

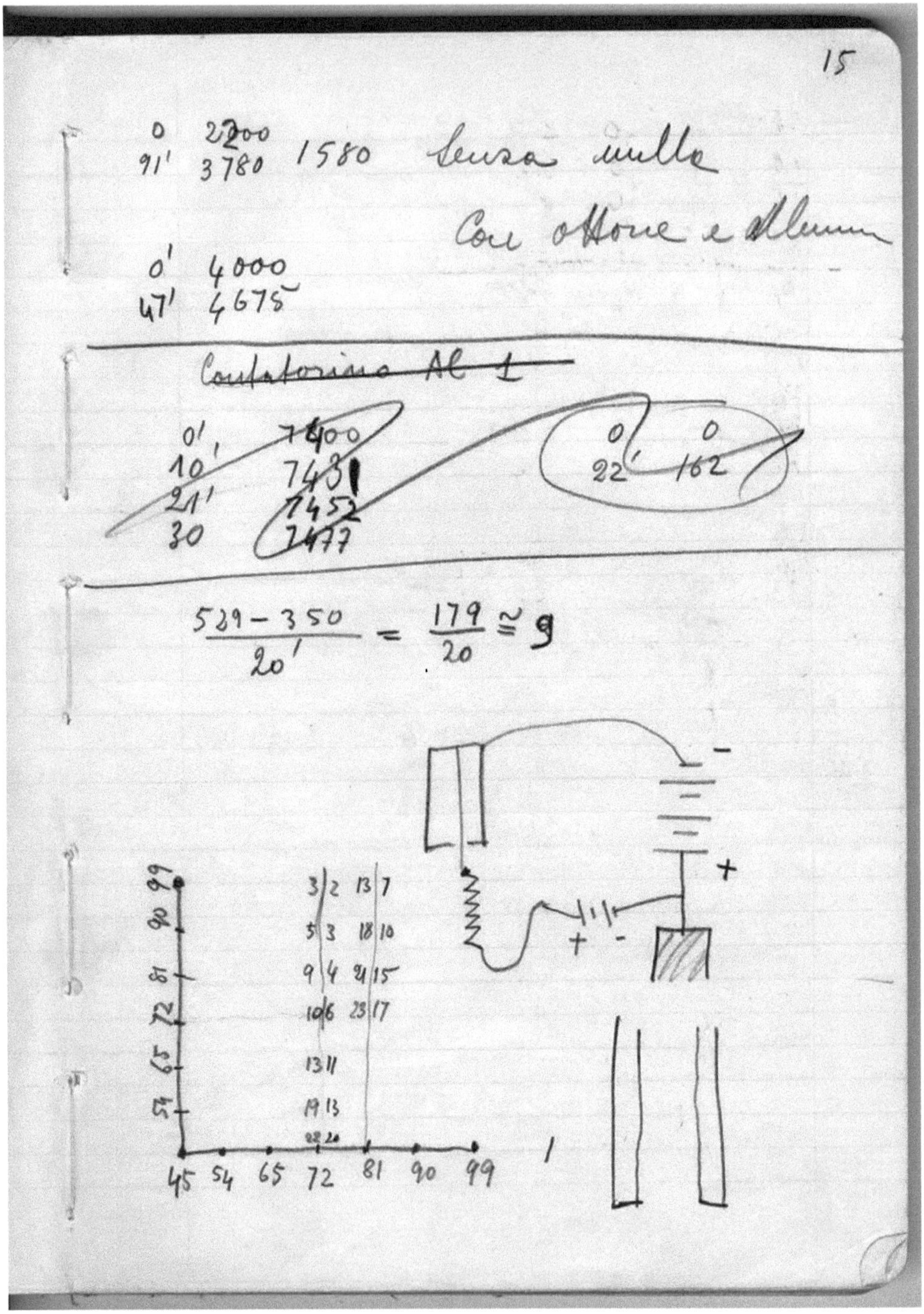

Q15

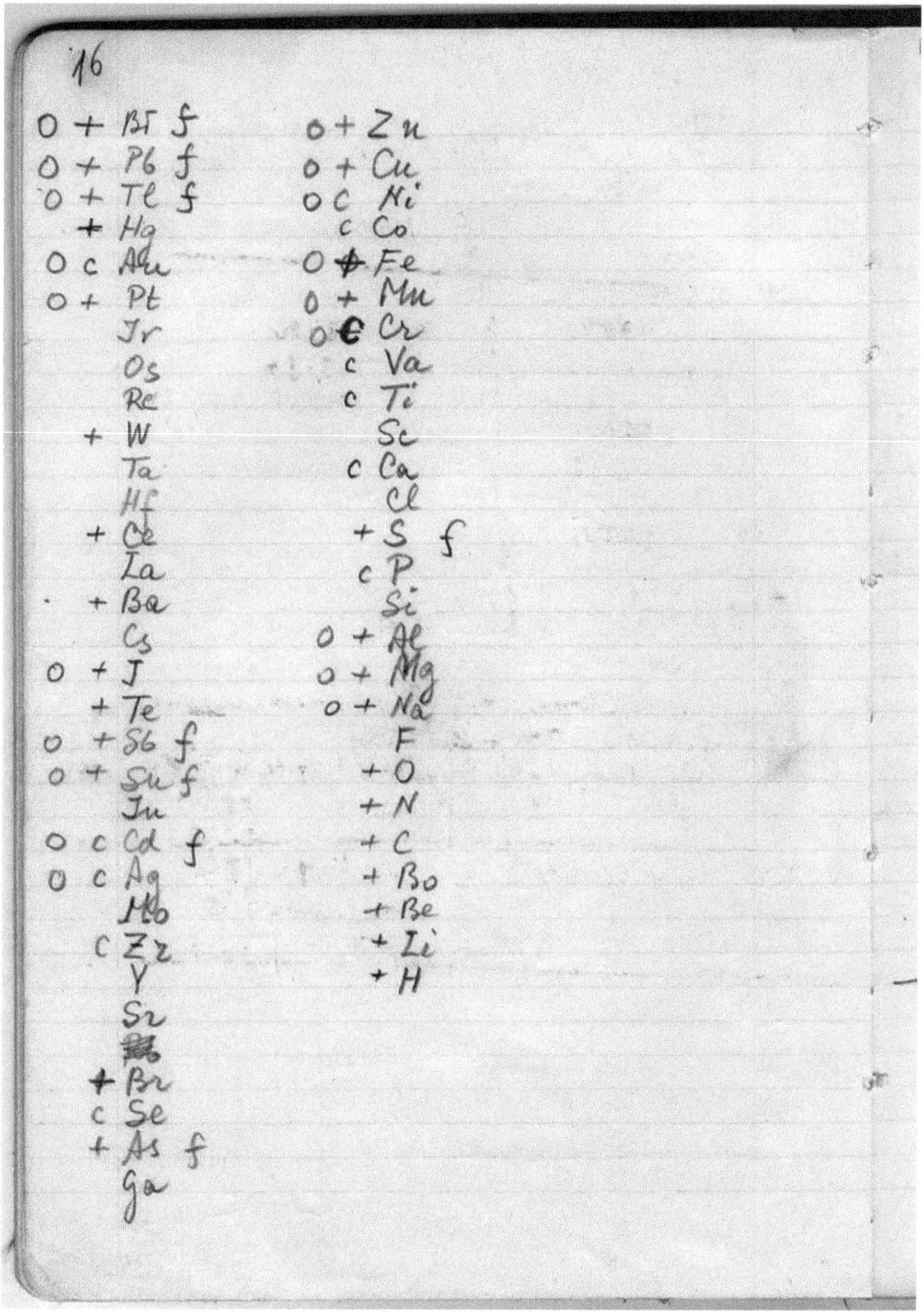
16

o + Bi f
o + Pb f
o + Tl f
+ Hg
o c Au
o + Pt
Ir
Os
Re
+ W
Ta
Hf
+ Ce
La
+ Ba
Cs
o + J
+ Te
o + Sb f
o + Sn f
In
o c Cd f
o c Ag
Mo
c Zr
Y
Sr
~~Rb~~
+ Br
c Se
+ As f
Ga

o + Zn
o + Cu
o c Ni
c Co
o + Fe
o + Mn
o c Cr
c Va
c Ti
Sc
c Ca
Cl
+ S f
c P
Si
o + Al
o + Mg
o + Na
F
+ O
+ N
+ C
+ Bo
+ Be
+ Li
+ H

Q16

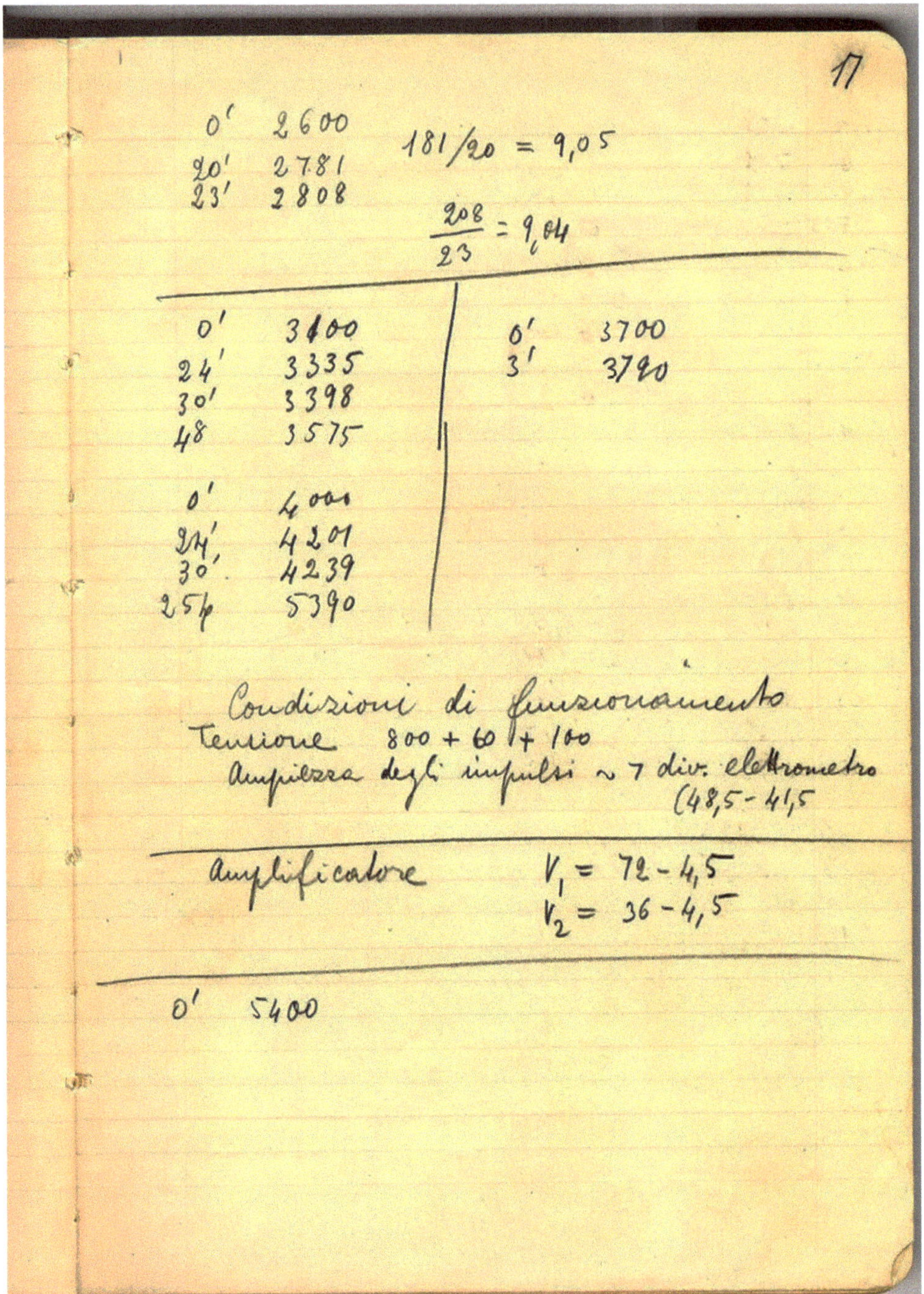

17

0' 2600
20' 2781
23' 2808

181/20 = 9,05

$\frac{208}{23}$ = 9,04

0' 3100
24' 3335
30' 3398
48 3575

0' 3700
3' 3720

0' 4000
24' 4201
30' 4239
254 5390

Condizioni di funzionamento
Tensione 800 + 60 + 100
Ampiezza degli impulsi ~ 7 div. elettrometro
(48,5 - 41,5

amplificatore
V_1 = 72 - 4,5
V_2 = 36 - 4,5

0' 5400

Q17

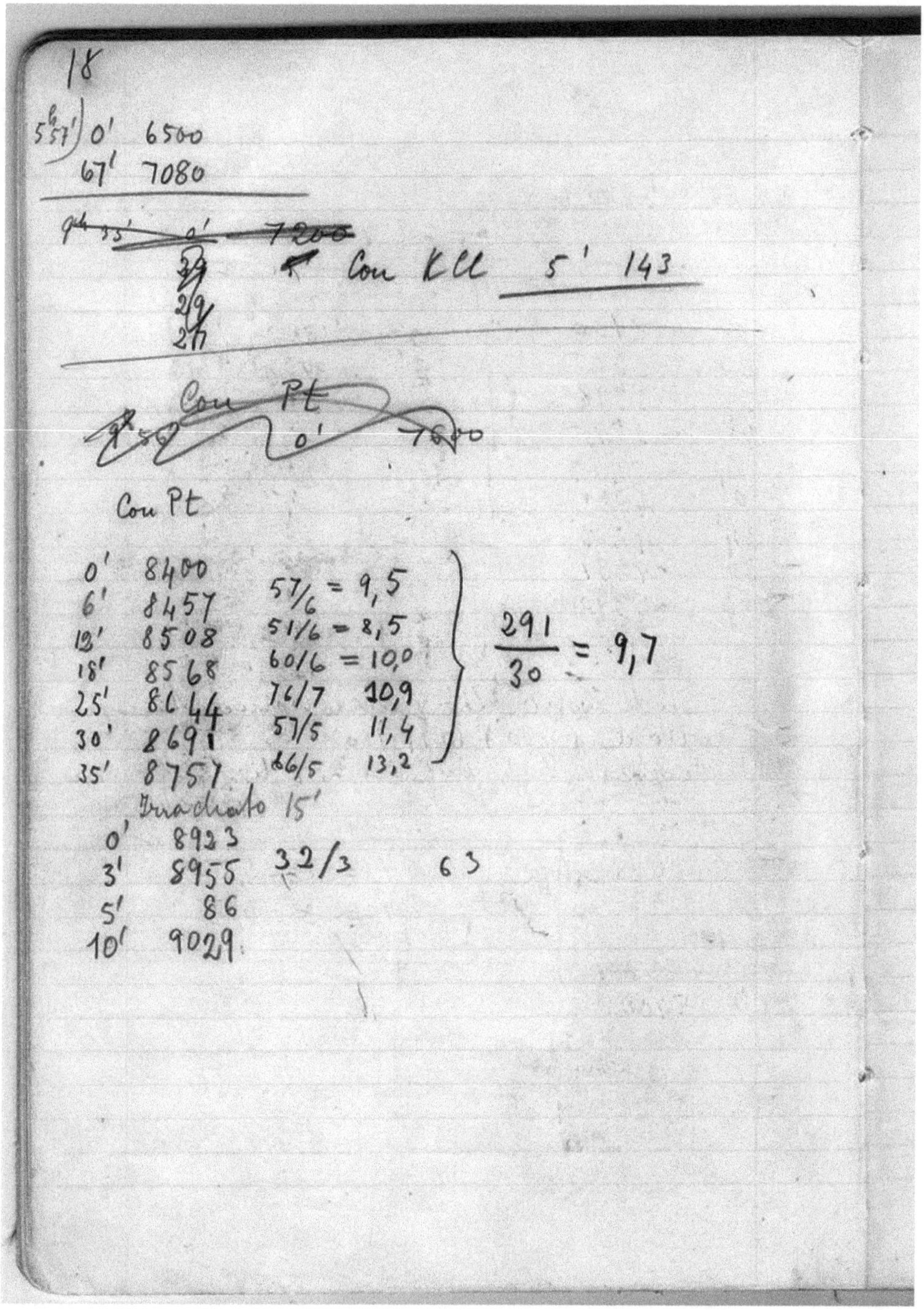
18

5h 57′) 0′ 6500
67′ 7080

Con Kll 5′ 143

Con Pt

0′	8400		
6′	8457	57/6 = 9,5	
12′	8508	51/6 = 8,5	
18′	8568	60/6 = 10,0	
25′	8644	76/7	10,9
30′	8691	57/5	11,4
35′	8757	66/5	13,2

$\frac{291}{30} = 9,7$

Irradiato 15′

0′	8923		
3′	8955	32/3	63
5′	86		
10′	9029		

Q18

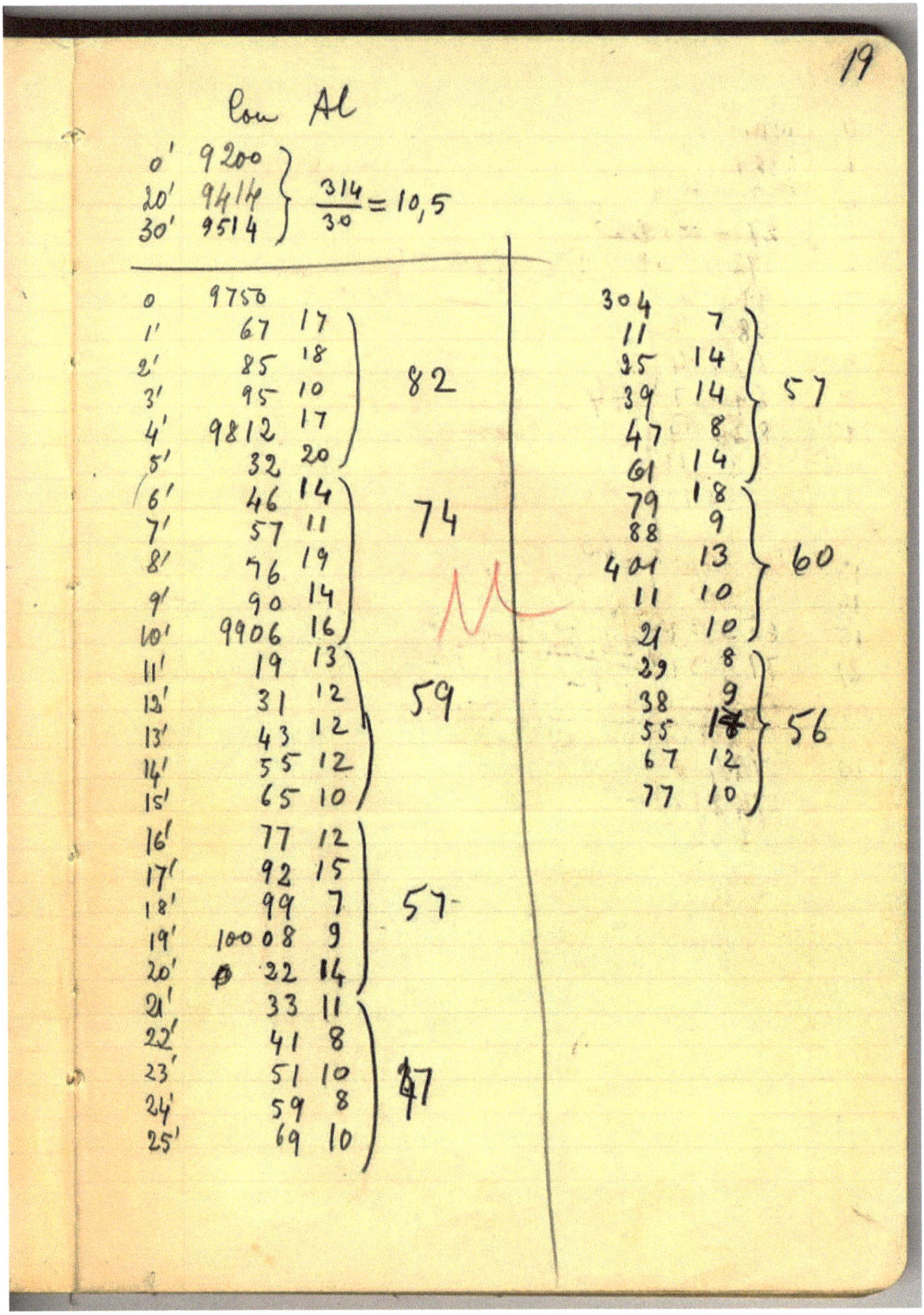

19

Con Al

0' 9200
20' 9414
30' 9514

$\frac{314}{30} = 10,5$

0	9750		
1'	67	17	
2'	85	18	
3'	95	10	82
4'	9812	17	
5'	32	20	
6'	46	14	
7'	57	11	74
8'	76	19	
9'	90	14	
10'	9906	16	
11'	19	13	
12'	31	12	59
13'	43	12	
14'	55	12	
15'	65	10	
16'	77	12	
17'	92	15	
18'	99	7	57
19'	10008	9	
20'	22	14	
21'	33	11	
22'	41	8	
23'	51	10	47
24'	59	8	
25'	69	10	

304		
11	7	
25	14	
39	14	57
47	8	
61	14	
79	18	
88	9	
401	13	60
11	10	
21	10	
29	8	
38	9	
55	17	56
67	12	
77	10	

Q19

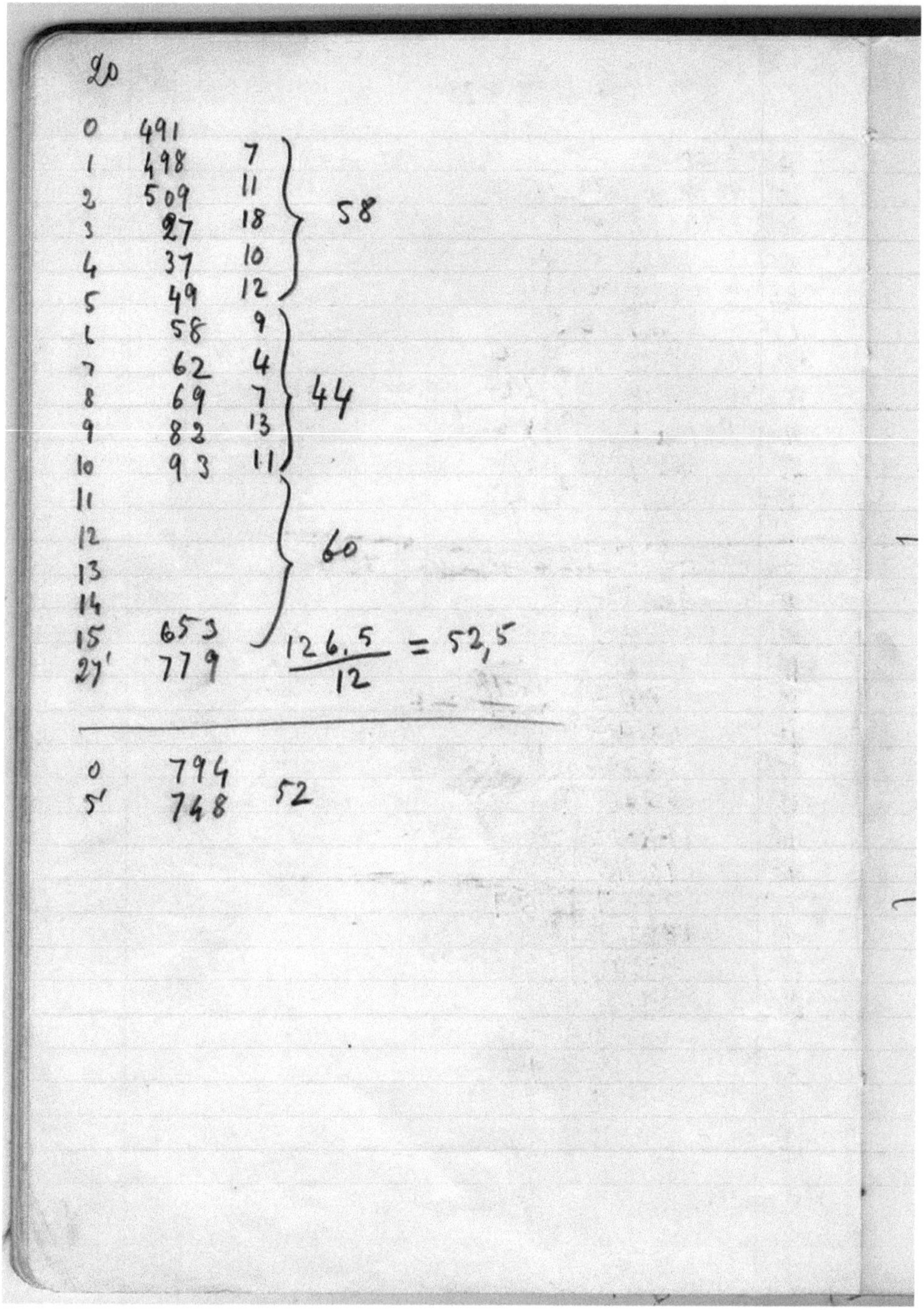

20

0	491		
1	498	7	
2	509	11	
3	27	18	58
4	37	10	
5	49	12	
6	58	9	
7	62	4	
8	69	7	44
9	82	13	
10	93	11	
11			
12			60
13			
14			
15	653		
27'	779		

$\frac{126,5}{12} = 52,5$

0	794	
5'	748	52

Q20

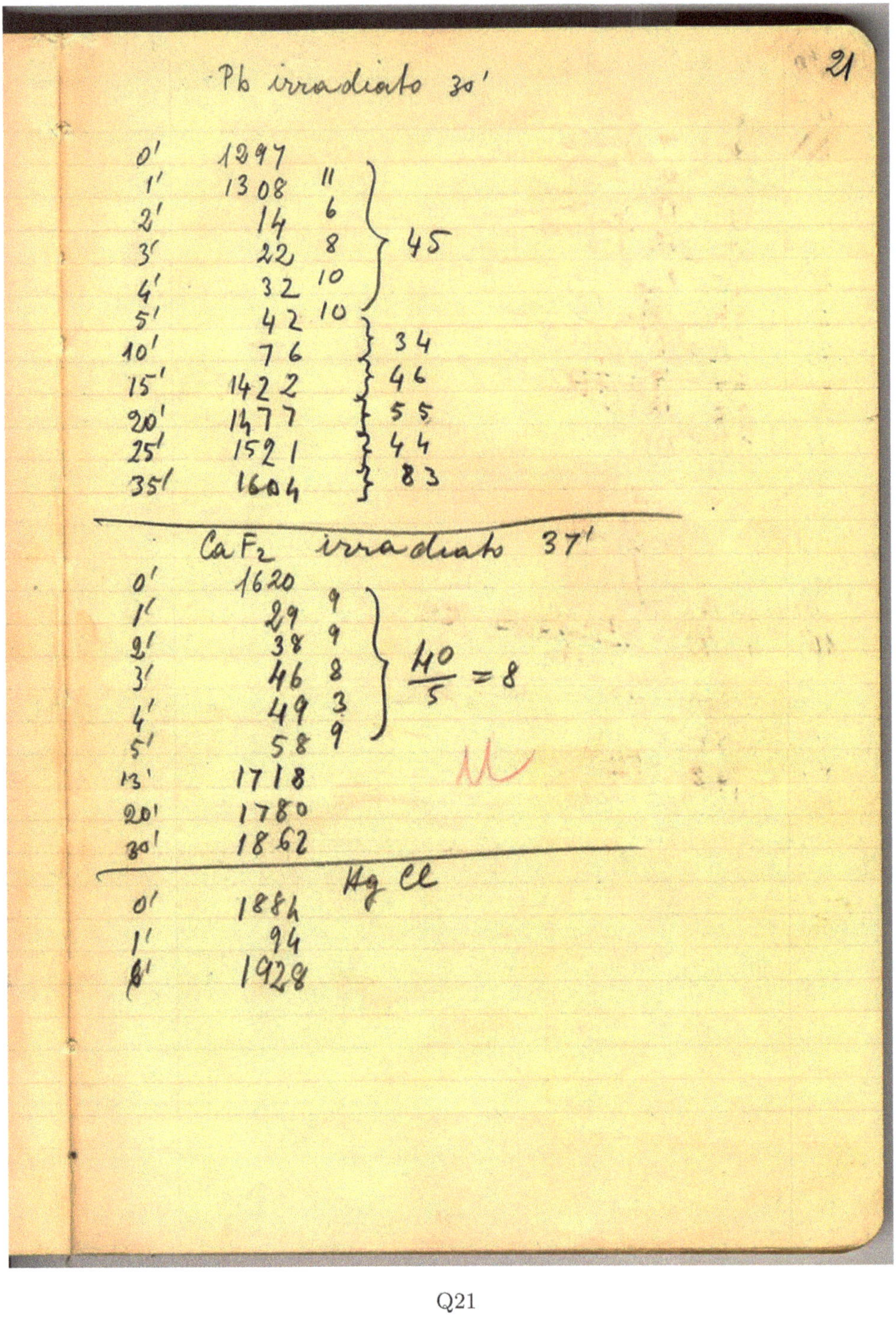

21

Pb irradiato 30'

0'	1297		
1'	1308	11	
2'	14	6	
3'	22	8	45
4'	32	10	
5'	42	10	
10'	76		34
15'	1422		46
20'	1477		55
25'	1521		44
35'	1604		83

CaF$_2$ irradiato 37'

0'	1620		
1'	29	9	
2'	38	9	
3'	46	8	$\frac{40}{5} = 8$
4'	49	3	
5'	58	9	
13'	1718		
20'	1780		
30'	1862		

Hg Cl

0'	1884
1'	94
8'	1928

Q21

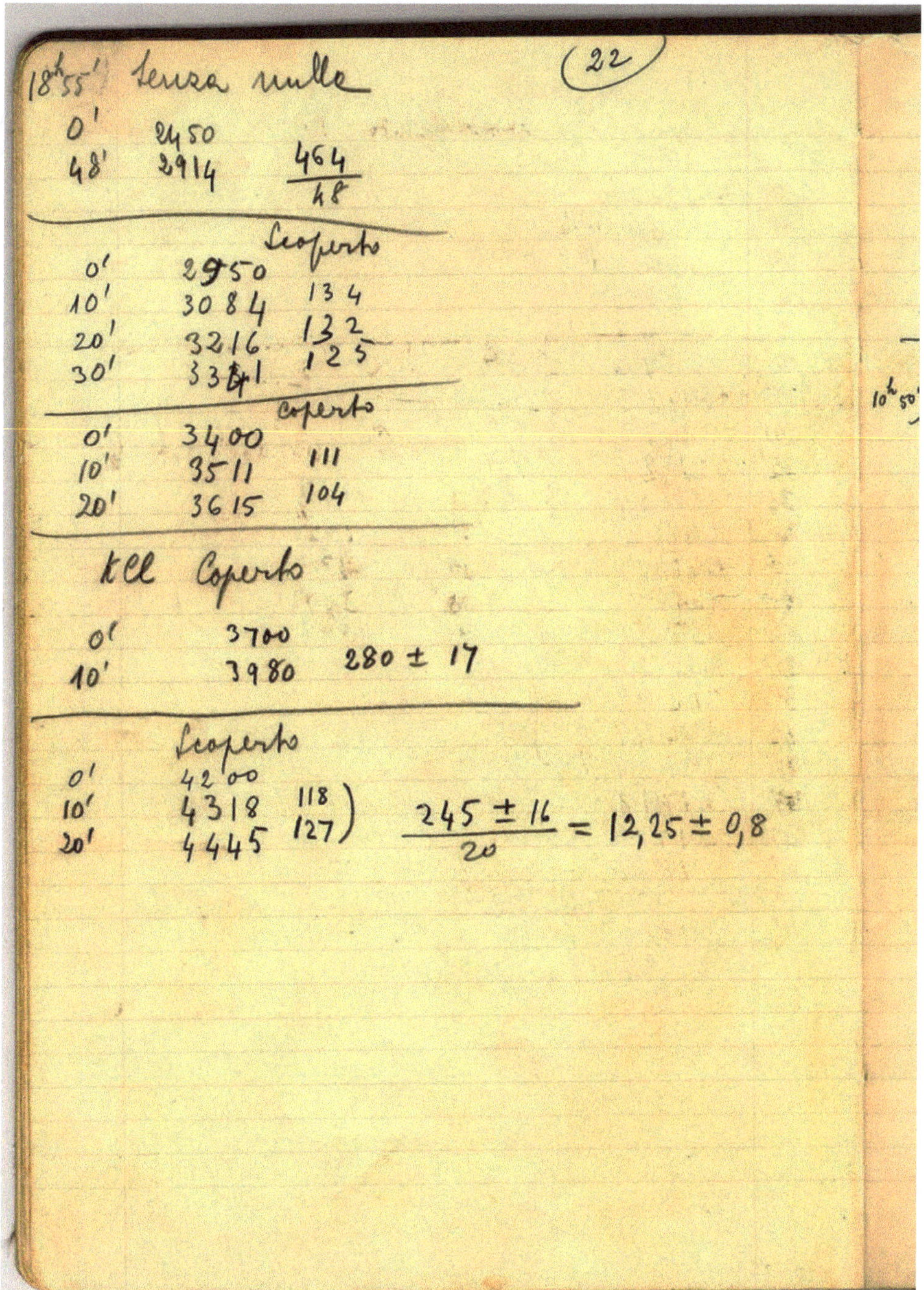

18h55' senza nulla (22)

0' 2450
48' 2914 464/48

Scoperto

0' 2950
10' 3084 134
20' 3216 132
30' 3341 125

Coperto

0' 3400
10' 3511 111
20' 3615 104

KCl Coperto

0' 3700
10' 3980 280 ± 17

Scoperto

0' 4200
10' 4318 118
20' 4445 127

$\frac{245 \pm 16}{20} = 12{,}25 \pm 0{,}8$

Q22

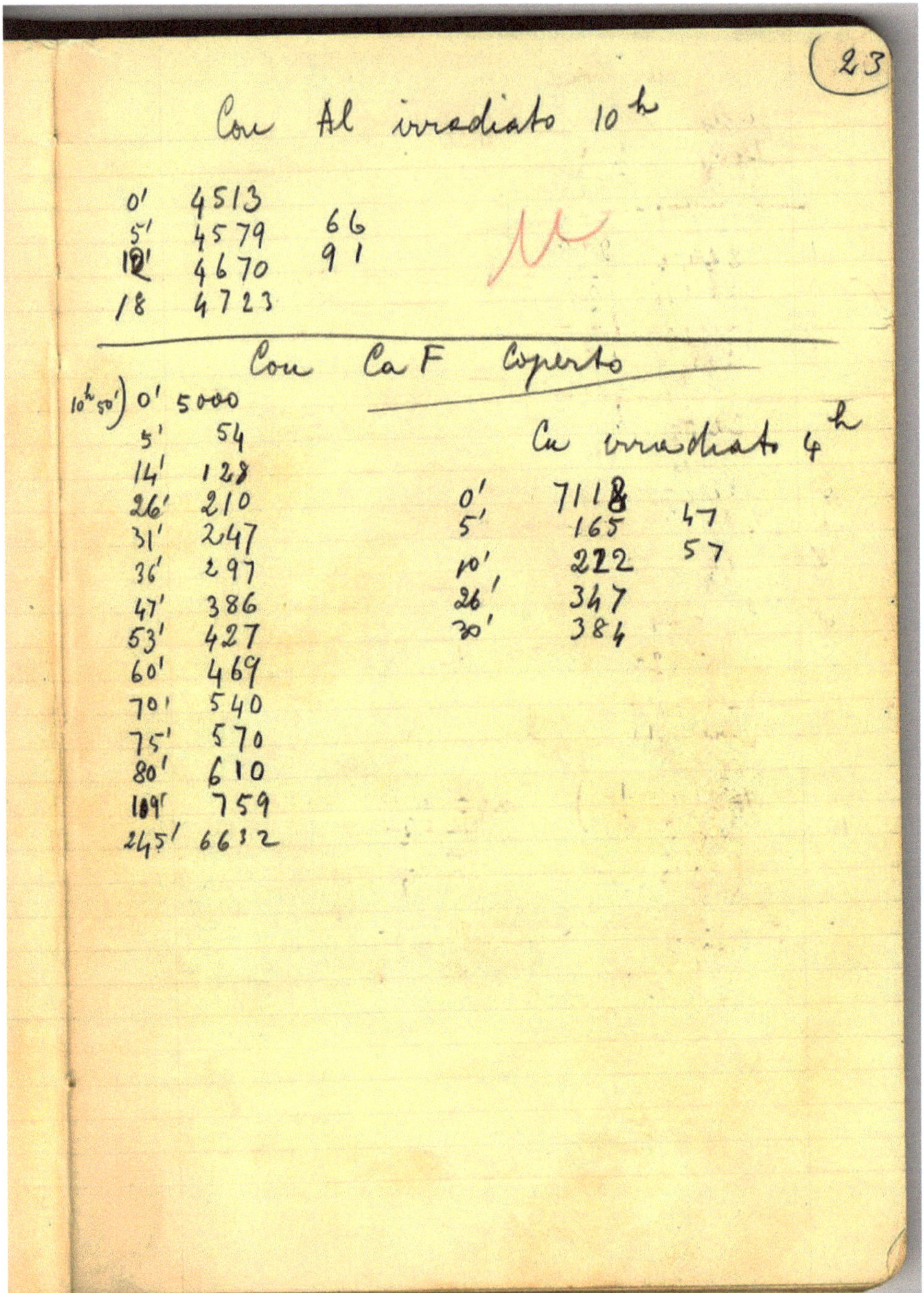

23

Con Al irradiato 10^h

0' 4513
5' 4579 66
12' 4670 91
18 4723

Con CaF coperto

($10^h 50'$) 0' 5000
5' 54
14' 128
26' 210
31' 247
36' 297
47' 386
53' 427
60' 469
70' 540
75' 570
80' 610
109' 759
245' 6632

Cu irradiato 4^h

0' 7118
5' 165 47
10' 222 57
26' 347
30' 384

Q23

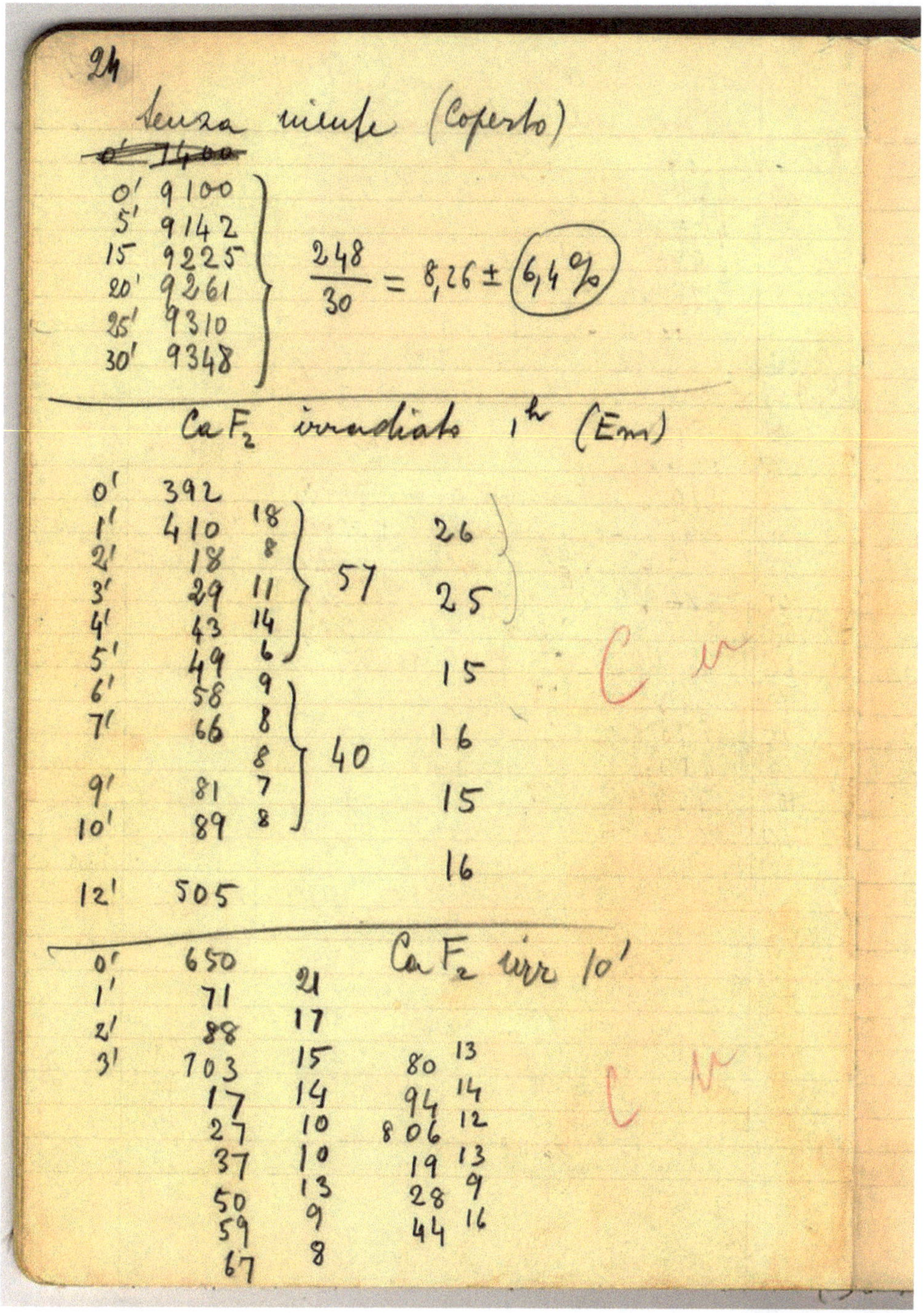
24

Senza niente (Coperto)

~~0' 7400~~

0'	9100	
5'	9142	
15	9225	$\frac{248}{30} = 8{,}26 \pm$ (6,4 %)
20'	9261	
25'	9310	
30'	9348	

CaF_2 irradiato 1^h (Em)

0'	392			
1'	410	18		26
2'	18	8		
3'	29	11	57	25
4'	43	14		
5'	49	6		15
6'	58	9		
7'	66	8		16
		8	40	
9'	81	7		15
10'	89	8		
				16
12'	505			

CaF_2 irr 10'

0'	650			
1'	71	21		
2'	88	17		
3'	703	15	80	13
	17	14	94	14
	27	10	806	12
	37	10	19	13
	50	13	28	9
	59	9	44	16
	67	8		

Q24

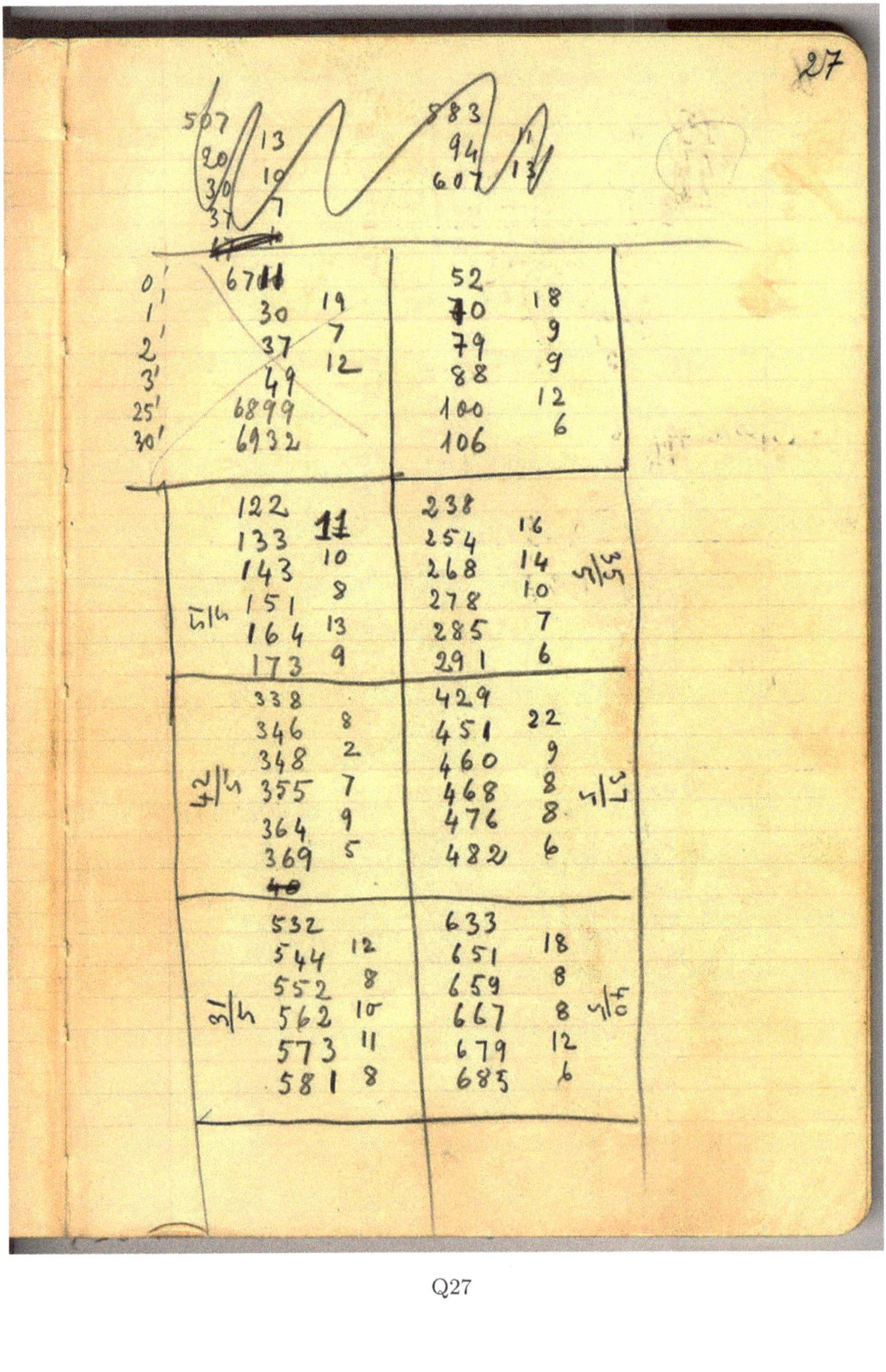

Q27

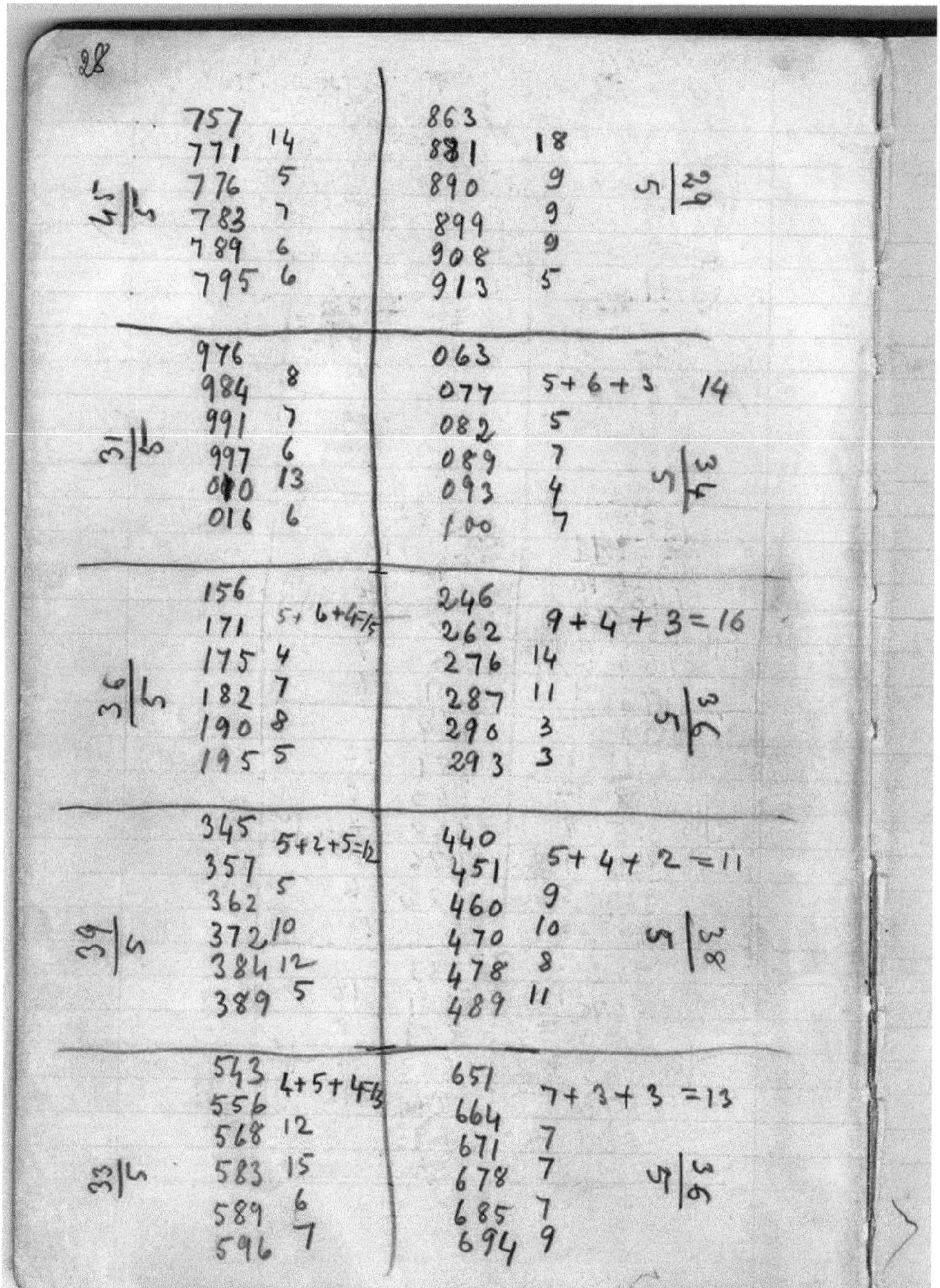

Q28

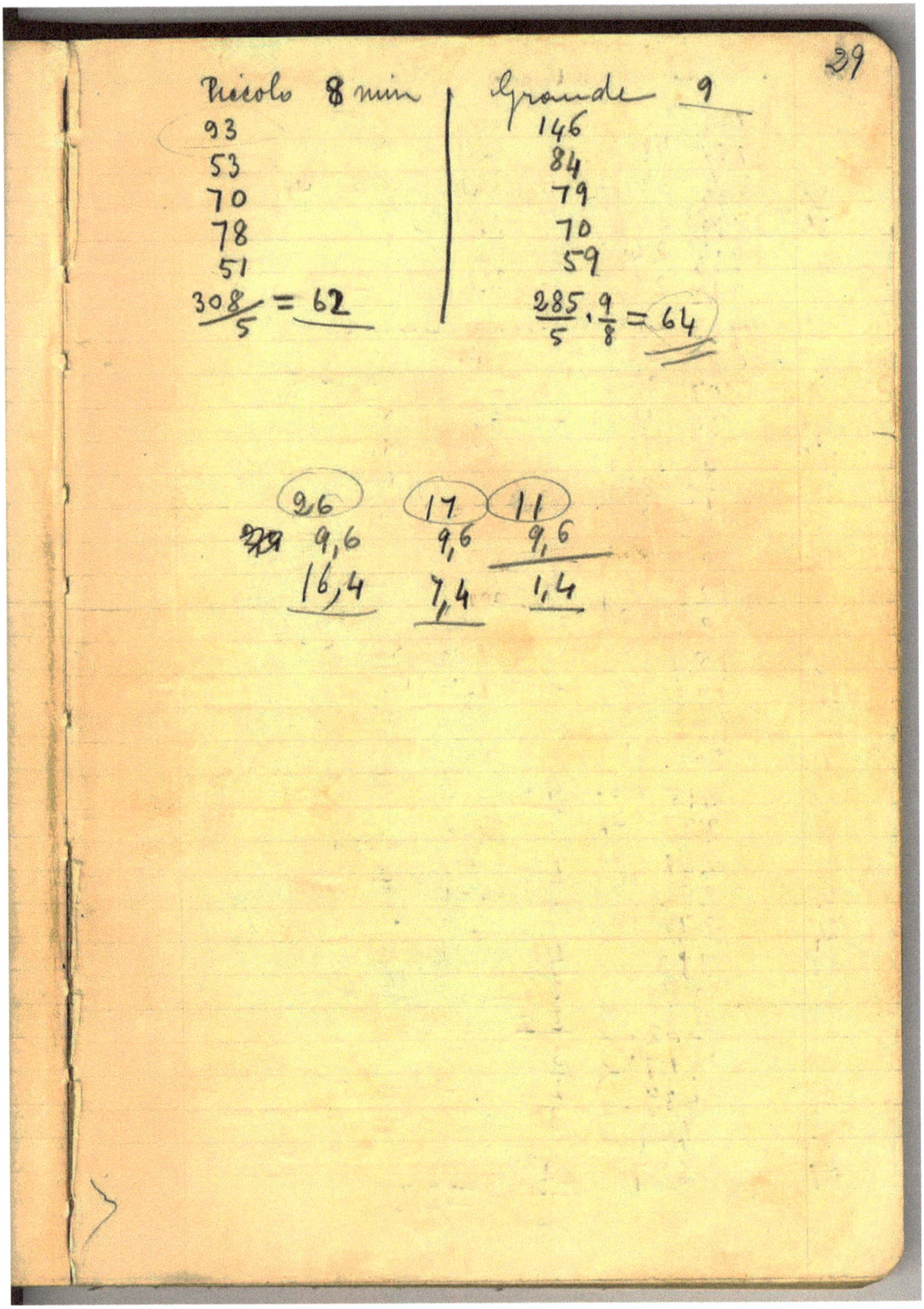

29

Piccolo 8 min	Grande 9
93	146
53	84
70	79
78	70
51	59
$\frac{308}{5} = 62$	$\frac{285}{5} \cdot \frac{9}{8} = 64$

26	17	11
9,6	9,6	9,6
16,4	7,4	1,4

Q29

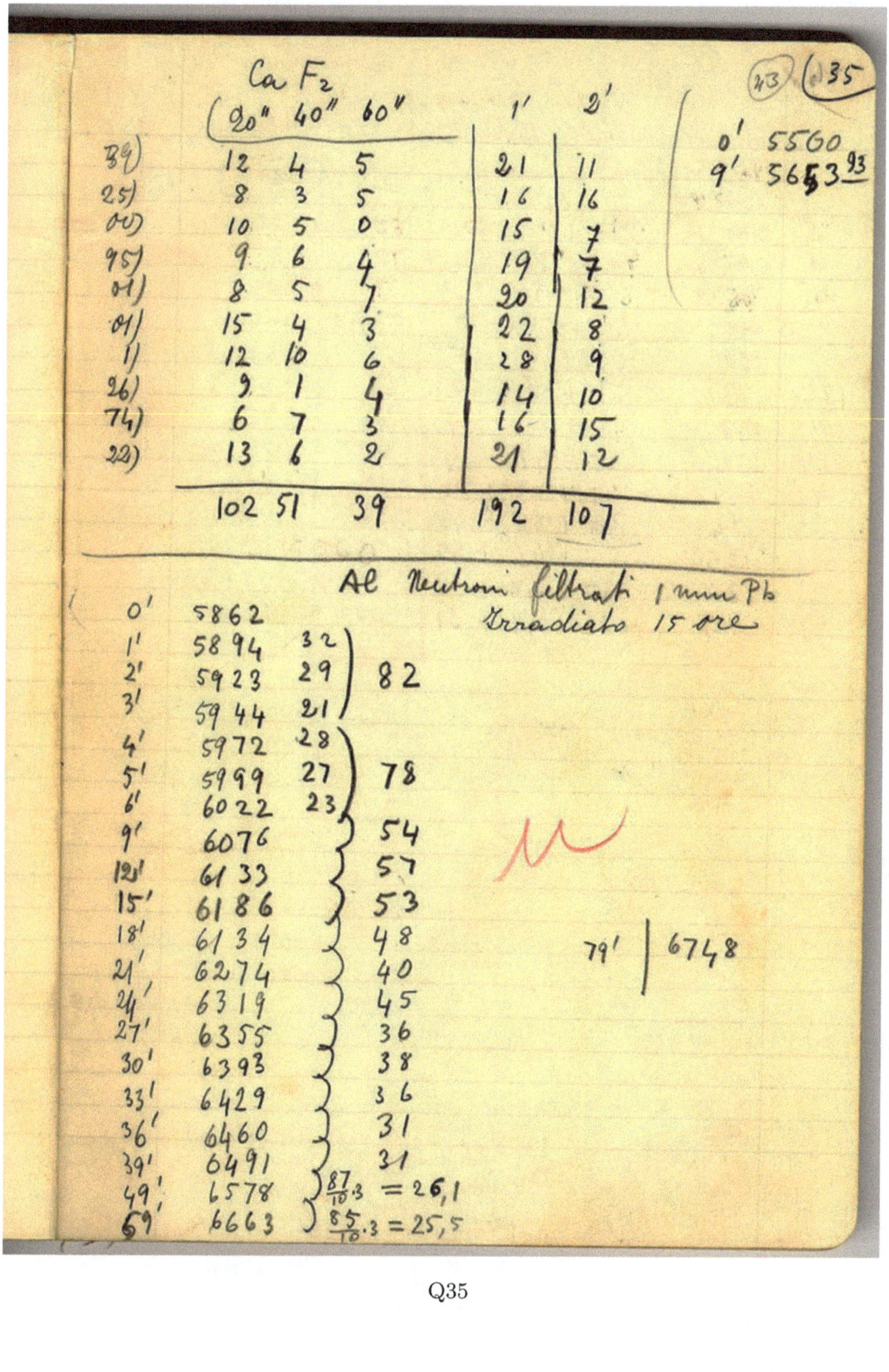

Ca F_2

	20″	40″	60″	1′	2′
39)	12	4	5	21	11
25)	8	3	5	16	16
00)	10	5	0	15	7
95)	9	6	4	19	7
01)	8	5	7	20	12
01)	15	4	3	22	8
1)	12	10	6	28	9
26)	9	1	4	14	10
74)	6	7	3	16	15
22)	13	6	2	21	12
	102	51	39	192	107

(43) Q35

0′ 5560
9′ 5653 93

Al Neutroni filtrati 1 mm Pb
Irradiato 15 ore

0′	5862		
1′	5894	32	
2′	5923	29	82
3′	5944	21	
4′	5972	28	
5′	5999	27	78
6′	6022	23	
9′	6076		54
12′	6133		57
15′	6186		53
18′	6134		48
21′	6274		40
24′	6319		45
27′	6355		36
30′	6393		38
33′	6429		36
36′	6460		31
39′	6491		31
49′	6578		$\frac{87}{10 \cdot 3} = 26,1$
59′	6663		$\frac{85}{10 \cdot 3} = 25,5$

79′ | 6748

Q35

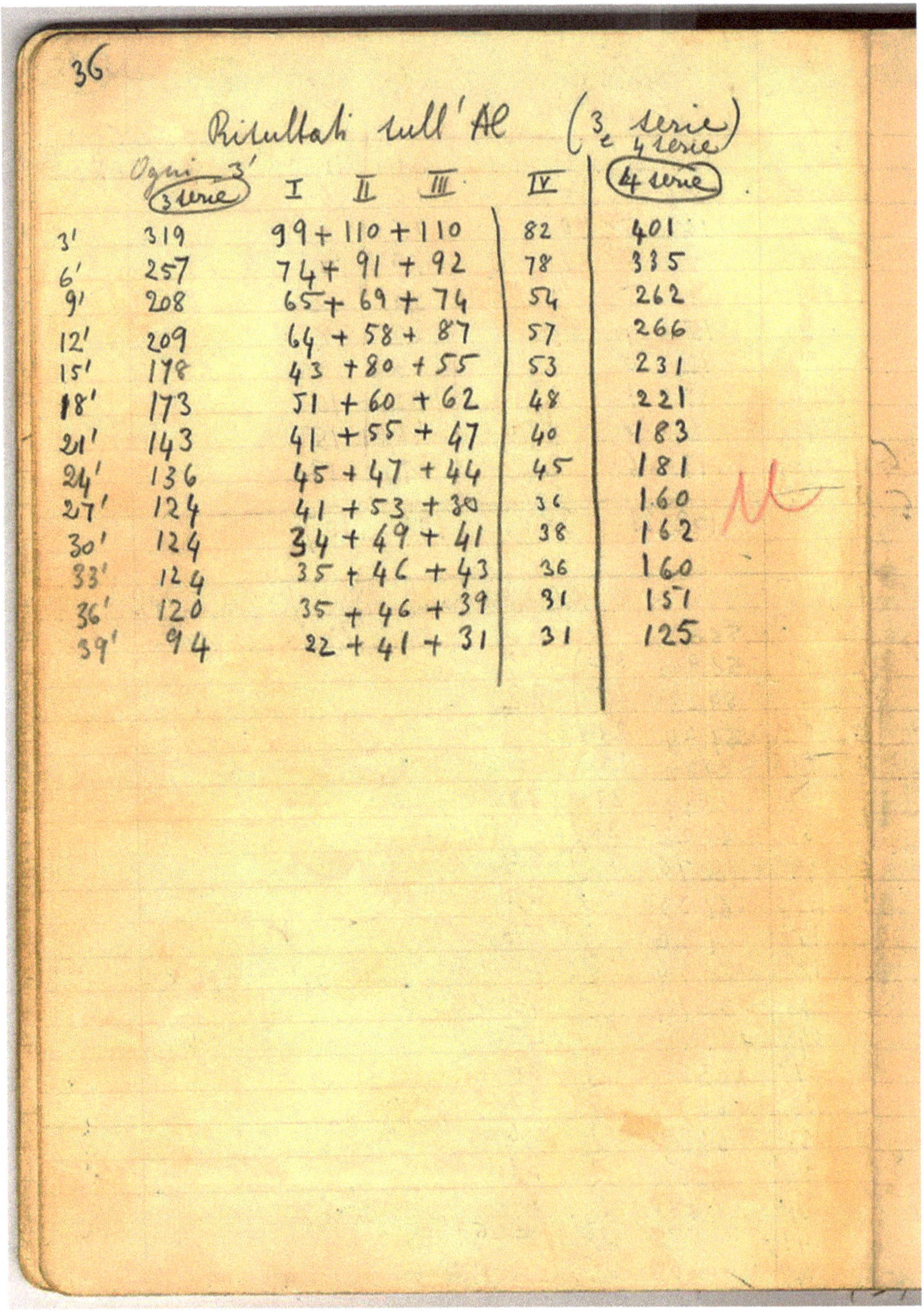
36

Risultati sull'Al (3 e 4 serie)

Ogni 3'	3 serie	I + II + III	IV	4 serie
3'	319	99 + 110 + 110	82	401
6'	257	74 + 91 + 92	78	335
9'	208	65 + 69 + 74	54	262
12'	209	64 + 58 + 87	57	266
15'	178	43 + 80 + 55	53	231
18'	173	51 + 60 + 62	48	221
21'	143	41 + 55 + 47	40	183
24'	136	45 + 47 + 44	45	181
27'	124	41 + 53 + 30	36	160
30'	124	34 + 49 + 41	38	162
33'	124	35 + 46 + 43	36	160
36'	120	35 + 46 + 39	31	151
39'	94	22 + 41 + 31	31	125

Q36

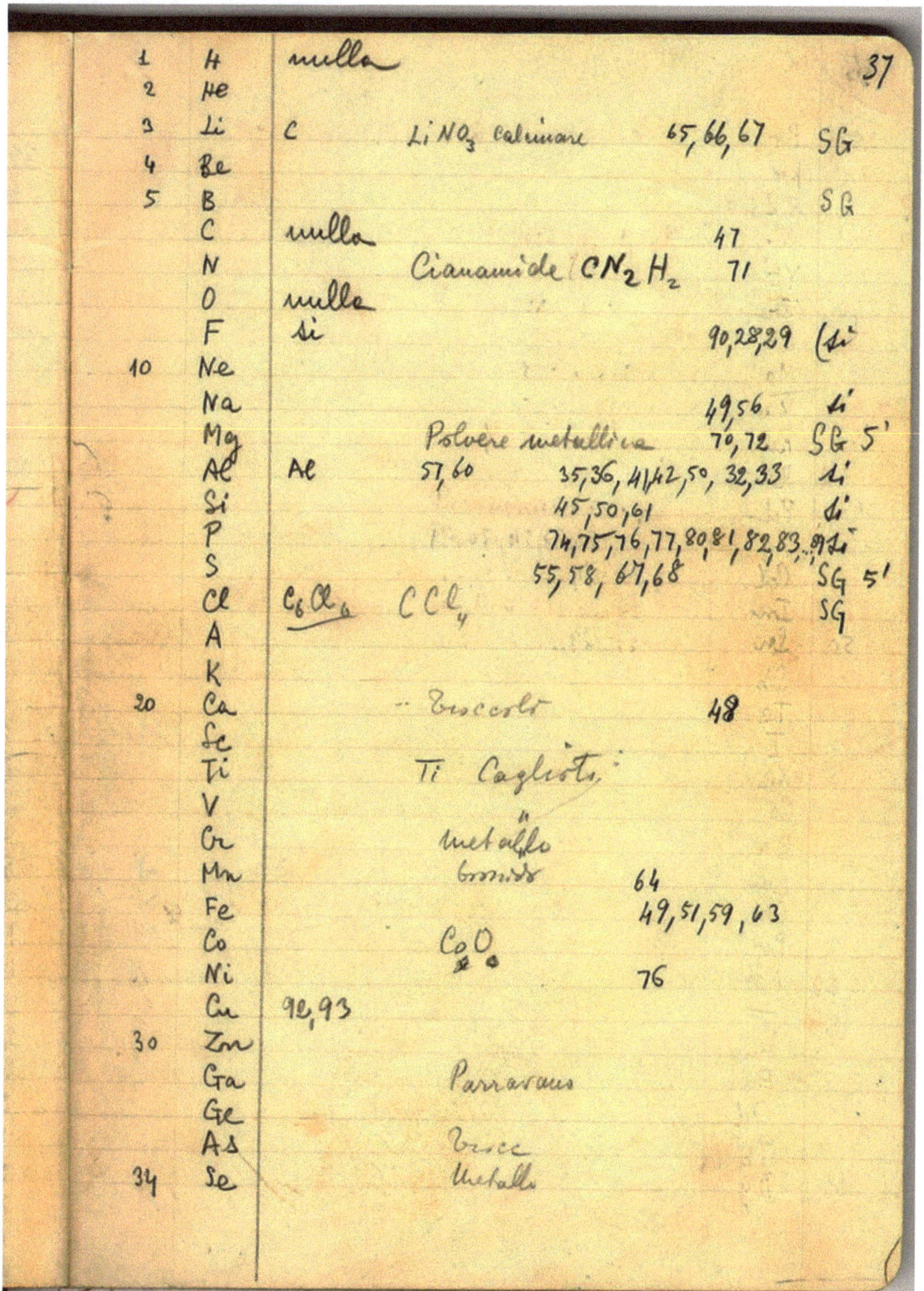

37

1	H	nulla			
2	He				
3	Li	C	$LiNO_3$ calcinare	65, 66, 67	SG
4	Be				
5	B				SG
	C	nulla		47	
	N		Cianamide CN_2H_2	71	
	O	nulla			
	F	si		90, 28, 29	(si
10	Ne				
	Na			49, 56	si
	Mg		Polvere metallica	70, 72	SG 5'
	Al	Al	57, 60	35, 36, 41, 42, 50, 32, 33	si
	Si			45, 50, 61	si
	P			74, 75, 76, 77, 80, 81, 82, 83, 91	si
	S			55, 58, 67, 68	SG 5'
	Cl	C_6Cl_6	CCl_4		SG
	A				
	K				
20	Ca		[illegible]	48	
	Sc				
	Ti		Ti Cagliotti		
	V				
	Cr		metallo		
	Mn		grosso	64	
	Fe			49, 51, 59, 63	
	Co		CoO		
	Ni			76	
	Cu	92, 93			
30	Zn				
	Ga		Parravano		
	Ge				
	As		Tracce		
34	Se		metallo		

Q37

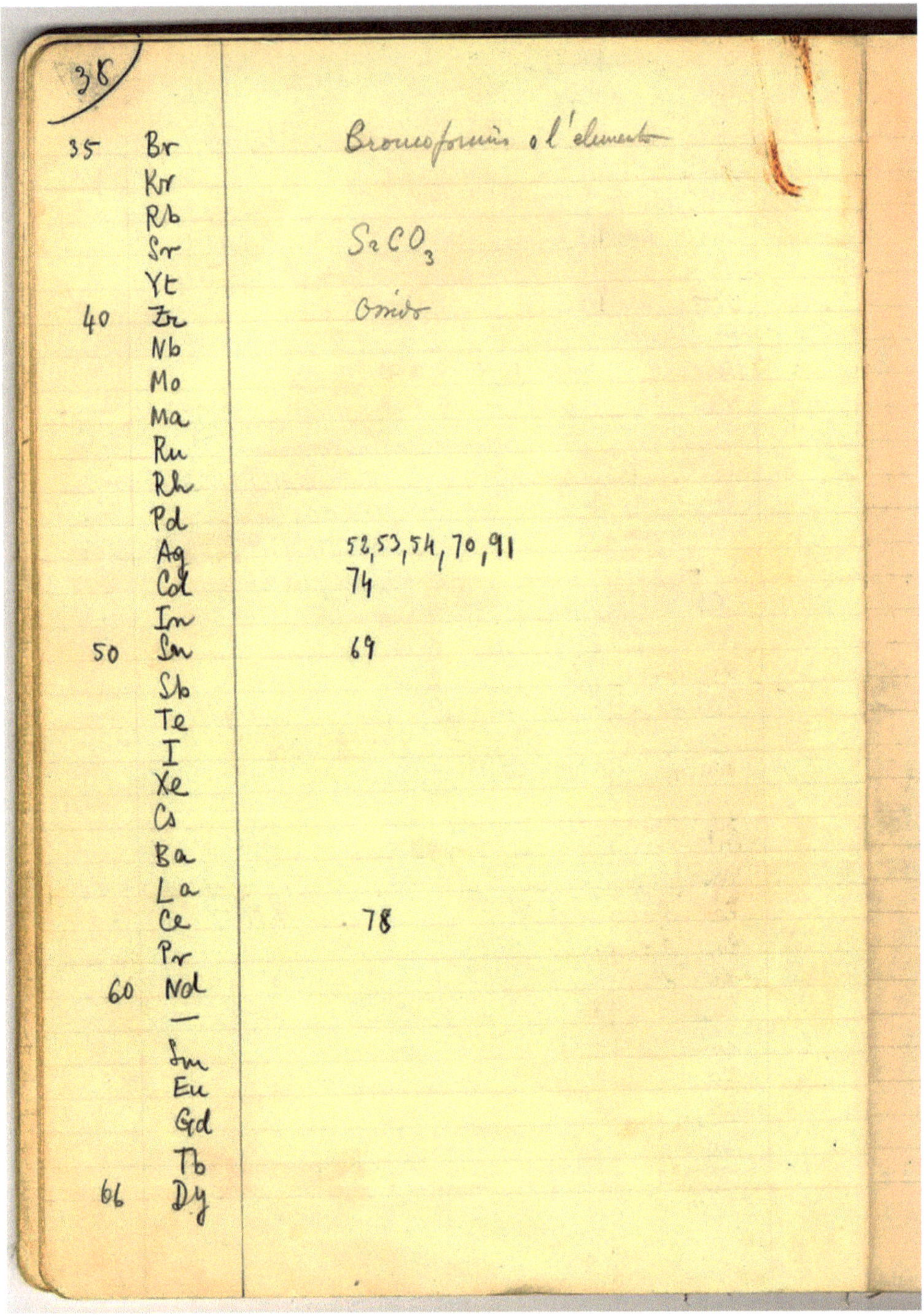

38

35	Br	Bromoformio o l'elemento
	Kr	
	Rb	
	Sr	$SrCO_3$
	Yt	
40	Zr	Ossido
	Nb	
	Mo	
	Ma	
	Ru	
	Rh	
	Pd	
	Ag	52, 53, 54, 70, 91
	Cd	74
	In	
50	Sn	69
	Sb	
	Te	
	I	
	Xe	
	Cs	
	Ba	
	La	
	Ce	78
	Pr	
60	Nd	
	—	
	Sm	
	Eu	
	Gd	
	Tb	
66	Dy	

Q38

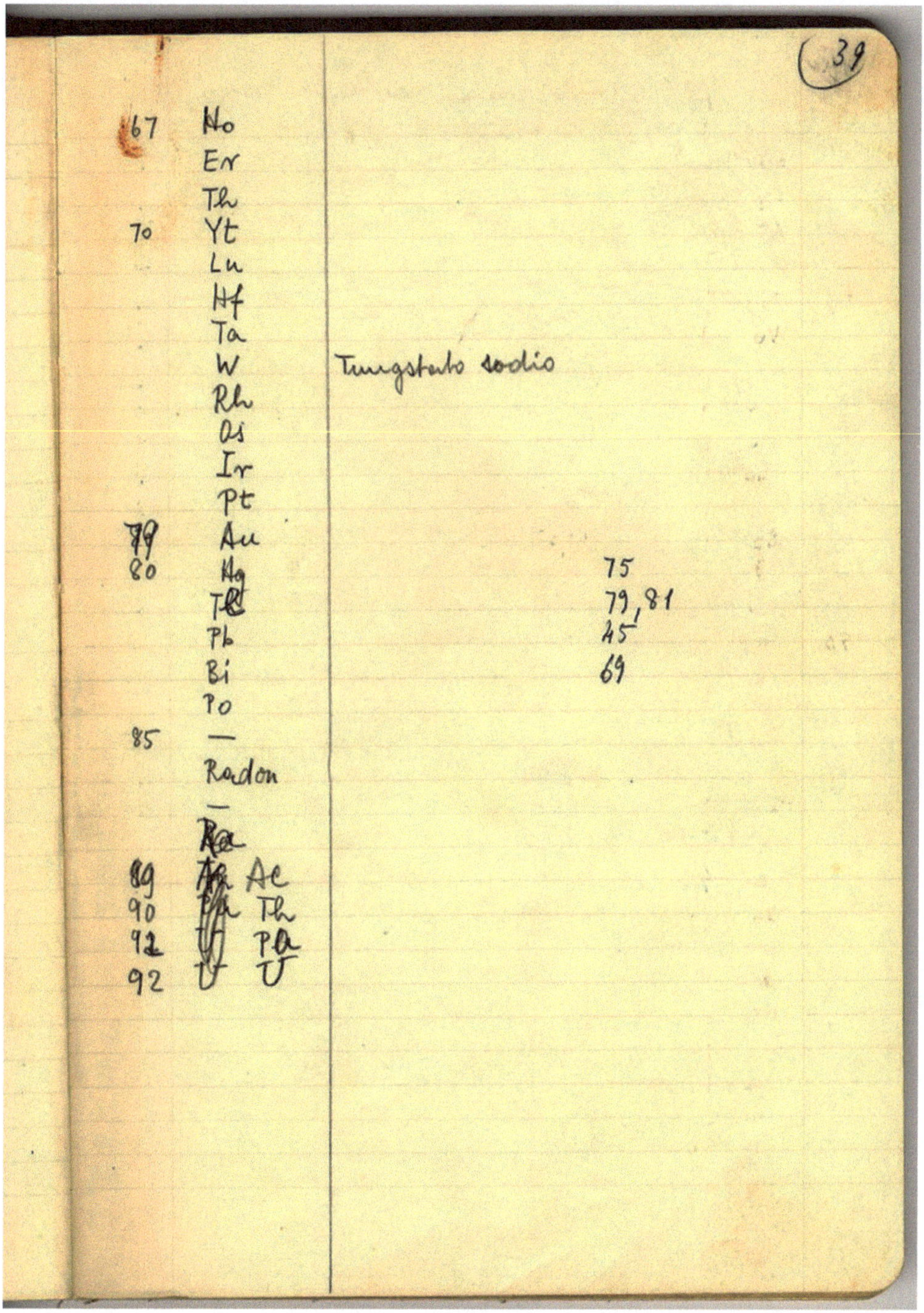
39

67 Ho
Er
Th
70 Yt
Lu
Hf
Ta
W Tungstato sodio
Rh
Os
Ir
Pt
79 Au
80 Hg 75
Tl 79, 81
Pb 45
Bi 69
Po
85 —
Radon
—
Ra
89 Ac Ac
90 Th Th
92 U Pa
92 U U

Q39

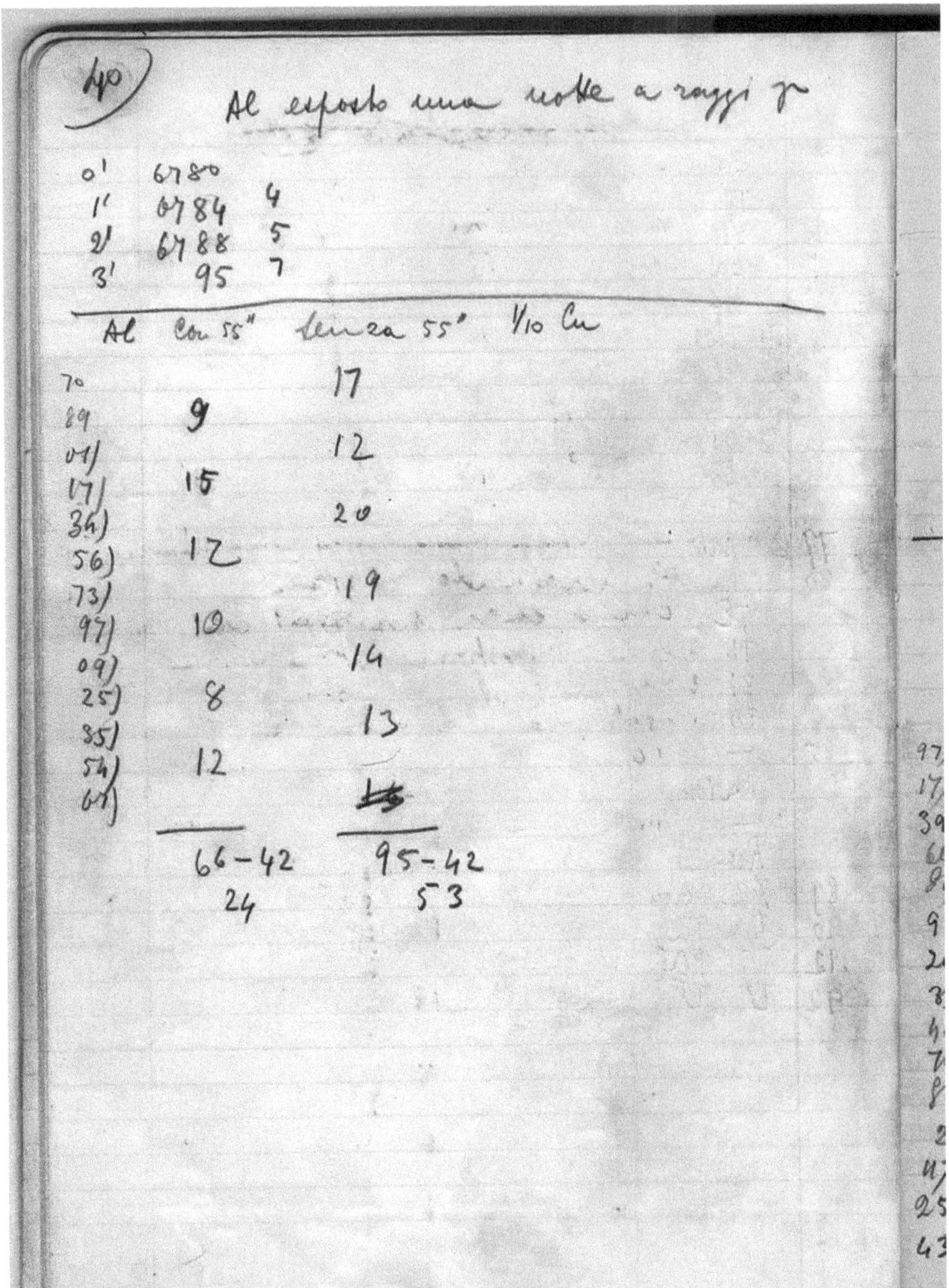

40

Al esposto una notte a raggi γ

0'	6780	
1'	6784	4
2'	6788	5
3'	95	7

Al	Con 55"	Senza 55" 1/10 Cu
70		17
89	9	
01)		12
17)	15	
34)		20
56)	12	
73)		19
97)	10	
09)		14
25)	8	
35)		13
54)	12	
67)		~~15~~
	66 − 42	95 − 42
	24	53

Q40

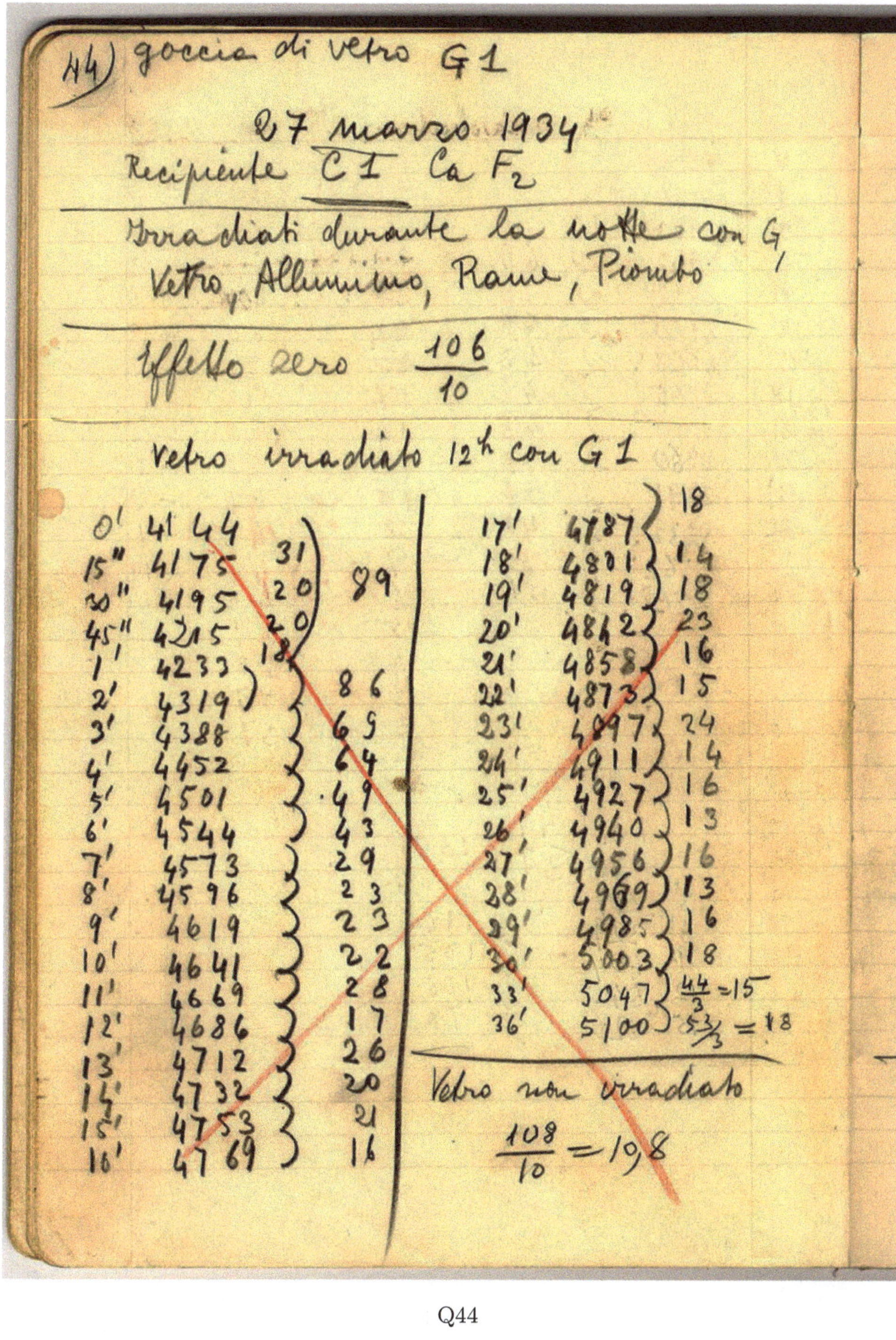

44) goccia di vetro G1

27 marzo 1934

Recipiente C1 CaF_2

Irradiati durante la notte con G, Vetro, Alluminio, Rame, Piombo

Effetto zero $\frac{106}{10}$

vetro irradiato 12^h con G1

0'	4144		
15"	4175	31	
30"	4195	20	89
45"	4215	20	
1'	4233	18	
2'	4319		86
3'	4388		69
4'	4452		64
5'	4501		49
6'	4544		43
7'	4573		29
8'	4596		23
9'	4619		23
10'	4641		22
11'	4669		28
12'	4686		17
13'	4712		26
14'	4732		20
15'	4753		21
16'	4769		16

		18
17'	4787	
18'	4801	14
19'	4819	18
20'	4842	23
21'	4858	16
22'	4873	15
23'	4897	24
24'	4911	14
25'	4927	16
26'	4940	13
27'	4956	16
28'	4969	13
29'	4985	16
30'	5003	18
33'	5047	$\frac{44}{3}=15$
36'	5100	$\frac{53}{3}=18$

Vetro non irradiato

$\frac{108}{10}=10,8$

Q44

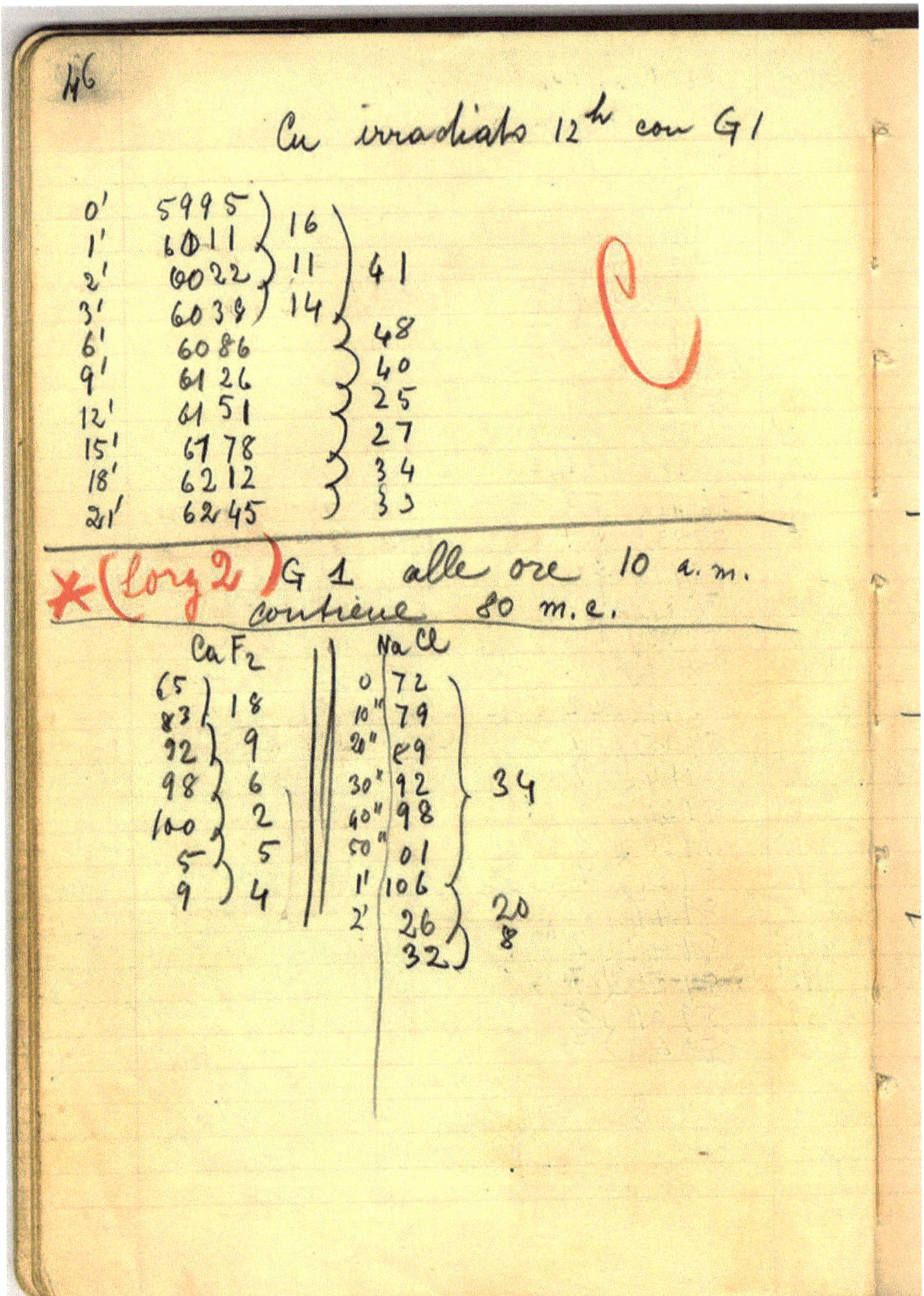

46

Cu irradiato 12^h con G1

0'	5995	16	41
1'	6011	11	
2'	6022	14	
3'	6038		48
6'	6086		40
9'	6126		25
12'	6151		27
15'	6178		34
18'	6212		33
21'	6245		

C

* (long 2) G 1 alle ore 10 a.m. contiene 80 m.c.

CaF_2			NaCl		
65	18	0	72	34
83	9	10"	79	
92	6	20"	89	
98	2	30"	92	
100	5	40"	98	
5	4	50"	01	
9		1'	106	20
		2'	26	8
			32	

Q46

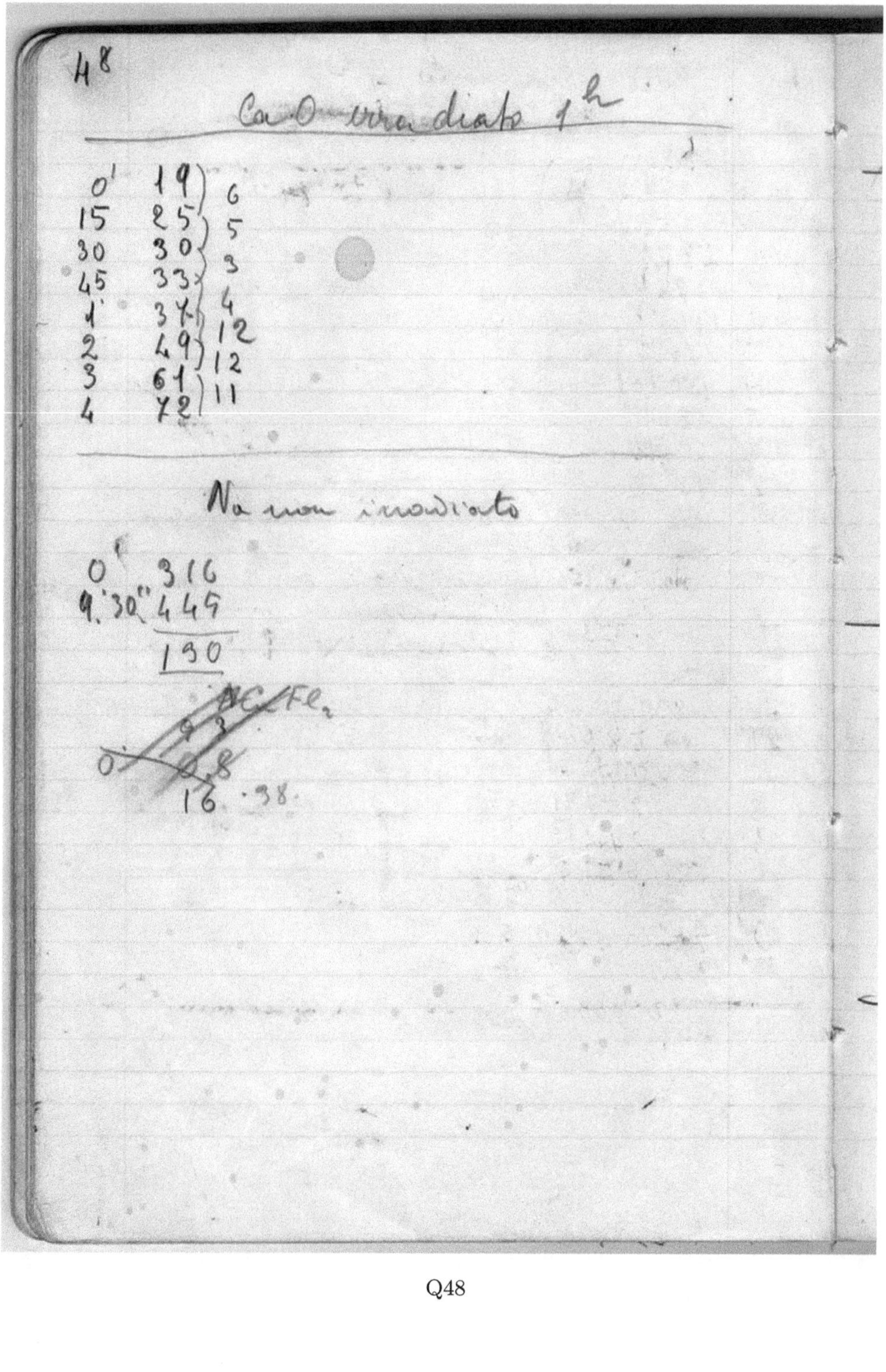

48

Ca O irradiato 1h

0'	19	
		6
15	25	
		5
30	30	
		3
45	33	
		4
1'	37	
		12
2	49	
		12
3	61	
		11
4	72	

Na non irradiato

0' 316
9'30" 445
190

Q48

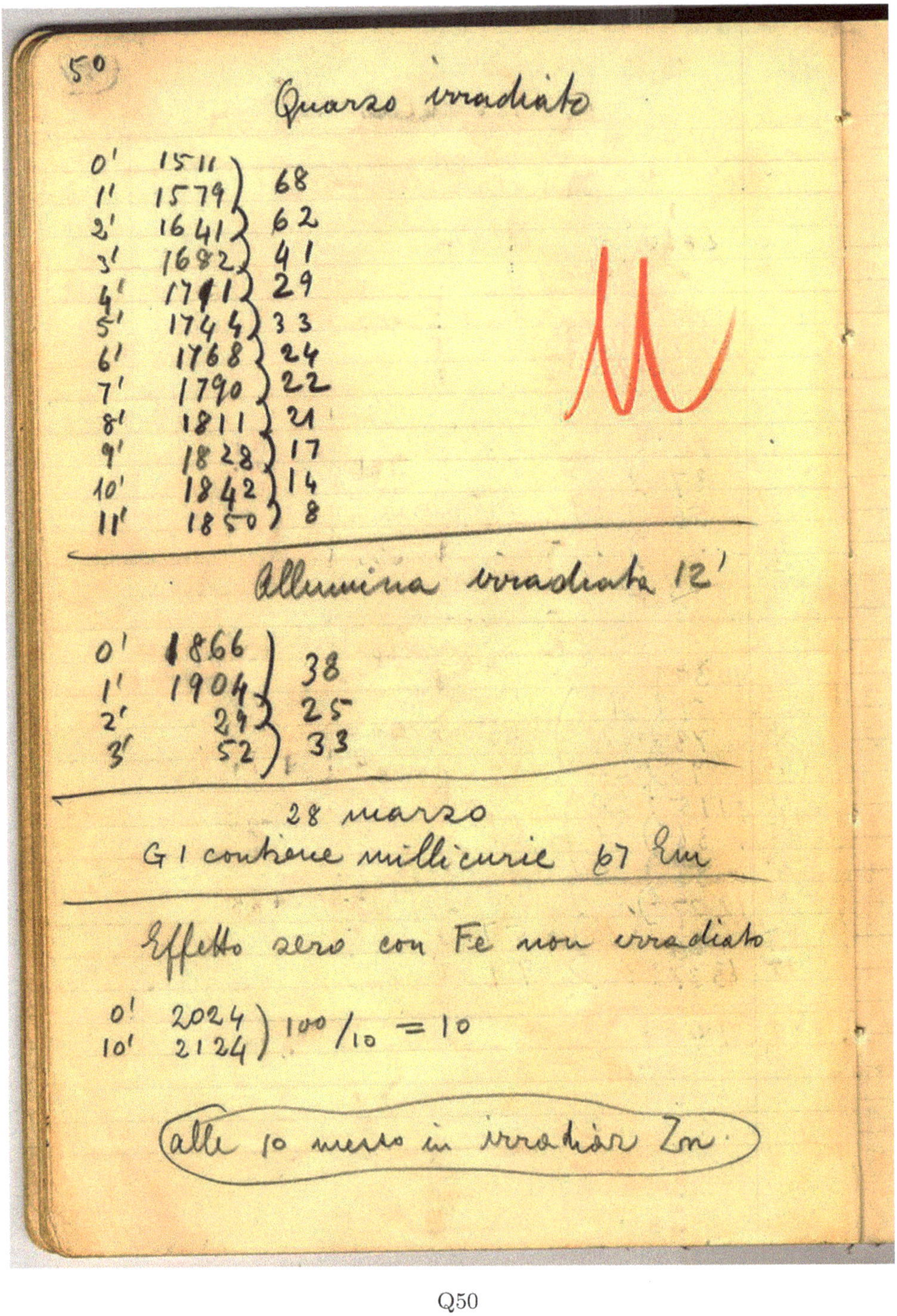

50

Quarzo irradiato

0'	1511	
		68
1'	1579	
		62
2'	1641	
		41
3'	1682	
		29
4'	1711	
		33
5'	1744	
		24
6'	1768	
		22
7'	1790	
		21
8'	1811	
		17
9'	1828	
		14
10'	1842	
		8
11'	1850	

Allumina irradiata 12'

0'	1866	
		38
1'	1904	
		25
2'	29	
		33
3'	52	

28 marzo

G1 contiene millicurie 67 Em

Effetto zero con Fe non irradiato

0'	2024	
		100/10 = 10
10'	2124	

alle 10 messo in irradiaz Zn.

Q50

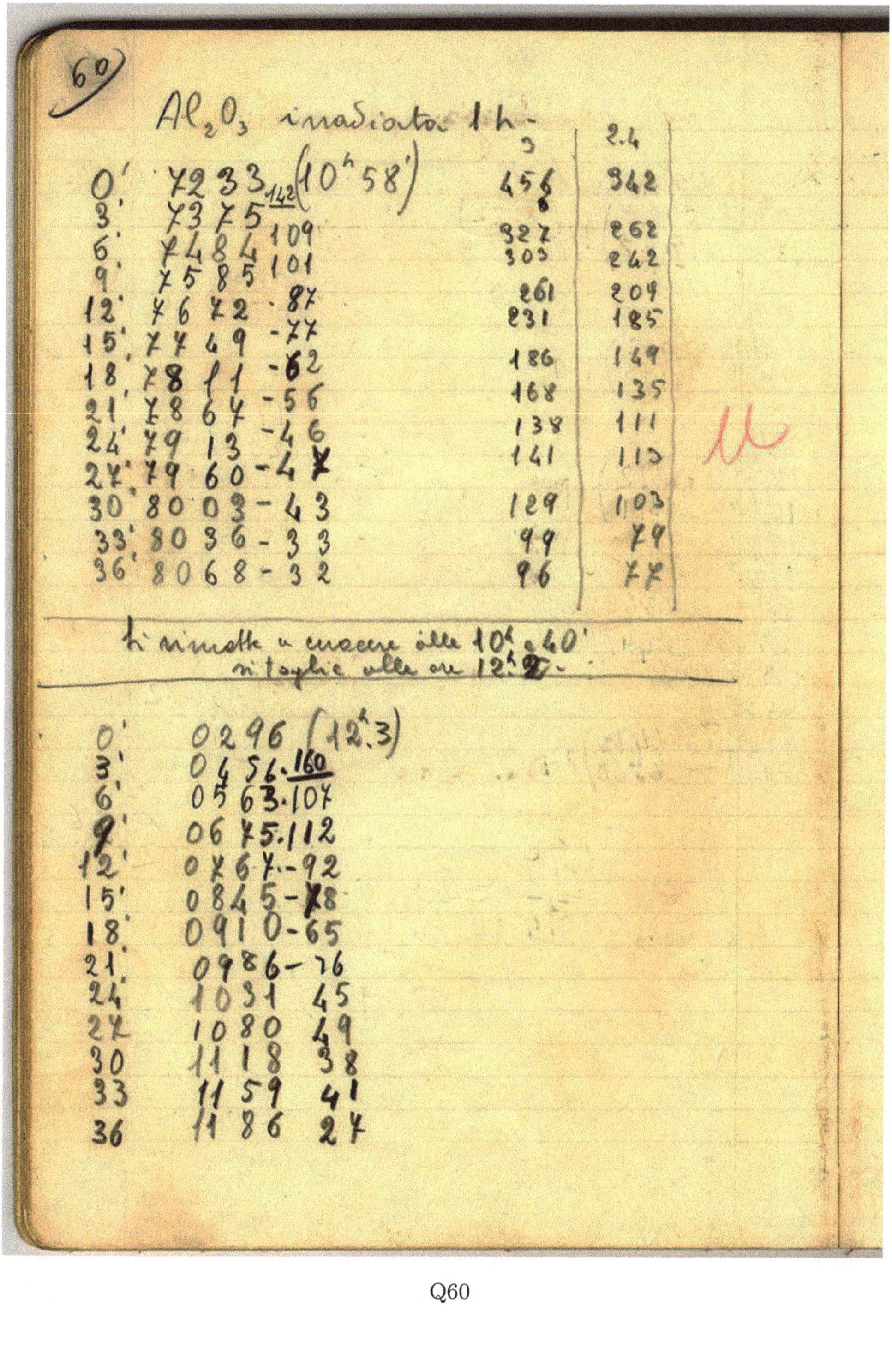
60

Al_2O_3 irradiato 1 h -

			3	2.4
0'	7233 (10h 58')	142	456	342
3'	7375			
6'	7484	109	327	262
9'	7585	101	303	242
12'	7672	87	261	204
15'	7749	77	231	185
18'	7811	62	186	149
21'	7867	56	168	135
24'	7913	46	138	111
27'	7960	47	141	113
30'	8003	43	129	103
33'	8036	33	99	79
36'	8068	32	96	77

si rimette a crescere alle 10h e 40'
si toglie alle ore 12h 2'-

0'	0296 (12h 3)	
3'	0456	160
6'	0563	107
9'	0675	112
12'	0767	92
15'	0845	78
18'	0910	65
21'	0986	76
24	1031	45
27	1080	49
30	1118	38
33	1159	41
36	1186	27

Q60

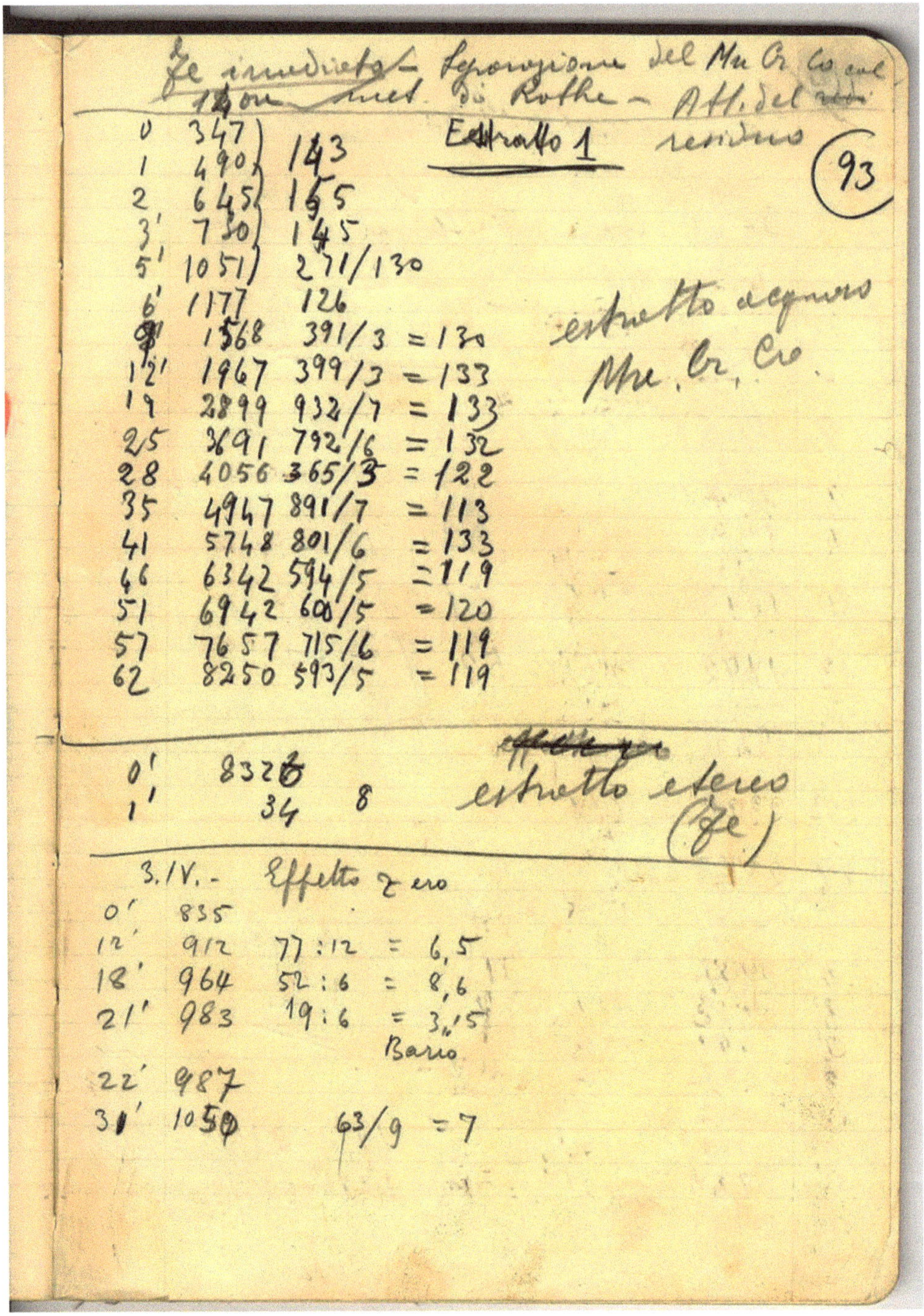

Fe immediato – separazione del Mn Cr Co col 1° metodo di Rothe – Attività del residuo

Estratto 1

93

0	347		
1	490	143	
2	645	155	
3′	790	145	
5′	1051	271/130	
6′	1177	126	
9′	1568	391/3	= 130
12′	1967	399/3	= 133
19	2899	932/7	= 133
25	3691	792/6	= 132
28	4056	365/3	= 122
35	4947	891/7	= 113
41	5748	801/6	= 133
46	6342	594/5	= 119
51	6942	600/5	= 120
57	7657	715/6	= 119
62	8250	593/5	= 119

estratto acquoso Mn, Cr, Co.

0′	8326	
1′	34	8

estratto etereo (Fe)

3.IV. – Effetto zero

0′	835		
12′	912	77 : 12	= 6,5
18′	964	52 : 6	= 8,6
21′	983	19 : 6	= 3,15

Bario

22′	987		
31′	1050	63/9	= 7

Q93

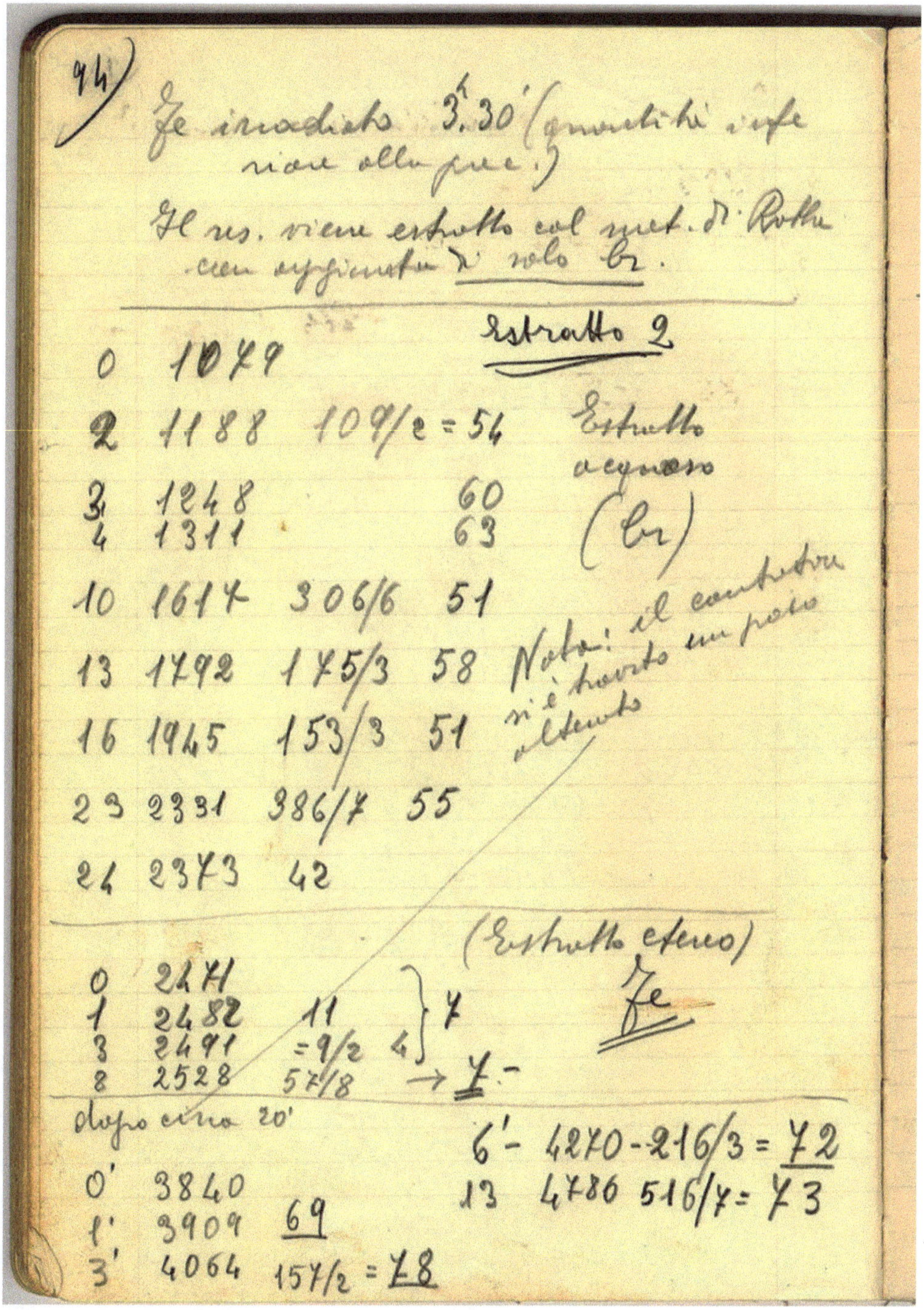

94)

Fe irradiato 3h 30' (quantità infe riore alla prec.)

Il res. viene estratto col met. di Rotha con aggiunta di solo Br.

Estratto 2

0 1079

2 1188 109/2 = 54 Estratto acquoso

3 1248 60 (Br)

4 1311 63

10 1617 306/6 51

13 1792 175/3 58 Nota: il contatore si è trovato un poco alterato

16 1945 153/3 51

23 2331 386/7 55

24 2373 42

(Estratto etereo) Fe

0 2471

1 2482 11

3 2491 = 9/2 4 } 7

8 2528 57/8 → 7.-

dopo circa 20'

0' 3840

1' 3909 69

3' 4064 157/2 = 78

6'- 4270 - 216/3 = 72

13 4786 516/7 = 73

Q94

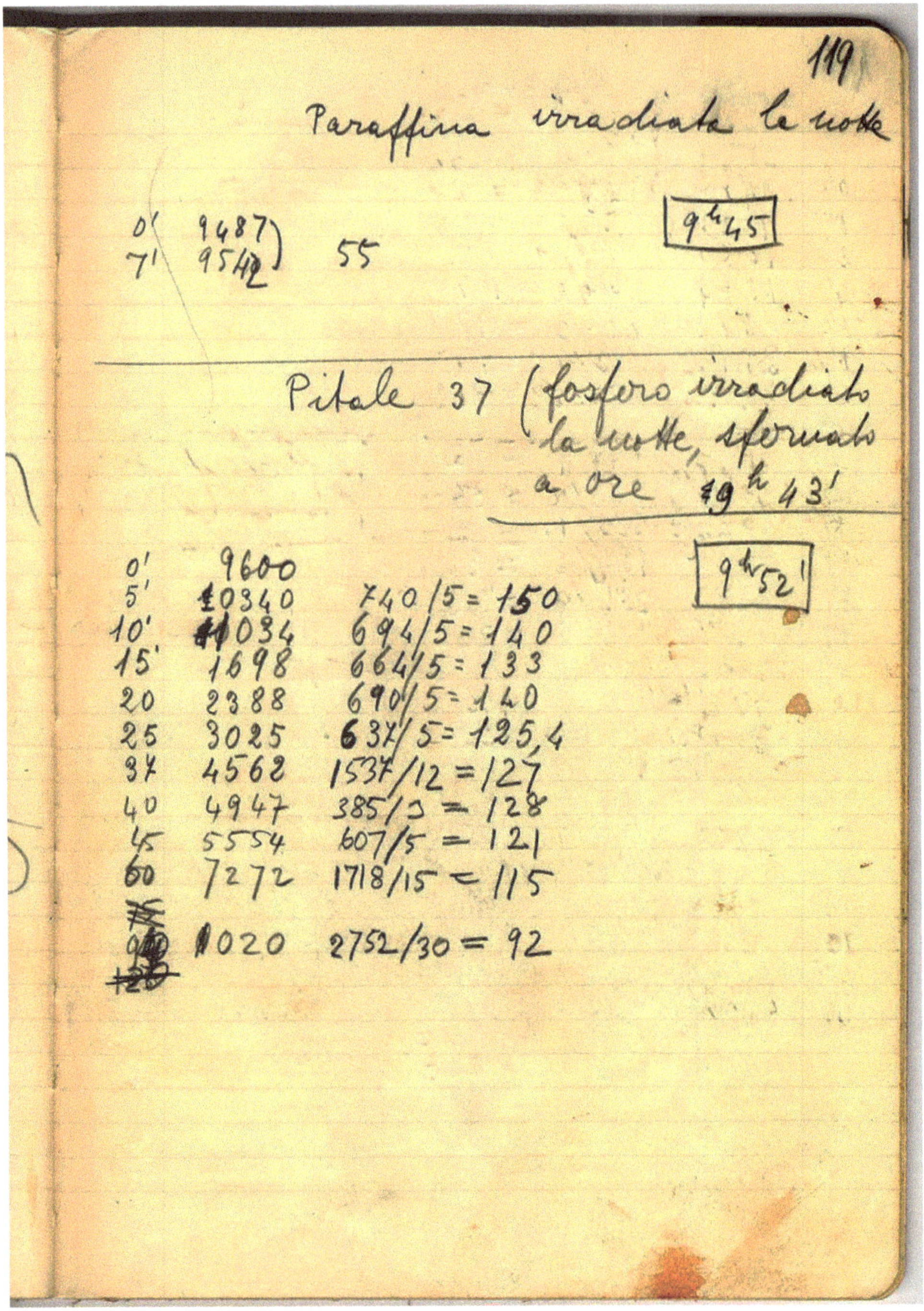

119

Paraffina irradiata la notte

0' 9487
7' 9542 } 55

9h45

Pitale 37 (fosforo irradiato la notte, sfornato a ore 19h 43'

9h52'

0'	9600	
5'	10340	740/5 = 150
10'	11034	694/5 = 140
15'	1698	664/5 = 133
20	2388	690/5 = 140
25	3025	637/5 = 125,4
37	4562	1537/12 = 127
40	4947	385/3 = 128
45	5554	607/5 = 121
60	7272	1718/15 = 115
90	1020	2752/30 = 92

Q119

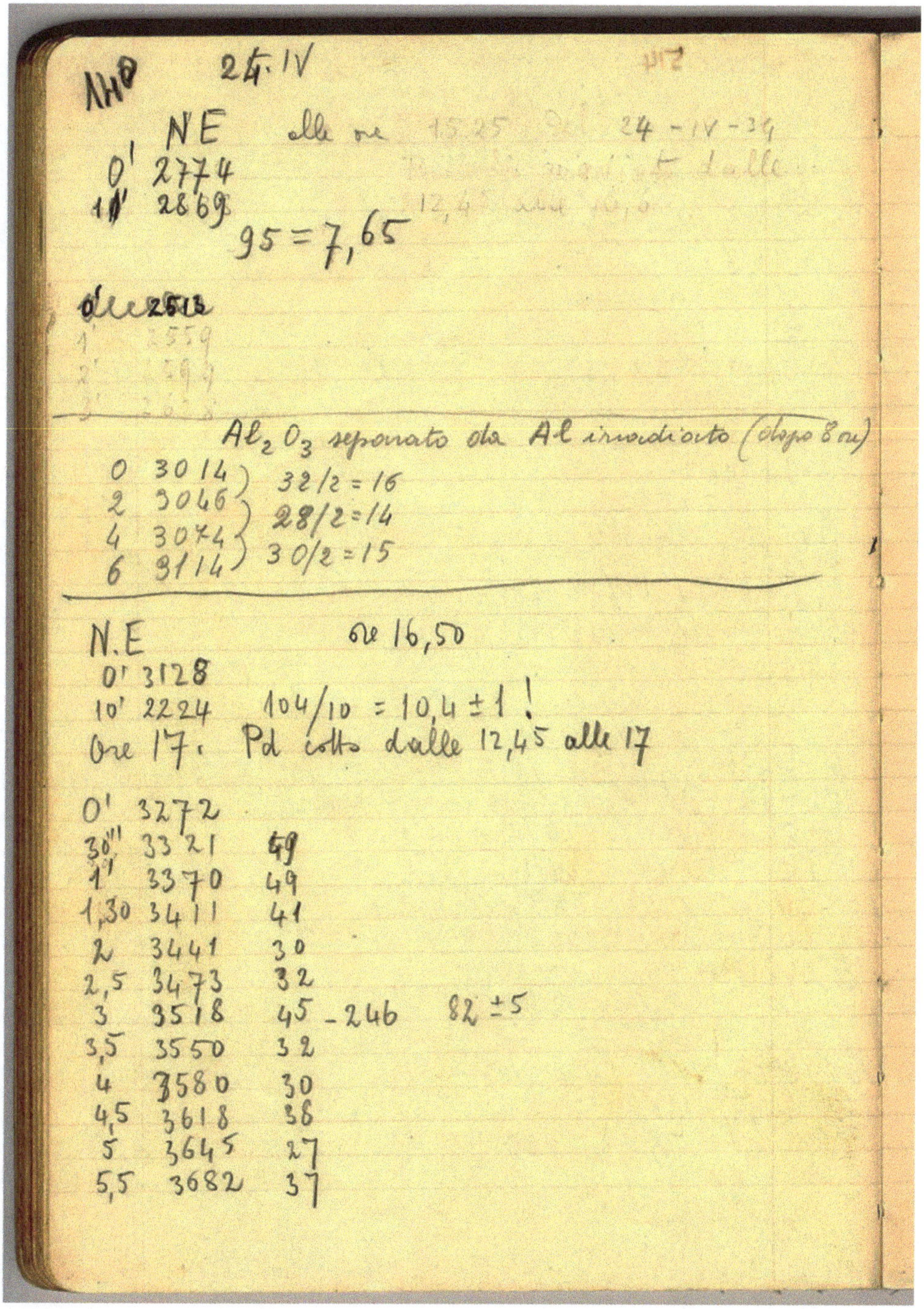

AHP 24. IV

NE alle ore 15,25 del 24 - IV - 34

0' 2774

10' 2869

95 = 7,65

Al₂O₃ separato da Al irradiato (dopo 8 ore)

0	3014	32/2 = 16
2	3046	28/2 = 14
4	3074	30/2 = 15
6	3114	

N.E ore 16,50

0' 3128

10' 2224 104/10 = 10,4 ± 1 !

Ore 17. Pd cotto dalle 12,45 alle 17

0'	3272		
30''	3321	49	
1'	3370	49	
1,30	3411	41	
2	3441	30	
2,5	3473	32	
3	3518	45 - 246	82 ± 5
3,5	3550	32	
4	3580	30	
4,5	3618	38	
5	3645	27	
5,5	3682	37	

Q140

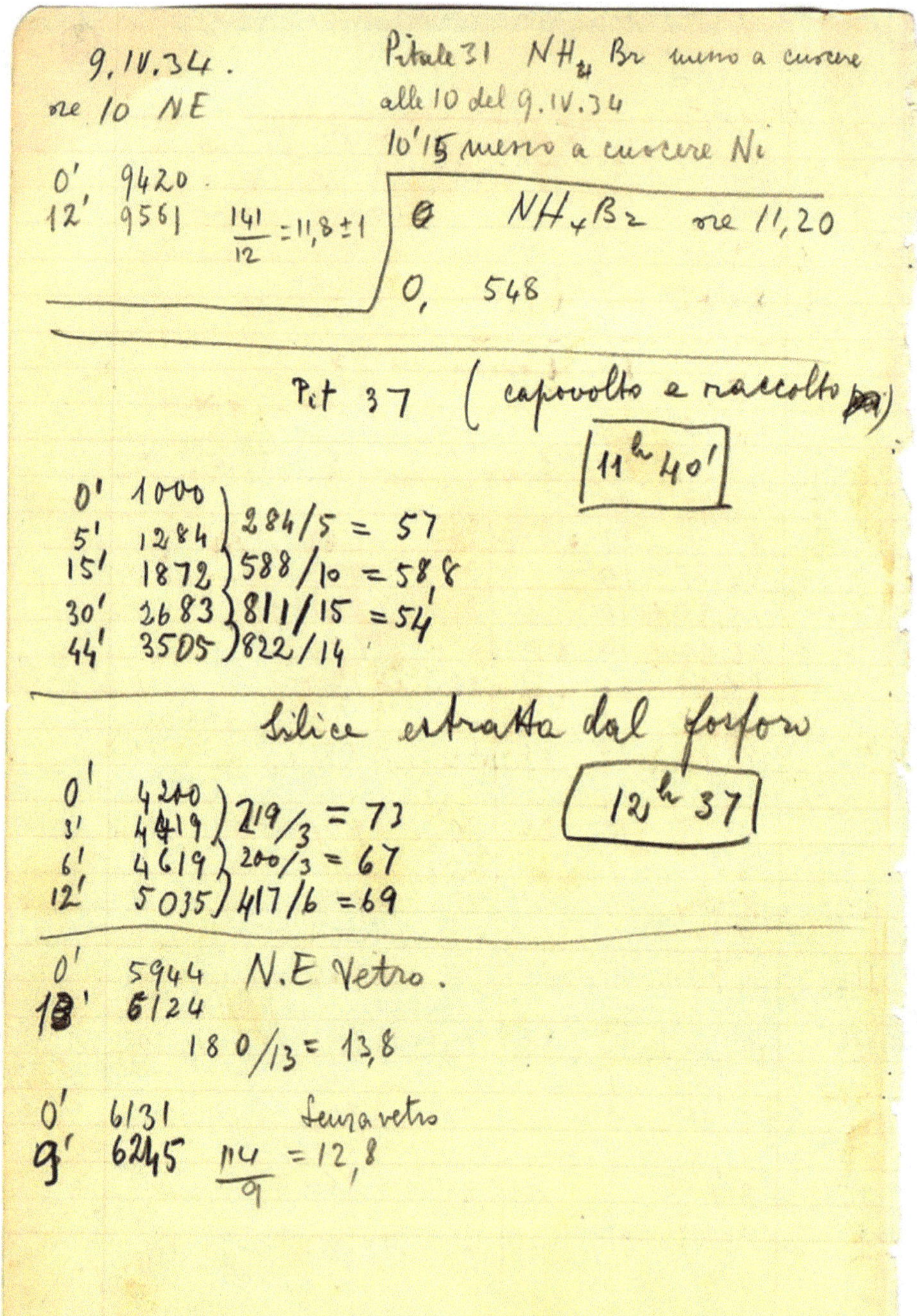

9.IV.34.
ore 10 NE
Pitale 31 NH_4Br messo a cuocere
alle 10 del 9.IV.34
10'15 messo a cuocere Ni

0' 9420
12' 9561 $\frac{141}{12}$ = 11,8 ± 1

NH_4Br ore 11,20
0, 548

Pit 37 (capovolto e raccolto)
11h 40'

0' 1000
5' 1284 284/5 = 57
15' 1872 588/10 = 58,8
30' 2683 811/15 = 54
44' 3505 822/14

Silice estratta dal fosforo
12h 37

0' 4200
3' 4419 219/3 = 73
6' 4619 200/3 = 67
12' 5035 417/6 = 69

0' 5944 N.E Vetro.
13' 6124
180/13 = 13,8

0' 6131 senza vetro
9' 6245 $\frac{114}{9}$ = 12,8

S3a

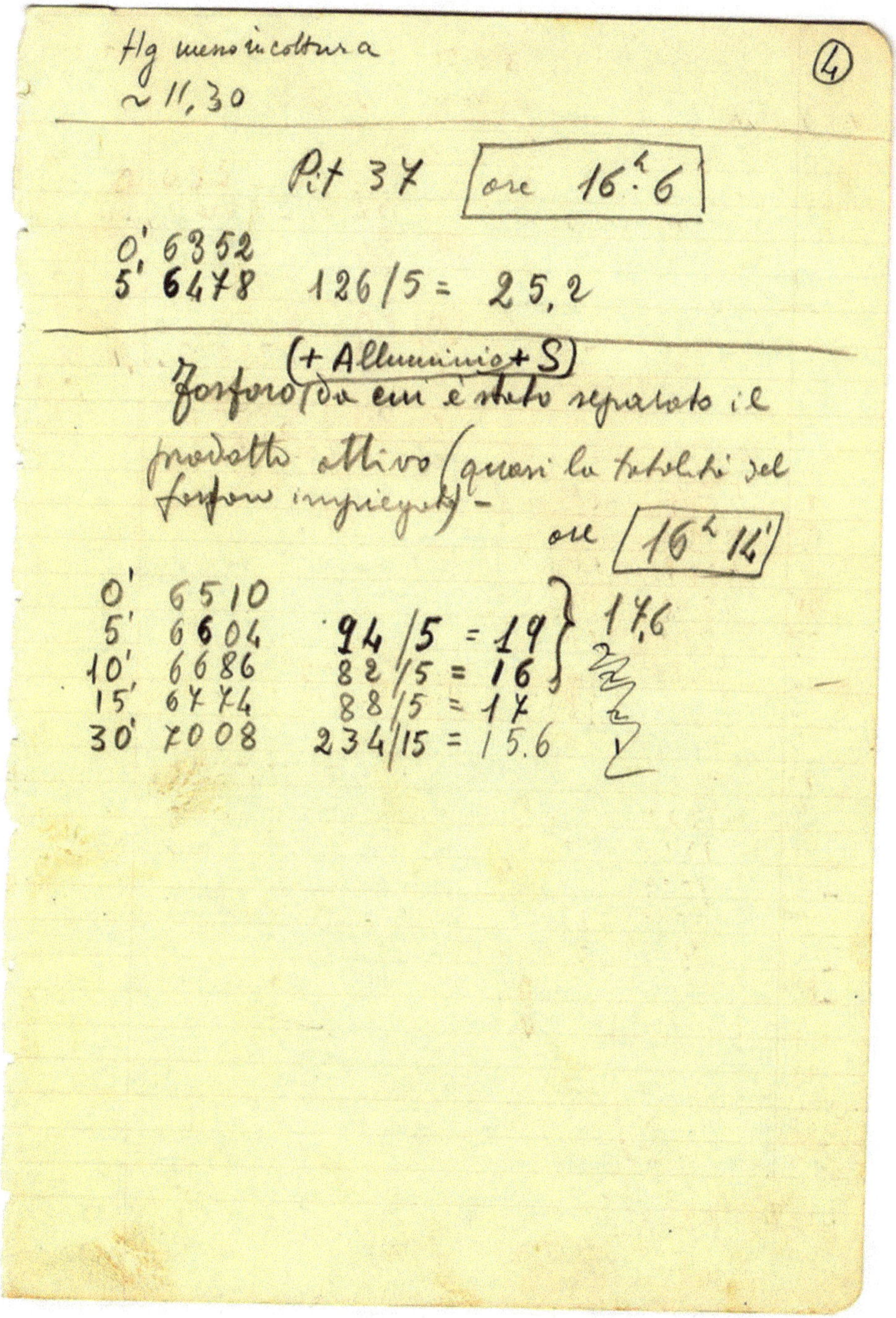

Hg meno in coltura a
~ 11,30

(4)

Pit 37 ore 16h 6

0' 6352
5' 6478 126/5 = 25,2

Fosforo (+ Alluminio + S) da cui è stato separato il prodotto attivo (quasi la totalità del fosforo impiegato) –

ore 16h 14'

0'	6510		
5'	6604	94/5 = 19	17,6
10'	6686	82/5 = 16	
15'	6774	88/5 = 17	
30'	7008	234/15 = 15.6	

S4

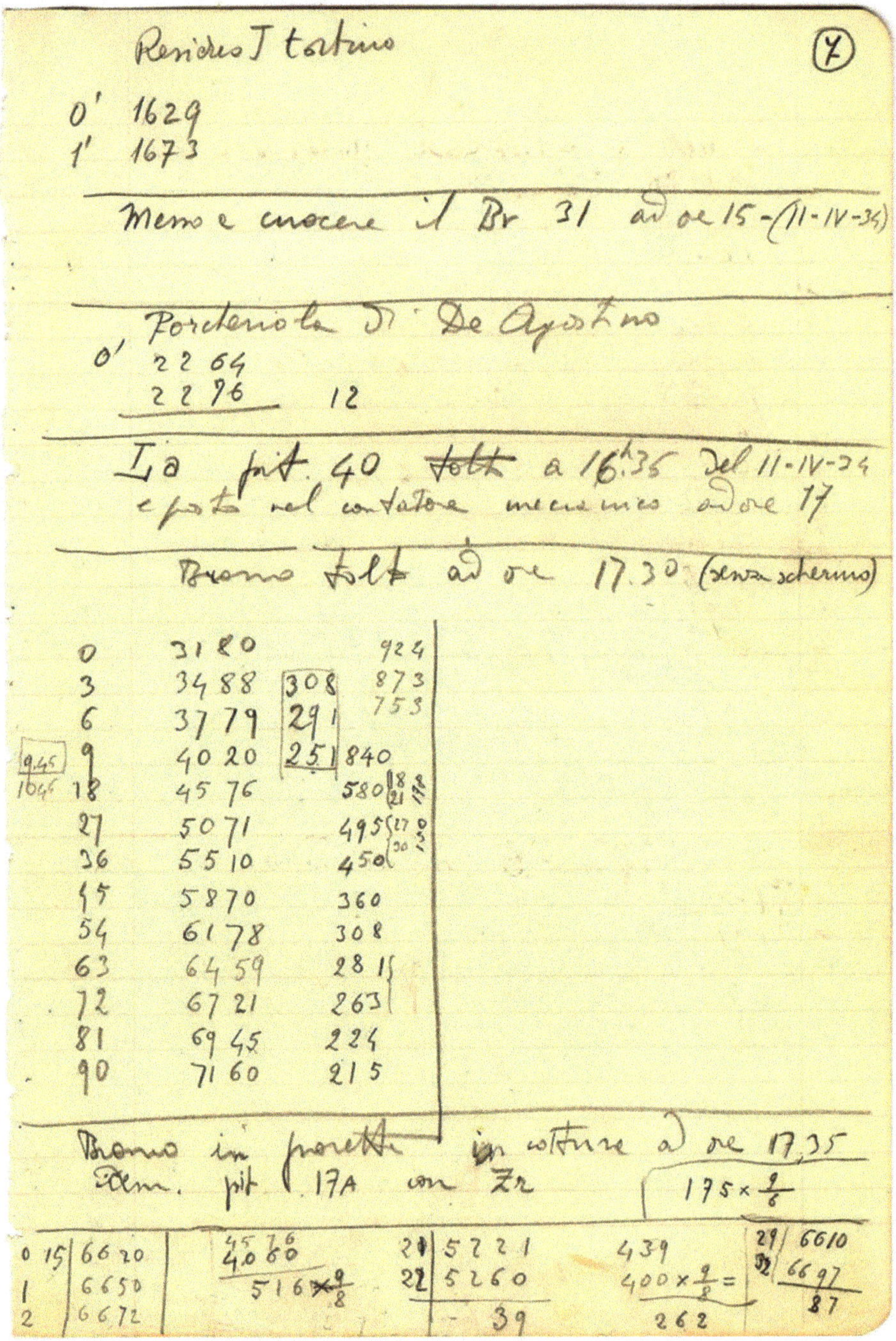
Residuo I tartino (7)

0' 1629
1' 1673

Messo a cuocere il Br 31 ad ore 15 – (11-IV-34)

Porcheriola di De Agostino
0' 2264
2276 12

Ia pit. 40 ~~tolta~~ a 16h.36 del 11-IV-34
e posto nel contatore meccanico ad ore 17

Bromo tolt ad ore 17.30 (senza schermo)

0	3180		924
3	3488	308	873
6	3779	291	753
9	4020	251	840
18	4576		580
27	5071		495
36	5510		450
45	5870		360
54	6178		308
63	6459		281
72	6721		263
81	6945		224
90	7160		215

Bromo in provette in cottura ad ore 17,35
Elem. pit 17A con Zr 175 × 9/6

0 15 6620 | 4576 4060 516 × 9/8 | 21 5221 22 5260 39 | 439 400 × 9/8 = 262 | 29 6610 32 6697 87
1 6650
2 6672

S7

12-IV-34 (8)

S' messo in prova ad ore 10.20 (15' 30' di attesa)
(pit 29)

0'	8630	
6'	9016	374
12'	9415	399
18'	9795	380
24'	10200	405
30'	10597	397
	1967	

0'	1235
1'	1332
2'	1431
3'	1526
8'	1975
15'	2518
20'	2896

ore 12,47

Bromuro di ammonio (dopo la ~~separazione~~ dei selenio e dell'arsenico) dopo 15' dal termine della irradiazione

→ 1661/20 = 83,0

Se + As separati da NH_4Br irradiato

– non sono ~~attivi~~ attivi –

ore 14.30

0 - 123		2924	
10	133	3491	567/10 = 56.7
23	146	4162	671/13 = 51.8
36	159	4721	559/13 = 43.
45	168	5110	389/9

il contatore non funzionava bene

ore 15,38 seguito bromuro di ammonio di stamani

0' 6213
18' 7143

N.E. 0' 7212
4' 7250

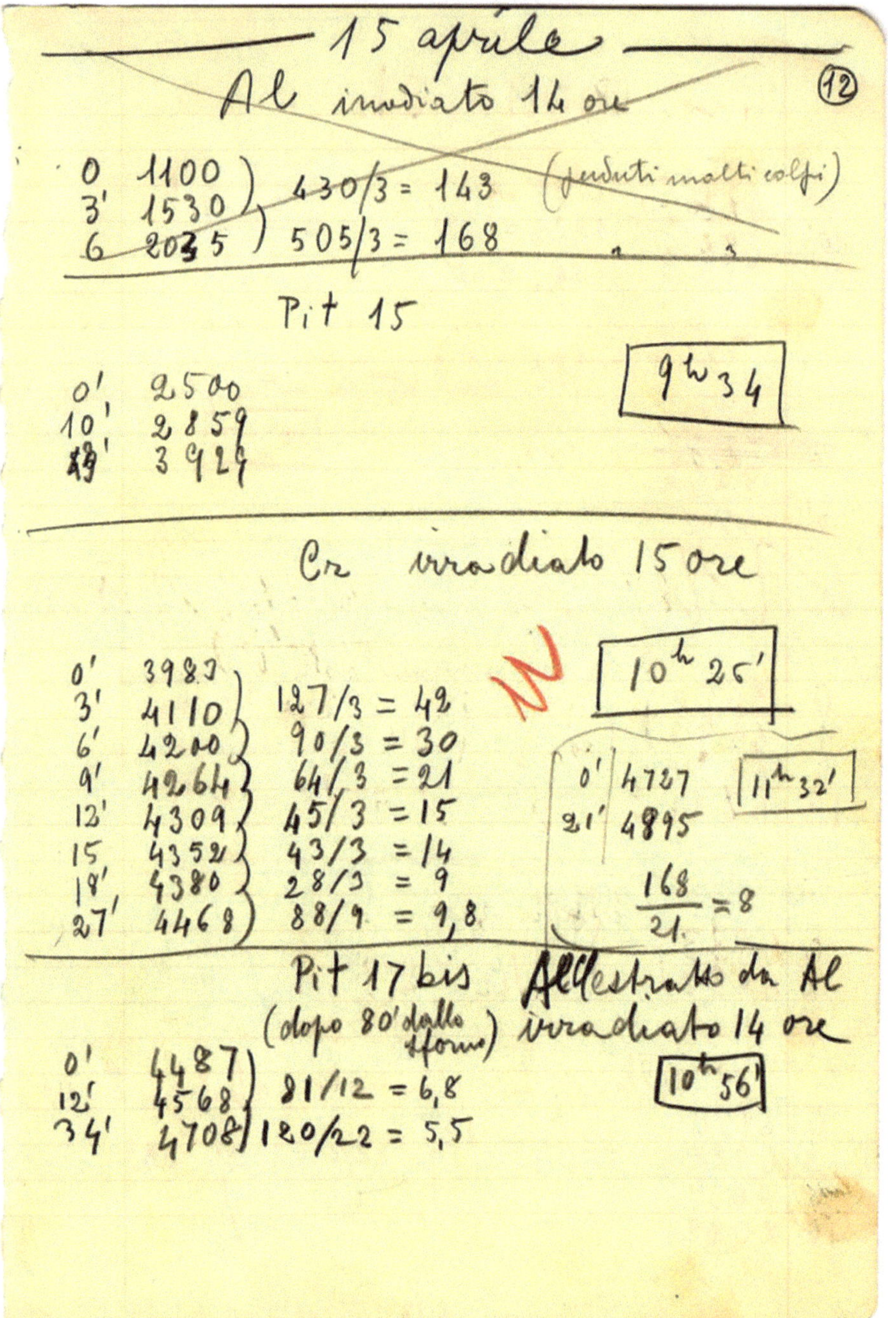

15 aprile

~~Al irradiato 14 ore~~ (12)

~~0 1100~~
~~3' 1530~~ ~~430/3 = 143~~ ~~(perduti molti colpi)~~
~~6 2035~~ ~~505/3 = 168~~

Pit 15

9h 34

0' 2500
10' 2859
13' 3929

Cr irradiato 15 ore

10h 25'

0' 3983
3' 4110 127/3 = 42
6' 4200 90/3 = 30
9' 4264 64/3 = 21
12' 4309 45/3 = 15
15 4352 43/3 = 14
18' 4380 28/3 = 9
27' 4468 88/9 = 9,8

0' 4727 11h 32'
21' 4895

$\frac{168}{21} = 8$

Pit 17 bis (dopo 80' dallo sforno)

Allestratto da Al irradiato 14 ore

10h 56'

0' 4487
12' 4568 81/12 = 6,8
34' 4708 120/22 = 5,5

(15)

465 8666) 2152/43 = 50 ore 18
510 0920 2254/45 = 50
ore 18.45

Silice estratta dallo Z di cui al Pit. 100 alle ore 18.45

0'.0956) 20/3 = 6,6) 115/15 = 7,8
3 0976
15 1071

P- estratto dallo Z di cui al Pit. 100 alle ore 19.22

0 1450) 282/3 = 94) 1504/15 = 100
3 1732
15 2954
28 4050 3600 = 128

18-IV-

0 P estratto dallo Z. di cui al Pit. 100 ore 8.30

0 4320) 329/3 = 109 | 0 - 6010) 299/3 = 100
3 4649 | 3 6309
~~4 4759~~ | 15 7504
~~0 5808~~ ~~7 5808~~ ~~6847~~

S15

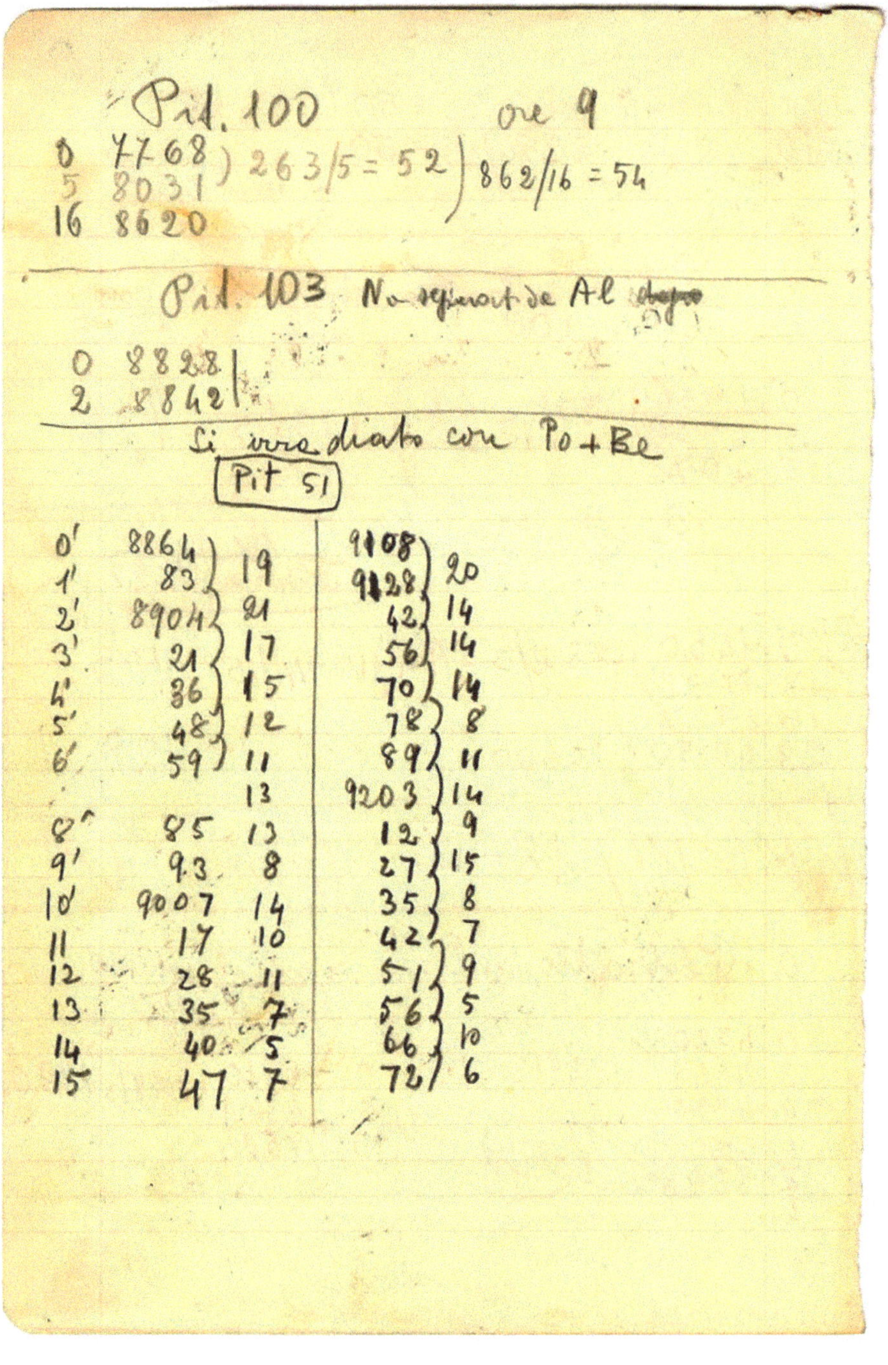

Pit. 100 ore 9

0 7768) 263/5 = 52) 862/16 = 54
5 8031
16 8620

Pit. 103 Non separato da Al ~~dopo~~

0 8828
2 8842

Si irradiato con Po + Be

Pit 51

0'	8864	19	9108	20
1'	83	21	9128	14
2'	8904	17	42	14
3'	21	15	56	14
4'	36	12	70	8
5'	48	11	78	11
6'	59	13	89	14
			9203	9
8'	85	13	12	15
9'	93	8	27	8
10'	9007	14	35	7
11	17	10	42	9
12	28	11	51	5
13	35	7	56	10
14	40	5	66	6
15	47	7	72	

S15a

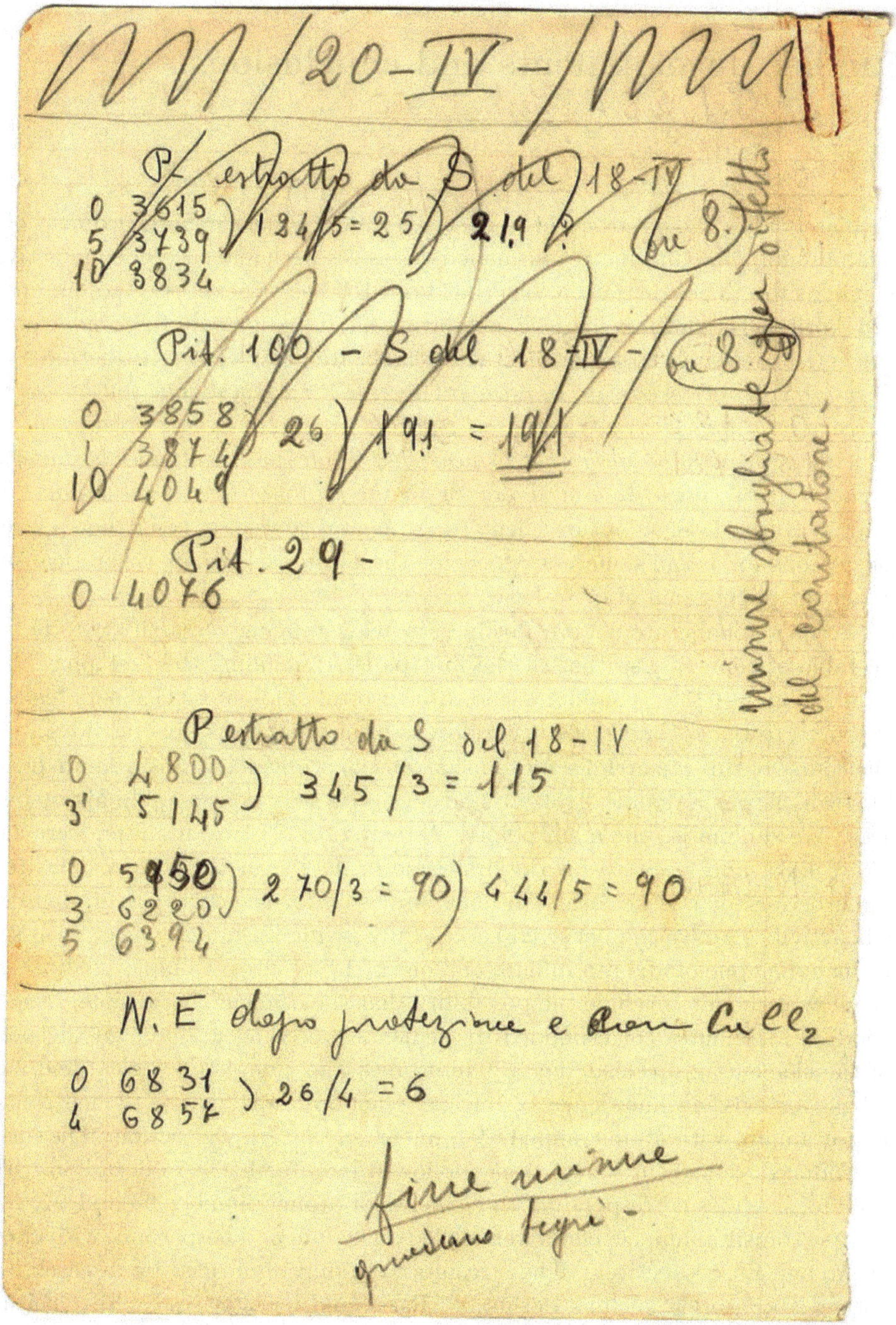
20-IV-

P/ estratto da S del 18-IV

0 3615) 124/5 = 25) 21.9 ?
5 3739
10 3834

su 8.

Pit. 100 - S del 18-IV - su 8.

0 3858) 26) 19.1 = 19.1
1 3874
10 4049

Pit. 29 -

0 4076

P estratto da S del 18-IV

0 4800) 345/3 = 115
3' 5145

0 5950) 270/3 = 90) 444/5 = 90
3 6220
5 6394

N. E dopo protezione e [illegible] $CuCl_2$

0 6831) 26/4 = 6
4 6857

fine misure

quaderno [illegible]

S16a

12 Further developments and conclusions

Immediately after the discovery of radioactivity induced by the bombardment of neutrons in aluminium and fluorine Fermi carried on his research and succeeded in activating other elements too. The congratulations that Lord Rutherford expressed to him in a letter on 23 April 1934, shown in Fig. 65, exhibiting his full approval of Fermi's transition from theoretical research to experimental research are significant. Rutherford wrote: *Dear Fermi,* [...] *Your results are of great interest,* [...] *I congratulate you on your successful escape from the sphere of theoretical physics! You seem to have struck a good line to start with. You may be interested to hear that Professor Dirac also is doing some experiments. This seems to be a good augury for the future of theoretical physics!*

Naturally Rutherford, as an excellent experimental physicist, could not appreciate that the reasons for Fermi's success rested for the most part on his theoretical understanding of the phenomena of beta decay.

Soon Fermi's collaborators were involved in wide ranging research over the whole periodic table of elements, so that in the first paper concluding the first phase of this systematic research [F135], as we have said, from a total of about 60 elements tested well over 40 proved to be activated producing dozens of new beta unstable nuclides.

In the course of this research Fermi's mishap on the problem of transuranium elements is well known. It was a double stroke of bad luck with very considerable historical consequences. We summarise the main points. Attempts to activate uranium were carried out in May 1934, and reported for the first time in [F123]. It was a very delicate kind of research because it was necessary to make provision for the effects of uranium's natural radioactivity. Activation occurred successfully giving an intense effect with several periods, including one of about a minute and one of 13 minutes. Complex radiochemical analysis allowed them to exclude the possibility that the element that disintegrated with the period of 13 minutes corresponded to elements preceding uranium in the periodic table of the elements up to lead. Fermi's interpretation, completely logical and natural on the basis of existing knowledge of nuclear reactions, was that with neutron bombardment uranium, with atomic number 92, underwent an (n, γ) reaction. The unstable nuclide produced underwent a beta decay, followed by a further beta decay, so with the production of nuclides corresponding to elements of atomic number 93 and 94, respectively. These transuranium elements were called ausonium and hesperium, with chemical symbols Ao and Es, respectively. These results were fully confirmed by research carried out in other laboratories, such as Berlin and Paris, and by later research by Fermi and his collaborators when the effects of slow neutrons were clear, and thus the efficiency of the neutron sources available could be increased.

F. Guerra and N. Robotti, *The Lost Notebook of ENRICO FERMI*,
https://doi.org/10.1007/978-3-319-69254-8_12

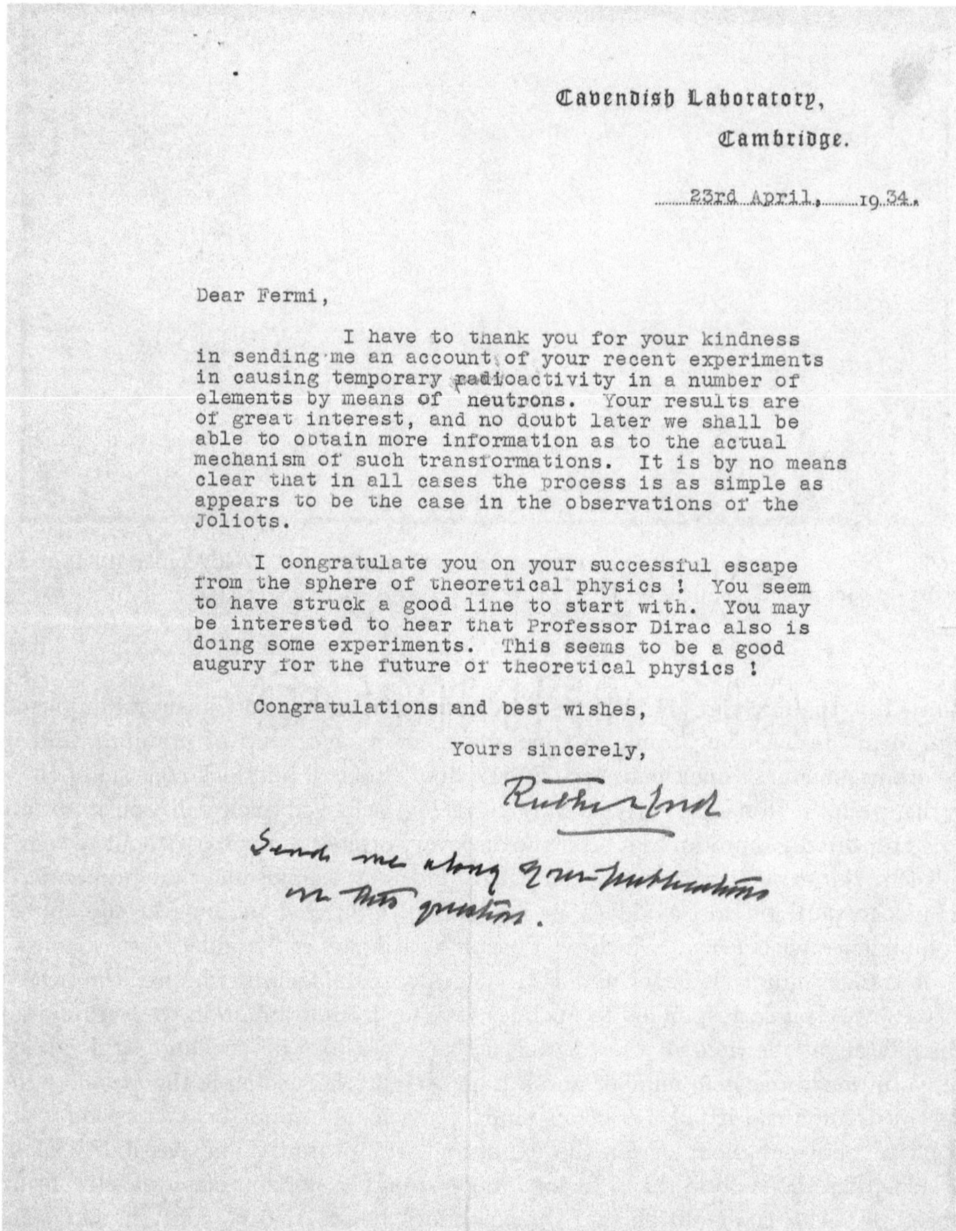

Cavendish Laboratory,
Cambridge.

23rd April, 1934.

Dear Fermi,

I have to thank you for your kindness in sending me an account of your recent experiments in causing temporary radioactivity in a number of elements by means of neutrons. Your results are of great interest, and no doubt later we shall be able to obtain more information as to the actual mechanism of such transformations. It is by no means clear that in all cases the process is as simple as appears to be the case in the observations of the Joliots.

I congratulate you on your successful escape from the sphere of theoretical physics ! You seem to have struck a good line to start with. You may be interested to hear that Professor Dirac also is doing some experiments. This seems to be a good augury for the future of theoretical physics !

Congratulations and best wishes,

Yours sincerely,

Rutherford

Send me along your publications on this question.

Fig. 65. – Letter of congratulations from Ernest Rutherford to Fermi (Cambridge, 23 April 1934) - FDG.

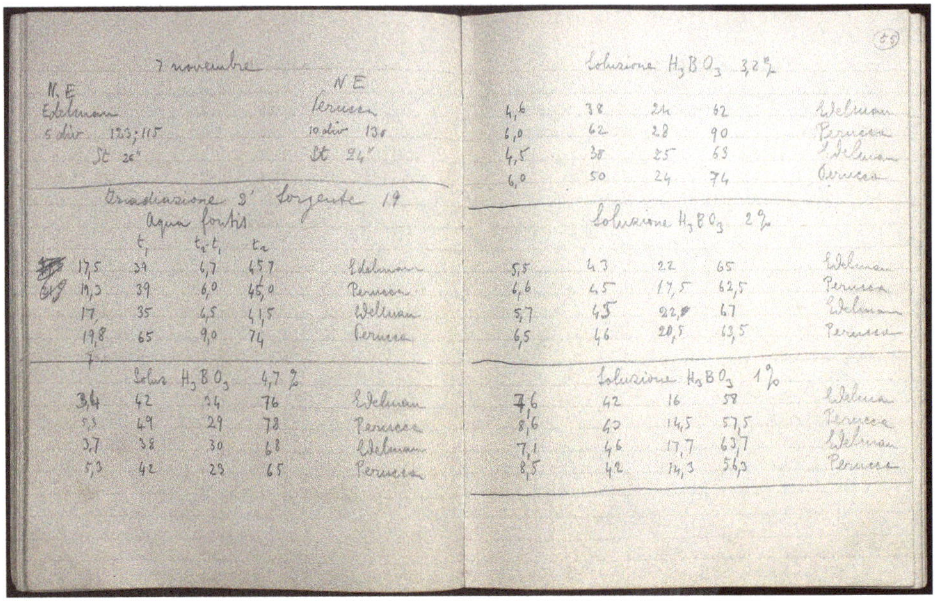

Fig. 66. – Two pages from a laboratory notebook of Fermi's, dated 7 November 1934, in which the words "aqua fontis" appear - FDG.

In reality, the nuclides found by Fermi did not correspond to transuranium elements but to elements with an atomic number much lower than that of uranium, and were therefore fragments of nuclear fission. So the first stroke of bad luck consists of the fact that the group in Rome, as early as May 1934, had achieved nuclear fission in uranium, and had repeatedly reproduced it after the discovery of slow neutrons, without recognising it as such. An explicit warning, which with hindsight seems somewhat prophetic, had been put forward by Ida Noddack in her paper [119], and perhaps in the course of correspondence with Fermi, which we do not have however. Noddack was against the idea of transuranium elements and had suggested establishing whether the activities observed by Fermi corresponded to nuclides very far from uranium in the periodic table. Unfortunately at the time all known nuclear reactions, both natural and artificial, led to small variations in atomic number and atomic weight. For example the atomic number varies by -2 for the expulsion of an alpha particle, $+1$ for a beta emission, -1 for a negative beta emission, or for the expulsion of a proton. The possibility of large variations, like those obtained in fission, was not on the horizon of physicists' intuition at the time. And it should be said that Noddack never tried to verify her hypothesis experimentally, not even in collaboration with someone in the best equipped laboratories.

It would be many years before Otto Hahn and Fritz Strassmann in Berlin, in December 1938, succeeded in demonstrating by complex radiochemical analysis that in the bombardment of uranium with slow neutrons radioactive barium was produced as a fission fragment.

This discovery happened immediately after the Nobel Prize ceremony for Enrico Fermi, who was forced to correct his draft Nobel Lecture to take into account the new discoveries. In any case Fermi recovered well from this "blow", also because perhaps with his intuition he had already glimpsed something. After his arrival in the United States he immediately inserted himself into fission research, and on 2 December 1942 he managed to create the first self-sustaining and controlled nuclear chain reaction, in an atomic pile with natural uranium as fuel, and graphite as moderator, due to the slowing down of the neutrons.

The second stroke of bad luck was that the hesperium hypothesised by Fermi, as the result of a sequence of two beta decays in irradiated uranium, actually exists, as element 94, and it is plutonium, a highly fissile element. Naturally the weak radon-beryllium neutron sources used by Fermi in 1934 and in the following years certainly produced plutonium (hesperium) but in absolutely negligible quantities which were impossible to detect. The radioactivity then seen by Fermi was due to fission fragments. Anyway, by a strange irony of fate, Fermi's nuclear reactor, as an extremely strong source of neutrons, was able to produce hesperium (plutonium) in considerable quantity, enough to build the first atom bombs for the Trinity test (16 July 1945) and Nagasaki (9 August 1945).

Therefore Fermi's overall research activity and the elusive, but very real, new transuranium element hesperium are strangely woven together in a very instructive way.

In continuing his research into neutron-induced radioactivity, Fermi arrived on 20 October 1934 at what was probably his most significant discovery in terms of applications. It was demonstrated that by appropriately slowing down the neutrons emitted by the source, by repeated collisions with nuclei of light elements, such as the hydrogen in paraffin, these neutrons were more effective in producing (n, γ)-type nuclear reactions in heavy nuclei. This discovery not only allowed the study of neutron-induced radioactivity to be expanded but is also the basis of the discovery of nuclear fission.

It is well known that according to a widespread tradition, derived from Laura Fermi's book [69], the water in the goldfish fountain in the garden of the Physics Institute in Via Panisperna was used to produce the slowing down of the neutrons, and thus increase their efficiency in nuclear reactions. Recently the fountain was declared a Historical Site by the European Physical Society. Possible documentary evidence for this tradition can be supplied, for example, by the records on 7 November 1934 of measurements of neutron source activity, in in which it is stated that the experiments were carried out in the medium "aqua fontis", as recorded in a laboratory notebook of Fermi's in the Domus Galilaeana, two pages of which are reproduced in Fig. 66.

Naturally, given the great sense of humour Fermi sometimes displayed, as we have seen on many occasions, one might think that the appellation "aqua fontis" chosen by Fermi simply denoted "tap water", compared to the solutions of boric acid H_3BO_3 in various concentrations used later. In any case the link between the goldfish fountain and the description "aqua fontis" is certainly appealing.

At the beginning of the summer holidays in 1935, the working group that had formed around Fermi on the problem of neutron induced radioactivity began to break up. Rasetti went to the United States, to Columbia University, to stay there for at least a year.

Segrè took up his post at the end of 1935 at the University of Palermo where in the meantime he had won the chair in Experimental Physics. D'Agostino, whose expertise in radiochemistry evidently was no longer required, would find a stable position at the Chemistry Institute of the C.N.d.R., to concern himself amongst other things with the study of aerosols and defence against chemical weapons. So in Rome, at the end of the summer, only Fermi, Amaldi and Pontecorvo were left. But very soon, in April 1936, Pontecorvo would also leave Rome and transfer to Paris.

Fermi abandoned radiochemical research and concentrated on the study of the diffusion, slowing down and absorption of neutrons, arriving at results of great significance which would then be fully used in building the nuclear pile.

The conferment of the Nobel Prize for Physics on Enrico Fermi, in the ceremony on 10 December 1938 in Stockholm, crowned an extraordinary research cycle, which originated with the formulation of the theory of beta decay, immediately arrived at the discovery of radioactivity induced by neutron bombardment, and continued with the discovery of the effects due to slow neutrons, concluding with the study of the properties of diffusion, slowing down and absorption of neutrons in matter.

It was an all Italian Nobel Prize, which in some aspects is almost miraculous.

In this book of ours we have analysed in the national and international context the rise of Nuclear Physics in Rome in the early Thirties and the discovery of neutron-induced radioactivity by Enrico Fermi, making use of the extraordinary mine of information of the Irpinia notebook for a complete description of Fermi's strategic and tactical choices and of the apparatus and procedures used.

We have seen that Fermi's methods were based first of all on his extraordinary physics intuition in the framework of a solid theoretical understanding of the phenomena involved. Moreover, his experimental success was due to a harmonious and bold use of all the best possible procedures, in a context where the limited means available required him to resort to ingenuity.

A significant episode is told which happened in an important research centre where one of the first cyclotrons was operating. Since some anomalies had been observed in the functioning of the detecting counters after the cyclotron was switched off it was decided to turn off the detectors together with the cyclotron. In reality the "anomalies" were none other than the signals of induced radioactivity produced by bombardment with protons. The excess of zeal in turning off the detectors together with the cyclotron prevented a great discovery being made which the Joliot-Curies would arrive at many months later and with great effort.

Enrico Fermi, who was very proud of proceeding C.I.F. (Con Intuito Formidabile - With Formidable Intuition), would never have made such a mistake. In any case maybe he was lucky to have only a modest radon-beryllium neutron source available that "could be hidden in a fist" instead of a powerful cyclotron.

Erratum to: The Lost Notebook of ENRICO FERMI

Erratum to:
F. Guerra and N. Robotti, *The Lost Notebook of ENRICO FERMI*, https://doi.org/10.1007/978-3-319-69254-8

In the original version of the book, the translation texts "Translated from Italian by Christine V. Pennison" and "Translation from the Italian language edition: *Enrico Fermi e il quaderno ritrovato* by F. Guerra and N. Robotti, © SIF 2015. All rights reserved" have to be included in the Copyright page of Frontmatter. The erratum book has been updated with the change.

The updated original online version for this book can be found at
https://doi.org/10.1007/978-3-319-69254-8

F. Guerra and N. Robotti, *The Lost Notebook of ENRICO FERMI*,
https://doi.org/10.1007/978-3-319-69254-8_13

Fermi's Publications in his Italian period

Journals

F1 E.F., "Sull'elettrostatica di un campo gravitazionale uniforme e sul peso delle masse elettromagnetiche", Nuovo Cimento **22**, 176-188 (1921).

F2 E.F., "Sulla dinamica di un sistema rigido di cariche elettriche in moto traslatorio", Nuovo Cimento **22**, 199-207 (1921).

F3 E.F., "Un teorema di calcolo delle probabilità ed alcune sue applicazioni", Tesi di Abilitazione della Scuola Normale Superiore, Pisa, 20 giugno 1922.

F4 E.F., "Studi sopra i raggi Röntgen", Tesi di Laurea in Fisica, Regia Università di Pisa, Pisa, 21 giugno 1922.

F5 E.F., "Sopra i fenomeni che avvengono in vicinanza di una linea oraria", Rendiconti Accademia Lincei **31**, 21-23, 51-52, 101-103 (1922).

F6 E.F., "Über einen Widerspruch zwischen der elektrodynamischen und der relativistischen Theorie der elektromagnetischen Masse", Physikalische Zeitschrift **23**, 340-344 (1922).

F7 E.F., "Correzione di una grave discrepanza tra la teoria delle masse elettromagnetiche e la teoria della relatività. Inerzia e peso dell'elettricità", Rendiconti Accademia Lincei **31**, 184-187 (1922).

F8 E.F., "Correzione di una grave discrepanza tra la teoria elettrodinamica e quella relativistica delle masse elettromagnetiche. Inerzia e peso dell'elettricità", Rendiconti Accademia Lincei **31**, 306-309 (1922).

F9 E.F., "I raggi Röntgen", Nuovo Cimento **24**, 133-163 (1922).

F10 E.F., "Formazione di immagini coi raggi Röntgen", Nuovo Cimento **25**, 63-68 (1923).

F11 E.F., "Correzione di una contraddizione tra la teoria elettrodinamica e quella relativistica delle masse elettromagnetiche", Nuovo Cimento **25**, 159-170 (1923).

F12 E.F., Il principio delle adiabatiche ed i sistemi che non ammettono coordinate angolari, Nuovo Cimento **25**, 171-175 (1923).

F. Guerra and N. Robotti, *The Lost Notebook of ENRICO FERMI*,
https://doi.org/10.1007/978-3-319-69254-8

F13 E.F., "Dimostrazione che in generale un sistema meccanico normale è quasi ergodico", Nuovo Cimento **25**, 267-269 (1923).

F14 E.F., "Alcuni teoremi di meccanica analitica importanti per la teoria dei quanti", Nuovo Cimento **25**, 271-285 (1923).

F15 E.F., "Le masse nella teoria della relatività", in *I fondamenti della relatività Einsteiniana*, R. Contu, T. Bembo, (ed.), Hoepli, Milano 1923, pag. 342-344.

F16 E.F., "Sul peso dei corpi elastici", Memorie Accademia Lincei **14**, 114-124 (1923).

F17 E.F., "Sul trascinamento del piano di polarizzazione da parte di un mezzo rotante", Rendiconti Accademia Lincei **32**, 115-118 (1923).

F18 E.F., A. Pontremoli, "Sulla massa della radiazione in uno spazio vuoto", Rendiconti Accademia Lincei **32**, 162-164 (1923).

F19 E.F., "Beweis dass ein mechanisches Normalsystem im allgemeinen quasi-ergodisch ist", Physikalische Zeitschrift **24**, 261-265 (1923).

F20 E.F., "Sulla teoria statistica di Richardson dell'effetto fotoelettrico", Nuovo Cimento **26**, 97-104 (1923).

F21 E.F., "Generalizzazione del teorema di Poincaré sopra la non esistenza di integrali uniformi di un sistema di equazioni canoniche normali", Nuovo Cimento **26**, 105-113 (1923).

F22 E.F., "Sopra la teoria di Stern della costante assoluta dell'entropia di un gas perfetto monoatomico", Rendiconti Accademia Lincei **32**, 395-398 (1923).

F23 E.F., "Sulla probabilità degli stati quantici", Rendiconti Accademia Lincei **32**, 493-495 (1923).

F24 E.F., "Nuova Meccanica Quantistica", Atti della Società Italiana per il Progresso delle Scienze, **XV**, 552-554 (1926).

F25 E.F.,"Über die Wahrscheinlichkeit der Quantenzustände", Zeitschrift für Physik **26**, 54-56 (1924).

F26 E.F., "Über die Existenz quasi-ergodischer Systeme", Physikalische Zeitschrift **25**, 166-167 (1924).

F27 E.F., "Sopra la riflessione e la diffusione di risonanza", Rendiconti Accademia Lincei **33**, 90-93 (1924).

F28 E.F., "Considerazioni sulla quantizzazione dei sistemi che contengono degli elementi identici", Nuovo Cimento **1**, 145-152 (1924).

F29 E.F., "Sull'equilibrio termico di ionizzazione", Nuovo Cimento **1**, 153-158 (1924).

F30 E.F., "Berekeningen over de intensiteiten van spektraallijnen", Physica **4**, 340-343 (1924).

F31 E.F., "Über die Theorie des Stosses zwischen Atomen und elektrisch geladenen Teilchen", Zeitschrift für Physik **29**, 315-327 (1924).

F32 E.F., "Sopra l'intensità delle righe multiple", Rendiconti Accademia Lincei **1**, 120-124 (1925).

F33 E.F., "Sui principi della teoria dei quanti", Rendiconti Seminario Matematico Università di Roma **8**, 7-12 (1925).

F34 E.F., "Sulla teoria dell'urto tra atomi e corpuscoli elettrici", Nuovo Cimento **2**, 143-158 (1925).

F35 E.F., "Sopra l'urto tra atomi e nuclei di idrogeno", Rendiconti Accademia Lincei **1**, 77-80 (1925).

F36 E.F., "Una relazione tra le costanti delle bande infrarosse delle molecole triatomiche", Rendiconti Accademia Lincei **1**, 386-387 (1925).

F37 E.F., F. Rasetti, "Effect of an Alternating Magnetic Field on the Polarisation of the Resonance Radiation of Mercury Vapour", Nature **115**, 764 (1925).

F38 E.F., F. Rasetti, "Über den Einfluss eines wechselnden magnetischen Feldes auf die Polarisation der Resonanzstrahlung", Zeitschrift für Physik **33**, 246-250 (1925).

F39 E.F., F. Rasetti, "Effetto di un campo magnetico alternato sopra la polarizzazione della luce di risonanza", Rendiconti Accademia Lincei **1**, 716-722 (1925).

F40 E.F., F. Rasetti, "Ancora dell'effetto di un campo magnetico alternato sopra la polarizzazione della luce di risonanza", Rendiconti Accademia Lincei **2**, 117-120 (1925).

F41 E.F., "Sopra la teoria dei corpi solidi", Periodico di Matematiche **5**, 264-274 (1925).

F42 E.F., "Sulla quantizzazione del gas perfetto monoatomico", Rendiconti Accademia Lincei **3**, 145-149 (1926).

F43 E.F., "Zur Quantelung des idealen einatomigen Gases", Zeitschrift für Physik **36**, 902-912 (1926).

F44 E.F., "Sopra l'intensità delle righe proibite nei campi magnetici intensi", Rendiconti Accademia Lincei **3**, 478-483 (1926).

F45 E.F., "Argomenti pro e contro la ipotesi dei quanti di luce", Nuovo Cimento **3**, XLVII-LIV (1926).

F46 E.F., "Problemi di chimica, nella fisica dell'atomo", Periodico di Matematiche **6**, 19-26 (1926).

F47 F. Rasetti, E.F., "Sopra l'elettrone rotante", Nuovo Cimento **3**, 226-235 (1926).

F48 E.F., "Zur Wellenmechanik des Stossvorganges", Zeitschrift für Physik **40**, 399-402 (1926).

F49 E.F., E. Persico, "Il principio delle adiabatiche e la nozione di forza viva nella nuova meccanica ondulatoria", Rendiconti Accademia dei Lincei **4**, 452-457 (1926).

F50 E.F., "Sopra una formula di calcolo delle probabilità", Nuovo Cimento **3**, 313-318 (1926).

F51 E.F., "Quantum Mechanics and the Magnetic Moment of Atoms", Nature **118**, 876 (1926).

F52 E.F., F. Rasetti, "Eine Messung des Verhältnisses h/k durch die anomale Dispersion des Thalliumdampfes", Zeitschrift für Physik **43**, 379-383 (1927).

F53 E.F., F. Rasetti, "Una misura del rapporto h/k per mezzo della dispersione anomala del tallio", Rendiconti Accademia Lincei **5**, 566-570 (1927).

F54 E.F., "Gli effetti elettro e magnetoottici e le loro interpretazioni", Fascicolo speciale dell'"Energia Elettrica", nel 1° centenario della morte di A. Volta, Uniel, Roma 1927, pag. 109-120.

F55 E.F., "Sul meccanismo dell'emissione nella meccanica ondulatoria", Rendiconti Accademia Lincei **5**, 795-800 (1927).

F56 E.F., "Un metodo statistico per la determinazione di alcune proprietà dell'atomo", Rendiconti Accademia Lincei **6**, 602-607 (1927).

F57 E.F., "Sulla deduzione statistica di alcune proprietà dell'atomo. Applicazione alla teoria del sistema periodico degli elementi", Rendiconti Accademia Lincei 7 (1928), 342-346.

F58 E.F., "Sulla deduzione statistica di alcune proprietà dell'atomo. Calcolo della correzione di Rydberg per i termini s", Rendiconti Accademia Lincei **7**, 726-730 (1928).

F59 E.F., "Anomalous Groups in the Periodic System of Elements", Nature **121**, 502 (1928).

F60 E.F., "Eine statistische Methode zur Bestimmung einiger Eigenschaften des Atoms und ihre Anwendung auf die Theorie des periodischen Systems der Elemente", Zeitschrift für Physik **48**, 73-79 (1928).

F61 E.F., "Statistische Berechnung der Rydbergkorrektionen der s-Terme", Zeitschrift für Physik **49**, 550-554 (1928).

F62 E.F., "Über die Anwendung der statistischen Methode auf die Probleme des Atombaues" Falkenhagen, *Quantentheorie und Chemie*, Leipziger Vorträge (1928), Hirzel, Leipzig 1928, pag. 95-111.

F63 E.F., "Sopra l'elettrodinamica quantistica", Rendiconti Accademia Lincei **9**, 881-887 (1929).

F64 E.F., "Sul moto di un corpo di massa variabile", Rendiconti Accademia Lincei **9**, 984-986 (1929).

F65 E.F., "Sulla teoria quantistica delle frange di interferenza", Rendiconti Accademia Lincei **10**, 72-77 (1929).

F66 E.F., "Sul complesso 4d della molecola di elio", Rendiconti Accademia Lincei, **10**, 515-517 (1929).

F67 E.F., "Problemi attuali della fisica", Annali dell'Istruzione Media **5**, 424-428 (1929).

F68 E.F., "I fondamenti sperimentali delle nuove teorie fisiche", Atti della Società Italiana per il Progresso delle Scienze, **XVIII**, vol. 1, 365-371 (1929).

F69 E.F., "Affinità elettronica e teoria statistica dell'atomo", Nuovo Cimento, **6**, XIII-XIV (1929).

F70 E.F. "L'interpretazione del fenomeno dell'irradiazione e dell'assorbimento nella attuale teoria dei quanti", Nuovo Cimento, **6**, XVI-XVII (1929).

F71 E.F., "Fondamenti sperimentali delle nuove teorie fisiche", Nuovo Cimento, **6**, CLIII-CLVI (1929).

F72 E.F., "Magnetic Moments of Atomic Nuclei", Nature **125**, 16 (1930).

F73 E.F., "Über die magnetischen Momente der Atomkerne", Zeitschrift für Physik **60**, 320-333 (1930).

F74 E.F., "Sui momenti magnetici dei nuclei atomici, Memorie Accademia d'Italia **1** (Fis.), 139-148 (1930).

F75 E.F., "Sulla teoria quantistica delle frange di interferenza", Nuovo Cimento **7**, 153-158 (1930).

F76 E.F., "Sul complesso 4d della molecola di elio", Nuovo Cimento **7**, 159-161 (1930).

F77 E.F., "Über das Intensitätsverhältnis der Dublettkomponenten der Alkalien", Zeitschrift für Physik **59**, 680-686 (1930).

F78 E.F., "Sul rapporto delle intensità nei doppietti dei metalli alcalini", Nuovo Cimento **7**, 201-207 (1930).

F79 E.F., "L'interpretazione del principio di causalità nella meccanica quantistica", Rendiconti Accademia dei Lincei **11**, 980-985 (1930).

F80 E.F., "L'interpretazione del principio di causalità nella meccanica quantistica", Nuovo Cimento **7**, 361-366 (1930).

F81 E.F., "Atomi e stelle", Atti della Società Italiana per il Progresso delle Scienze, **XIX**, vol. 1, 228-235 (1930).

F82 E.F., "I fondamenti sperimentali della nuova meccanica atomica", Periodico di Matematiche **10**, 71-84 (1930).

F83 E.F., "La fisica moderna", Nuova Antologia **65**, 137-45 (1930).

F84 E.F., "Sul calcolo degli spettri degli ioni", Memorie Accademia d'Italia **1** (Fis.), 149-156 (1930).

F85 E.F., "Sopra l'elettrodinamica quantistica", Rendiconti Accademia Lincei **12**, 431-435 (1930).

F86 E.F., "Sul momento magnetico del nucleo", Atti della Società Italiana per il Progresso delle Scienze, **XIX**, 96 (1930), solo il titolo.

F87 E.F., "Sur les moments magnétiques des noyaux", in *Le Magnétisme, Rapports et discussions du Sixième Conseil de Physique*, Bruxelles 20-25 Octobre 1930, Gauthier-Villars, Paris, 1932.

F88 E.F., "Sull'effetto Raman delle molecole poliatomiche", Atti della Società Italiana per il Progresso delle Scienze, **XX**, vol. 2, 149-150 (1930).

F89 E.F., "Le masse elettromagnetiche nella elettrodinamica quantistica", Nuovo Cimento **8**, 121-132 (1931).

F90 E.F., "Sul calcolo degli spettri degli ioni", Nuovo Cimento **8**, 7-14 (1931).

F91 E.F., "Sur la théorie de la radiation", Annales de l'Institut H. Poincaré **1**, 53-74(1931).

F92 E.F., "Über den Ramaneffekt des Kohlendioxids", Zeitschrift für Physik **71**, 250-259 (1931).

F93 E.F., F. Rasetti, "Über den Ramaneffekt des Steinsalzes", Zeitschrift für Physik **71**, 689-695 (1931).

F94 E.F., "Sopra la teoria quantistica dei campi elettromagnetici", Nuovo Cimento **8**, XCIII (1931).

F95 E.F., "Nuclei ed elettròni", Gerarchia **9**, 879-885 (1931).

F96 E.F., "Quantum Theory of Radiation", Review of Modern Physics **4**, 87-132 (1932).

F97 H. Bethe, E.F., "Über die Wechselwirkung von zwei Elektronen", Zeitschrift für Physik **77**, 296-306 (1932).

F98 E.F., "L'effetto Raman nelle molecole e nei cristalli", Memorie Accademia d'Italia **3** (Fis.), 239-256 (1932).

F99 E.F., "La physique du noyau atomique", in R. de Valbreuze, ed., *Comptes Rendus du Congrès International d'Électricité, Paris 1932, Première Section*, 709-807, Gauthier-Villars, Paris, 1933.

F100 E.F., "Lo stato attuale della fisica del nucleo atomico", Ricerca Scientifica **3** (2), 101-113 (1932).

F101 E.F., "Stato attuale della fisica del nucleo atomico", L'Elettricista **41**, 89-93 (1932).

F102 E.F., "Sulle bande di oscillazione e rotazione dell'ammoniaca", Rendiconti Accademia Lincei **16**, 179-185.

F103 E.F., "Sulle bande di oscillazione e rotazione dell'ammoniaca", Nuovo Cimento **9**, 277-283 (1932).

F104 E.F., "Lo spettro di rotazione e oscillazione dell'ammoniaca", Atti della Società Italiana per il Progresso delle Scienze, **XXI**, vol. 2, 177 (1932).

F105 E.F., B. Rossi, "Azione del campo magnetico terrestre sulla radiazione penetrante", Rendiconti Accademia Lincei **17**, 346-350 (1933).

F106 E.F., B. Rossi, "Azione del campo magnetico terrestre sulla radiazione penetrante", Nuovo Cimento, **10**, 333-338 (1933).

F107 E.F., E. Segrè, "Sulla teoria delle strutture iperfini", Memorie Accademia d'Italia **4** (Fis.), 131-158 (1933).

F108 E.F., E. Segrè, "Zur Theorie der Hyperfeinstruktur", Zeitschrift für Physik **82**, 729-749 (1933).

F109 E.F., G. Uhlenbeck, "Sulla ricombinazione di elettroni e positroni", Ricerca Scientifica **4** (2), 157-60 (1933).

F110 E.F., G. Uhlenbeck, "On the Recombination of Electrons and Positrons", Physical Review **44**, 510-511 (1933).

F111 E.F., F. Rasetti, "Uno spettrografo per raggi «gamma»a cristallo di bismuto", Ricerca Scientifica **4** (2), 299-302 (1933).

F112 E.F., "Le ultime particelle costitutive della materia, Atti della Società Italiana per il Progresso delle Scienze, **XXII**, vol. 2, 7-14 (1933).

F113 E.F., "Theory of hyperfine structures", public address at the Chicago Meeting of the American Physical Society, June 19-24, 1933, Physical Review **44**, 313 (1933), solo titolo.

F114 E.F., "La combinazione degli elettroni e dei positroni", Atti della Società Italiana per il Progresso delle Scienze, **XXII**, vol. 2, 217 (1933).

F115 E.F., "Tentativo di una teoria dell'emissione dei raggi «beta»", Ricerca Scientifica **4** (2), 491-495 (1933).

F116 E.F., "Tentativo di una teoria dei raggi β", Nuovo Cimento **11**, 1-19 (1934).

F117 E.F., "Versuch einer Theorie der β-Strahlen. I", Zeitschrift für Physik, **88**, 161-171 (1934)

F118 E.F., "Sopra lo spostamento per pressione delle righe elevate delle serie spettrali", Nuovo Cimento **11**, 157-166 (1934).

F119 E.F., "Le ultime particelle costitutive della materia", Scientia **55**, 21-28 (1934).

F120 E.F., "Radioattività indotta da bombardamento di neutroni", Ricerca Scientifica **5** (1), 283 (1934).

F121 E.F., "Radioattività provocata da bombardamento di neutroni", Ricerca Scientifica **5** (1), 330-331 (1934).

F122 E.F., "Radioactivity induced by neutron bombardment", Nature **133**, 757 (1934).

F123 E. Amaldi, O. D'Agostino, E. F., F. Rasetti, E. Segrè, "Radioattività "beta,, provocata da bombardamento di neutroni - III", Ricerca Scientifica **5** (1), 452-453 (1934).

F124 E. Amaldi, E. D'Agostino, E.F., F. Rasetti, E. Segrè, "Radioattività "beta" provocata da bombardamento di neutroni", Nuovo Cimento, **11** 332-333 (1934).

F125 E.F., "Zur Bemerkung von G. Beck und K. Sitte", Zeitschrift für Physik **89**, 522 (1934).

F126 E.F., E. Amaldi, "Le orbite ∞ s degli elementi", Memorie Accademia d'Italia **6** (1), 119-149 (1934).

F127 E.F., F. Rasetti, O. D'Agostino, "Sulla possibilità di produrre elementi di numero atomico maggiore di 92", Ricerca Scientifica, **5** (1), 536-537 (1934).

F128 E. Amaldi, O. D'Agostino, E. F., F. Rasetti, E. Segrè, "Radioattività provocata da bombardamento di neutroni. - IV.", Ricerca Scientifica **5** (1), 652-653 (1934).

F129 E. Amaldi, O. D'Agostino, E. F., F. Rasetti, E. Segrè, "Radioattività provocata da bombardamento di neutroni - V.", Ricerca Scientifica **5** (2), 21-22 (1934).

F130 E.F., "Possible Production of Elements of Atomic Number Higher than 92", Nature **133**, 898-899 (1934).

F131 E.F., "Radioattività prodotta da bombardamento di neutroni", Nuovo Cimento **11**, 429-441 (1934).

F132 E. Amaldi, E. F., F. Rasetti, E. Segrè, "Nuovi radioelementi prodotti con bombardamento di neutroni", Nuovo Cimento **11**, 442-451 (1934).

F133 E.F., "Radioattività artificiale", Nuovo Cimento, **11**, 651-652 (1934) (Discorso a Classi Riunite alla XXIII Riunione della Società Italiana per il Progresso delle Scienze, Napoli, 1934, congiunta con la XVII Adunanza Generale della SIF).

F134 E.F., Commento al discorso sulla "Radioattività artificiale", Nuovo Cimento, **11**, 656-658 (1934) (Comunicazione di sezione alla XXVII Adunanza Generale della SIF, Napoli, 1934).

F135 E. F., E. Amaldi, O. D'Agostino, F. Rasetti, E. Segrè, "Artificial Radioactivity Produced by Neutron Bombardment", Proceedings of the Royal Society London **A146**, 483-500 (1934).

F136 E. F., "Artificial Radioactivity Produced by Neutron Bombardment", Nature **134**, 668 (1934).

F137 E. F., in *Conferencias. Facultad de Ciencias exactas Fisicas y Naturales*, Publicación 15, Buenos Aires (1934).

F138 E. F., "Artificial Radioactivity Produced by Neutron Bombardment", in *International Conference on Physics*, London 1934, vol. I. Nuclear Physics, 75-77, published by The Physical Society, printed at The University Press, Cambridge, 1935.

F139 E. F., Contribution to the "Discussion on Natural β-decay", in *International Conference on Physics*, London 1934, vol. I. Nuclear Physics, 67-68, published by The Physical Society, printed at The University Press, Cambridge, 1935.

F140 E. F., E. Amaldi, B. Pontecorvo, F. Rasetti, E. Segrè, "Azione di sostanze idrogenate sulla radioattività provocata da neutroni", Ricerca Scientifica **5** (2), 282-283 (1934).

F141 E. F., B. Pontecorvo, F. Rasetti, "Effetto di sostanze idrogenate sulla radioattività provocata da neutroni. II", Ricerca Scientifica **5** (2), 380-381 (1934).

F142 E. Amaldi, O. D'Agostino, E. F., B. Pontecorvo, F. Rasetti, E. Segrè, "Radioattività provocata da bombardamento di neutroni - VII.", Ricerca Scientifica **5** (2), 467-470 (1934).

F143 E. F., "La radioattività artificiale", Atti della Società Italiana per il Progresso delle Scienze, **XXIII**, vol. 1, 34-39 (1934).

F144 E. Amaldi, O. D'Agostino, E. F., B. Pontecorvo, F. Rasetti, E. Segrè, "Radioattività provocata da bombardamento di neutroni - VIII", Ricerca Scientifica **6** (1), 123-125 (1935).

F145 E. F., "Trasmutazione artificiale degli elementi", Sapere, Anno I, Volume I, N. 2, 31 gennaio 1935-XIII.

F146 E. Amaldi, O. D'Agostino, E. F., B. Pontecorvo, F. Rasetti, E. Segrè, Artificial Radioactivity Produced by Neutron Bombardment. II, Proceedings of the Royal Society London **A149**, 522-558 (1935).

F147 E. Amaldi, O. D'Agostino, E. F., B. Pontecorvo, E. Segrè, "Radioattività indotta da bombardamento di neutroni - IX", Ricerca Scientifica **6** (1), 435-437 (1935).

F148 E. Amaldi, O. D'Agostino, E. F., B. Pontecorvo, E. Segrè, "Radioattività indotta da bombardamento di neutroni. X", Ricerca Scientifica **6** (1), 581-584 (1935).

F149 E. F., F. Rasetti, "Ricerche sui neutroni lenti", Nuovo Cimento **12**, 201-210 (1935).

F150 E. F., "On the Velocity Distribution Law for the Slow Neutrons", *Zeeman Verhandelingen*, 128-130, Martinus Nijoff, the Hague, 1935.

F151 E. F., "On the Recombination of Neutrons and Protons", Physical Review **48**, 570 (1935).

F152 E. Amaldi, E. F., "Sull'assorbimento dei neutroni lenti", Ricerca Scientifica **6** (2), 344-347 (1935).

F153 E. F., "Recenti risultati della radioattività artificiale", Ricerca Scientifica **6** (2), 399-402 (1935).

F154 E. F., "Recenti risultati della radioattività artificiale", Atti della Società Italiana per il Progresso delle Scienze, **XXIV**, vol. 3, 116-120, (1935).

F155 E. F., E. Amaldi, "Sull'assorbimento dei neutroni lenti. - II", Ricerca Scientifica **6** (2), 443-447 (1935).

F156 E.F. "La radioattività artificiale" (Redazione a cura del dott. Ing. A. Giacomini), in *Conferenze di Fisica e di Matematica tenute negli anni accademici 1934-35 e 1935-36*, 7-13, Villarboito e Figli, Torino, 1936.

F157 E. Amaldi, E. F., "Sull'assorbimento dei neutroni lenti. - III", Ricerca Scientifica **7** (1), 56-59 (1936).

F158 E. Amaldi, E. F., "Sul cammino libero medio dei neutroni nella paraffina", Ricerca Scientifica **7** (1), 223-5 (1936).

F159 E. Amaldi, E. F., "Sui gruppi di neutroni lenti", Ricerca Scientifica **7** (1), 310-313 (1936).

F160 E. Amaldi, E. F., "Sulle proprietà di diffusione dei neutroni lenti", Ricerca Scientifica **7** (1), 393-395 (1936).

F161 E. Amaldi, E. F., "Sopra l'assorbimento e la diffusione dei neutroni lenti", Ricerca Scientifica **7** (1), 454-503 (1936).

F162 E.F., "Recenti risultati della radiottività artificiale" Nuovo Cimento, **13**, 131-132 (1936) (Discorso di sezione alla XXIV Riunione della Società Italiana per il Progresso delle Scienze, Palermo, 1936).

F163 E. Amaldi, E. F., "On the Absorption and the Diffusion of Slow Neutrons", Physical Review **50**, 899-928 (1936).

F164 E. F., "Sul moto dei neutroni nelle sostanze idrogenate", Ricerca Scientifica **7** (2), 13-52 (1936).

F165 E. F., "Un maestro: Orso Mario Corbino", Nuova Antologia **72**, 313-316 (1937).

F166 E. Amaldi, E. F., F. Rasetti, "Un generatore artificiale di neutroni", Ricerca Scientifica **8** (2), 40-43 (1937).

F167 E. F., "Tribute to Lord Rutherford", Nature **140**, 1052 (1937).

F168 E. F., "Neutroni lenti e livelli energetici nucleari", Nuovo Cimento **15**, 41-42 (1938).

F169 E. F., F. Rasetti "Azione del boro sui neutroni caratteristici dello iodio", Ricerca Scientifica **9** (2), 472-473 (1938).

F170 E. F., E. Amaldi, G.C. Wick, "On the Albedo of Slow Neutrons", Physical Review **53**, 493 (1938).

F171 E. F., "Prospettive di applicazioni della radioattività artificiale", Rendiconti dell'Istituto di Sanità Pubblica, vol. 1, 421-432 (1938).

F172 E. F., "Guglielmo Marconi e la propagazione delle onde elettromagnetiche nell'alta atmosfera", Società Italiana per il Progresso delle Scienze, *Collectanea Marconiana*, 1-5, Roma 1938.

F173 E. F., "Artificial Radioactivity Produced by Neutron Bombardment", Nobel Lecture, December 12, 1938, in *Nobel Lectures, Physics 1922-1941*, Elsevier Publishing Company, Amsterdam, 1965.

Other publications

Entries in the Enciclopedia Italiana di Scienze, Lettere ed Arti, signed E. F. The Enciclopedia was published between 1929 and 1937, in 35 volumes, together with an appendix volume in 1938. The first volume was printed in March 1929, the later volumes were published at a rate of one volume every three months.

Vol. I: Abbe, Ernst; Abraham, Max; *Accelerazione*; Aepinus, Ulrich Theodor.

Vol. II: Aldini, Giovanni; Alembert, Jean Baptiste Le Rond d'.; *Principio di D'Alembert*; *alfa* (α), *Particelle*; *Algoritmo*; Amagat, Émile Hilaire.

Vol.III: *Ampère*; *Amperometro*; *Analitica, Geometria*; *Anelli di Newton*; Angeli, Stefano degli; Ångström, Anders Jonas; *anisotropia*; *Anodo*; *Apertura.*

Vol. IV: *Archimede - Principio di Archimede*; *Area - Principio delle Aree*; Argand, François Pierre Aimé; Artom, Alessandro; *Asse - Asse nella ruota.*

Vol.V: Aston, Francis William; *Atomico, Calore*; *Atomico, Numero*; *Atomo - Teoria elettrica dell'atomo*; *Atomo-Grammo*; *Attrito*; Atwood, George; Avogadro, Amedeo - *Numero di Avogadro.*

Vol. VI: *Baria*; Barkla, Charles Glover; *Bartholin*; Becquerel, Alexandre-Edmond; Becquerel, Antoine-César; Becquerel, Henri.

Vol.VII: *Bisettrice*; Bohr, Niels Henrik David; *Bolometro*; Bragg, William Henry e William Lawrence

Vol.VIII: *Bussola - Bussola Giroscopica.*

Vol. IX: Carnot, Sadi-Nicolas-Léonard

Vol. XI: Compton, Arthur Holly; Colulomb, Charles-Augustin de - *Legge di Coulomb*; *Crioscopia.*

Vol. XIII: *Elettrone*; *Elettroottica.*

Vol. XX: Kerr, John - *Effetto Kerr.*

Vol. XXXII: *Statistica, Meccanica.*

Books

E. F., *Introduzione alla Fisica Atomica*, pp. 332, Zanichelli Bologna, 1928.

E. F., *Fisica ad uso dei Licei*, vol. I, pp. 240, vol. II, pp. 244, Zanichelli Bologna, 1929.

E. F., *Molecole e cristalli*, Trattato generale di Fisica a cura del Consiglio Nazionale delle Ricerche, 1, pp. 304, Zanichelli Bologna, 1934.

Appendix - Letters to Majorana

– Transcript of the postcard with the heading of the Physics Institute in Rome sent by Segrè to Majorana on 3 February 1933, Fig. 34.

Caro Ettorre,
dacchè sei partito non abbiamo più avuto tue notizie.
Spero che tu ti trovi bene e che ti sia già un pò ambientato coll'elemento locale. In ogni modo fatti vivo e raccontaci un pò di te e dei fisici lipsiensi.
Da noi non ci sono gran novità. Fermi e io scriviamo le strutture iperfini. Ado fa o cerca di fare onde corte, Rasetti cristalli di Bi e gli altri entsprechend.
C'è in giro parecchia influenza e in istituto ha preso a vari ma ora sono guariti. Non si sente che ci siano in giro notizie sensazionali nella fisica, a meno che non ce le forniate voi. Molti auguri di ameno soggiorno lipsiese e saluti affettuosi da tutti e in particolare da me. Scrivici
Emilio Segrè

– Translation of the postcard with the heading of the Physics Institute in Rome sent by Segrè to Majorana on 3 February 1933, Fig. 34.

Dear Ettorre,
since you left we haven't heard any news from you. I hope you are getting along and that you've already settled in with the local element. In any case get in touch and tell us a bit about you and the Leipzig physicists.
There are no great changes here Fermi and I are writing on hyperfine structures. Ado is making short waves or trying to, Rasetti Bi crystals and the others entsprechend.
There's a lot of flu going around and plenty in the institute have caught it but now they've recovered. There isn't any sensational physics news around, unless you can provide us with some. Best wishes for a pleasant stay in Leipzig and affectionate greetings from all and in particular from me. Write to us
Emilio Segrè

F. Guerra and N. Robotti, *The Lost Notebook of ENRICO FERMI*,
https://doi.org/10.1007/978-3-319-69254-8

– Transcript of the typed letter from Fermi to Majorana (11 February 1933), Figs. 35, 36.
Headed paper: Istituto di Fisica della R. Università di Roma - via Panisperna n. 89-A

Roma 11 febbraio 1933
Caro Majorana,
ti ringrazio della tua lettera, da cui mi pare di capire che ti sei già discretamente ambientato nell'ambiente lipsiese.
Ti manderò tra qualche giorno una copia del manoscritto sulle strutture iperfine, che Segrè e io stiamo faticosamente elaborando in questi giorni.
Qui non sta succedendo nulla di sensazionale. Non trovando nulla di sensato da fare nella teoria io sto mettendomi a lavorare sperimentalmente, e insieme a Rasetti cerchiamo vari metodi per peggiorare le tecniche esistenti sulle misure radioattive; in questo indirizzo non si può negare che abbiamo qualche successo.
Attendiamo da Peierls notizie sopra il convegno che avete in questi giorni.
Dovresti dire da parte mia a Bloch, se non glielo ha già detto Peierls, che non c'è nessuna difficoltà perchè egli venga qui dal primo ottobre, perchè io non ci sono di solito ancora, ma l'istituto è aperto. Che lo sconsiglio invece dal venire in settembre, perchè fa ancora piuttosto caldo e troverebbe l'istituto deserto.
Molti auguri e saluti. Saluta da parte mia anche Heisenberg, Debye, Hund, Bloch, ecc...
- firmato E. Fermi -
(Enrico Fermi)

– Translation of the typed letter from Fermi to Majorana (11 February 1933), Figs. 35, 36.
Headed paper: Istituto di Fisica della R. Università di Roma - via Panisperna n. 89-A

Rome 11 February 1933
Dear Majorana,
thank you for your letter. I seem to understand that you have already settled in pretty well in the Leipzig environment.
In a few days I will send you a copy of the manuscript on hyperfine structures that Segrè and I are laboriously refining these days.
Nothing sensational is happening here. Not finding anything sensible to do in theory I have started working experimentally and together with Rasetti we are trying out various methods to make existing techniques for radioactive measurements worse; there is no denying that we have had some success in this direction.
We are waiting for news from Peierls about the conference you have in these days.
You should tell Bloch from me, if Peierls hasn't already told him, that there is no problem with him coming here from the first of October because I am not usually here yet but the institute is open. I advise him not to come in September because it will still be pretty hot and he would find the institute deserted.
Best wishes and greetings. Say hello from me also to Heisenberg, Debye, Hund, Bloch, etc... - signed E. Fermi

– Transcript of the letter from Placzek to Majorana (23 February 1933), Fig. 33.

København, 23./2.33.
Illustre Inquisitore.
Al mio arrivo a Copenhagen ho trovato qui la Sua lettera. Mi sono recato immediatamente da Bohr, informandolo di Sua lodevole intenzione. Come la SS. inquisizione forsè avrà già previsto, Bohr La fa sapere, cha avrà molto piacere di vederLa qui. Durante il spazio di tempo progettato per il Suo soggiorno, Bohr si permetterà forsè di essere assente per ca. 10 giorni, ma si spera nemmeno che si troverà l'occasione per ampie discussioni nel tempo restante. Inutile d'aggiungere, che anche noi altri aspettiamo il Suo arrivo con molta gioia. Mi scriva per favore il giorno d'arrivo.
Tanti saluti a Lei e tutti i dignitari Lipsiani.
Suo aff^{mo}
G. Placzek

– Translation of the letter from Placzek to Majorana (23 February 1933), Fig. 33.

Copenhagen, 23./2.33.
Illustrious Inquisitor.
On my arrival in Copenhagen I found your letter here. I immediately went to Bohr informing him of your praiseworthy intention. As the Most Holy Inquisition may perhaps already have predicted Bohr sends word that he will be very glad to see you here. During the time planned for your stay Bohr may perhaps be absent for about 10 days but we hope that there will be the chance for ample discussions in the remaining time. It is useless to add that we too await your arrival with great joy. Please write and tell me the date of your arrival.
Many greetings to you and all the dignitaries in Leipzig.
Your most affectionate
G. Placzek

– Transcript of the letter from Bronstein to Majorana (21 April, 1935). Figs. 29, 30, 31, 32.

Leningrad, Physical-Technical Institute (Lessnoi, Sosnowka 2)
April 21st, 1935
Dear Signor Majorana,
I desire to inform you that we are going to have in our Institute a congress on nuclear physics about 20th-30th of this September. We had such a congress two years ago in the September 1933, when our Institute started its work on nuclear physics; several foreign physicists, among them M. Joliot, Mr. Dirac and many others, have attended this congress together with our scientists (Mr. Skobelzyn, Mr. Gamow and others) and have delivered reports on their work. During these two years we have made a considerable progress - I may mention for instance the work of Skobelzyn, Alichanow and Kurtschatov

(mainly on the emission of positrons by radioactive bodies, and several other points) — and we would be very glad to have the opportunity of discussing our work with foreign physicists and of hearing their reports about their recent work.
Now, Signor Majorana, would you come to this 2nd Congress? Many other foreign physicists are also invited: for instance Messrs Dirac and Blackett are expected to come and many others. (I mention only those whose probability to come is very near to 1). I am sorry to say that we are not able to repay you your travelling expenses from Italy to our frontier; but as soon as you cross the frontier you become our guest (together with Signora Majorana), and our Institute will pay for your hotels, your travelling fare and other expenses during your sojourn here. You will have opportunity not only of discussing physics, but also of seeing our country which is so unlike your own and our city which is one of the beautifullest in Europe. If you wish your trip to other cities of our Union (Moskow, Kharkow and others) can also be organized. You shall have a great deal of good time here. I dare say, all foreign physicists, who have been here once, become our friends and readily revisit us at first opportunity. (Dirac beats the record — he has been in U.S.S.R. six times).
Please don't delay your answer. If you answer YES, a formal invitation shall be sent you as soon as the list of members of Congress is worked out and approved by our administration.
We should be very obliged to you if you will speak also to Signor Enrico Fermi and Signor Rossi. If they are inclined to consider our invitation with a benevolent eye or even if you find them to hesitate about it, please write us, and we shall send them invitation immediately, together with yours. We should be very glad to hear about further brilliant successes of nuclear research in Italy.
Yours truly
M. Bronstein
Secretary of the Congress

Bibliographical references

[1] Convegno di Fisica Nucleare, Roma, 11-18 ottobre 1931, Atti dei Convegni, Reale Accademia d'Italia, Fondazione Alessandro Volta, Roma 1932.

[2] G. Acocella, "La Donazione Oscar D'Agostino", Atti del XXII Congresso Nazionale di Storia della Fisica e dell'Astronomia, Genova-Chiavari, giugno 2002.

[3] G. Acocella, F. Guerra, N. Robotti, "La scoperta della radioattività indotta da neutroni: il ritrovamento ad Avellino del primo quaderno di Enrico Fermi", Il Nuovo Saggiatore **19** (2003).

[4] G. Acocella, F. Guerra, N. Robotti, "Enrico Fermi's discovery of Neutron induced artificial radioactivity: the recovery of his first laboratory notebook", Physics in Perspective **6**, 29-41 (2004).

[5] E. Amaldi, E. Segrè, "Segno ed energia degli elettroni emessi da elementi attivati con neutroni", Nuovo Cimento **11**, 452-459 (1934).

[6] E. Amaldi, E. Persico, F. Rasetti, E. Segrè , ed., *Enrico Fermi, Note e Memorie (Collected Papers)*, Volume I, Italia 1921-1938, Accademia Nazionale dei Lincei - Roma, The University of Chicago Press, 1962.

[7] E. Amaldi, D. Bocciarelli, F. Rasetti, G. C. Trabacchi, "Generatore di neutroni a 1000 kilovolt", Ricerca Scientifica, **10**, 623-632 (1939).

[8] E. Amaldi, "The Fermi Manuscripts at the Domus Galilaeana", Physis **1**, 69-72 (1959).

[9] E. Amaldi, *La vita e l'opera di Ettore Majorana (1906-1938)*, Accademia Nazionale dei Lincei, Roma, 1966.

[10] C. D. Anderson, "Energies of Cosmic-Ray Particles", Physical Review **41**, 405-421 (1932).

[11] C. D. Anderson, "The apparent existence of easily deflectable positives", Science **76**, 238-239 (1932).

[12] C. D. Anderson, "The positive electron", Physical Revue **43**, 491 (1933).

[13] C. D. Anderson, S. H. Neddermeyer, "Positrons from Gamma-Rays", Physical Review **43**, 1034 (1933).

[14] P. Auger, G. Monod-Herzen, "Sur l'émission de neutrons par l'aluminium sous l'action des particules α", Comptes Rendues **196**, 543 (1933).

[15] G. Battimelli, M. G. Ianniello, "Fermi e dintorni. Due secoli di fisica a Roma (1748-1960)", Mondadori Università, 2013.

[16] G. Beck, "Conservation laws and β-emission", Nature 132, 967 (1933).

[17] G. Beck, K. Sitte, "Zur Theorie des β - Zerfalls", Zeitschrift für Physik **86**, 105-119 (1933).

[18] H. Becker, W. Bothe, "Die γ-Strahlung von Bor und Beryllium", Naturwissenschaften **20**, 349 (1932).

F. Guerra and N. Robotti, *The Lost Notebook of ENRICO FERMI*,
https://doi.org/10.1007/978-3-319-69254-8

[19] H. Becker, W. Bothe, "Die in Bor und Beryllium erregten γ-Strahlen", Zeitschrift für Physik **76**, 421-438 (1932).

[20] G. Bernardini, "Über die Anregung der Neutronen in Beryllium", Zeitschrift für Physik, 555-558 (1933).

[21] P. M. S. Blackett, "The Ejection of Protons from Nitrogen nuclei, Photographed by Wilson Method", Proceedings of the Royal Society London **A107**, 349-359 (1925).

[22] P. M. S. Blackett and G. P. S. Occhialini, "Some Photographs of the Tracks of Penetrating Radiation", Proceedings of the Royal Society London **A139**, 699-726 (1933).

[23] N. Bohr, " Chemistry and the quantum theory of atomic constitution - Faraday Lecture", Journal of the Chemical Society, 349-384 (1932).

[24] A. Bonetti, M. Mandò, eds., *Arcetri dagli anni '20 agli anni '30*, Arcetri, 4 dicembre, 1987.

[25] L. Bonolis, "Bruno Rossi and the Racial Laws of Fascist Italy", Physics in Perspective **13**, 58-90 (2011).

[26] U. Bordoni, "Sottocomitato per la fisica. Relazione sui lavori svolti dal cessato Comitato sul programma del Sottocomitato pel 1933", Ricerca Scientifica **4** (1), 304-307 (1933).

[27] W. Bothe, H. Geiger, "Über das Wesen des Compton Effects: eine Experimentelle Beitrag zur Theorie der Strahlung", Zeitschrift für Physik **32**, 639-663 (1925).

[28] W. Bothe, H. Becker, "Künstliche Erregung von Kern-γ-Strahlen", Zeitschrift für Physik **66**, 289-306 (1930).

[29] L. M. Brown and H. Rechenberg, *The Origin of the Concept of Nuclear Forces*, Institute of Physics Publishing, Bristol and Philadelphia, 1996.

[30] R. Brunetti, "Il momento magnetico del nucleo atomico", Nuovo Cimento **7**, CCV-CCXX (1930).

[31] J. Chadwick, "Possible existence of a neutron", Nature **129**, 312 (1932).

[32] J. Chadwick, "The existence of a neutron", Proceedings of the Royal Society London **136**, 692-708 (1932).

[33] J. Chadwick, P. M. S. Blackett, G. Occhialini, "New Evidence for the Positive Electron", Nature **131**, 473 (1933).

[34] J. Chadwick, "The Bakerian Lecture: The neutron", Proceedings of the Royal Society London **A142**, 1-25 (1933).

[35] J. Chadwick, "The Neutron and Its Properties", in *Nobel Lectures, Physics 1922-1941*, Elsevier Publishing Company, Amsterdam, 1965.

[36] Proceedings of the American Physical Society, Minutes of the Chicago Meeting, June 19-24, 1933, Physical Review **44**, 313-330 (1933).

[37] J. D. Cockroft, E. T. S. Walton, "Experiments with high velocity positive ions", Proceedings of the Royal Society London **A129**, 477-489 (1930).

[38] J. D. Cockcroft, E. T. S. Walton, "Disintegration of Lithium by Swift Protons", Nature **129**, 649-649 (1932).

[39] J. D. Cockcroft, E. T. S. Walton, "Experiments with High Velocity Positive Ions. (I) Further Developments in the Method of Obtaining High Velocity Positive Ions", Proceedings of the Royal Society London **A136**, 619-630 (1932).

[40] J. D. Cockcroft, E. T. S. Walton, "Experiments with High Velocity Positive Ions. II. The Disintegration of Elements by High Velocity Protons", Proceedings of the Royal Society London **A137**, 229-242 (1932).

[41] J. D. Cockcroft, C. W. Gilbert, and E. T. S. Walton, "Production of Induced Radioactivity by High Velocity Protons", Nature **133**, 328 (1934).

[42] O. M. Corbino, "I compiti nuovi della fisica sperimentale", Società Italiana per il Progresso delle Scienze **XVIII**, vol. 1, 157-168 (1929).

[43] O. M. Corbino, "Le nuove esperienze sulla disintegrazione degli atomi" Società Italiana per il Progresso delle Scienze **XXI**, vol. 2, 7-16 (1932).
[44] O. M. Corbino, "Prospettive e risultati della fisica moderna", Ricerca Scientifica **5** (1), 609-619 (1934).
[45] H. R. Crane and C. C. Lauritsen, "Radioactivity from Carbon and Boron Oxide Bombarded with Deutons and the Conversion of Positrons into Radiation", Physical Review **45**, 430-432 (1934).
[46] H. R. Crane, C. C. Lauritsen, "Further Experiments with Artificially Produced Radioactive Substances", Physical Review **45**, 497 (1934).
[47] M. Curie, "Les Rayons des corps radioactifs en relation avec la structure nucléaire", in R. de Valbreuze, ed., *Comptes Rendus du Congrès International d'Électricité, Paris 1932, Premiere Section*, 809-832, Gauthier-Villars, Paris, 1933.
[48] I. Curie, "Sur le rayonnement γ nucléaire excité dans le glucinium et dans le lithium par les particules du polonium", Comptes Rendues **193**, 1412-1414 (1931).
[49] I. Curie, F. Joliot, "Emission de protons de grande vitesse par les substances hydrogénées sous l'influence des rayons α très pénétrants, Comptes Rendues **194**, 273-275 (1932).
[50] I. Curie, F. Joliot, "Effet d'absorption de rayons γ de très haute fréquence par projection de noyaux légers", Comptes Rendues **194**, 708-711 (1932).
[51] I. Curie, F. Joliot, "Sur la nature du rayonnement pénétrant excité dans les noyaux légers par les particules α, Comptes Rendues **194**, 1229-1232 (1932).
[52] I. Curie, F. Joliot, "Projection de noyaux atomiques par un rayonnement très pénétrant. L'existence du neutron", Actualités scientifiques et industrielles, **32**, 422-437 (1932).
[53] I. Curie, F. Joliot, "New evidence for the neutron", Nature **130**, 692-693 (1932).
[54] I. Curie, F. Joliot, "Sur les conditions d'émission par action des particules α sur les éléments leger", Comptes Rendues **196**, 397-399 (1933).
[55] I. Curie, F. Joliot, "Contribution à l'étude des électrons positifs", Comptes Rendues **196**, 1105-1107 (1933).
[56] I. Curie, F. Joliot, "Sur l'origine des électrons positifs", Comptes Rendues **196**, 1581-1583 (1933).
[57] I. Curie, F. Joliot, "Electrons positifs de transmutations", Comptes Rendus **196**, 1885-1887 (1933).
[58] I. Curie, F. Joliot, "Electrons de Matérialisation et de transmutation", Journal de Physique **4**, 494-500 (1933).
[59] I. Curie, F. Joliot, "Un nouveau type de radioactivité", Comptes Rendues **198**, 254-256 (1934).
[60] I. Curie, F. Joliot, "Séparation chimique des nouveax radioéléments émetteurs d'électrons positifs", Comptes Rendues **198**, 559-561 (1934).
[61] O. D'Agostino, "Nuovi elementi radioattivi artificiali. I", Gazzetta Chimica italiana **LXIV**, 835-851 (1934).
[62] M. de Broglie, L. Leprince-Ringuet, "Sur la dispersion des neutrons du glucinium et l'existence de noyaux de recul provoqués par le lithium excité", Comptes Rendues **194**, 1616-1617 (1932).
[63] P. A. M. Dirac, "On the Theory of Quantum Mechanics", Proceedings of the Royal Society London **A112**, 661-677 (1926).
[64] P. A. M. Dirac, "The Quantum Theory of the Emission and Absorption of Radiation", Proceedings of the Royal Society London **A114**, 243-265 (1927).
[65] P. A. M. Dirac, "A theory of electrons and protons", Proceedings of the Royal Society London **A126**, 360-365 (1930).
[66] C. D. Ellis, W. A. Wooster, "The Average Energy of Disintegration of Radium E", Proceedings of the Royal Society London **A117**, 109-123 (1927).

[67] N. Feather, "The collisions of neutrons with nitrogen nuclei", Proceedings of the Royal Society London **A136**, 709-727 (1932).
[68] E. Fermi, "Conferenze di Fisica Atomica", Fondazione Donegani 1949, Accademia Nazionale dei Lincei, Roma, 1950.
[69] Laura Fermi, *Atoms in the Family: My Life with Enrico Fermi*, University of Chicago Press, 1954.
[70] A. P. French, "The Strange Case of Emil Rupp", Physics in Perspective **1**, 3-21 (1999).
[71] R. Frisch, O. Stern, "Über die magnetische Ablenkung von Wasserstoffmolekülen und das magnetische Moment des Protons. I.", Zeitschrift für Physik **85**, 4-16 (1933).
[72] O. R. Frisch, J. A. Wheeler, "The Discovery of Fission", Physics Today **20**, 43-48 (1967).
[73] G. Gamow, "Zur Quantentheorie des Atomkernes", Zeitschrift für Physik **51**, 204-212 (1928).
[74] G. Gamow e F. G. Houtermans, "Zur Quantenmechanik des Radioaktiven Kerns", Zeitschrift für Physik **52**, 496-509 (1929).
[75] G. Gamow, "Zur Quantentheorie der Atomzertrümmerung", Zeitschrift für Physik **52**, 510-515 (1929).
[76] G. Gamow, *Constitution of Atomic Nuclei and Radioactivity*, The International Series of Monographs on Physics, Oxford at the Claredon Press, 1931.
[77] H. Geiger, W. Müller, "Das Elektronenzahlrohr", Physikalische Zeitscrift **29**, 839-841 (1928).
[78] H. Geiger, W. Müller, "Elektronenzählrohr zur Messung schwächster Aktivitäten", Naturwissenschaften **16**, 617-618 (1928).
[79] G. Gentile, "Sulla teoria dei satelliti di Rutherford", Rendiconti dell'Accademia dei Lincei **7**, 346-349 (1928).
[80] L. H. Gray, G. T. P. Tarrant, "The Nature of the Interaction between Gamma-Radiation and the Atomic Nucleus" Proceedings of the Royal Society London **A136**, 662-691 (1932).
[81] F. Guerra, M. Leone, N. Robotti, "Enrico Fermi's Discovery of Neutron-Induced Artificial Radioactivity: Neutrons and Neutron Sources", Physics in Perspective **8**, 255-281 (2006).
[82] F. Guerra, N. Robotti, "Ettore Majorana's Forgotten Publication on the Thomas-Fermi Model", Physics in Perspective **10**, 56-57 (2008).
[83] F. Guerra, N. Robotti, *Ettore Majorana - Aspects of his Scientific and Academic Activity*, Edizioni della Normale, Pisa, 2008.
[84] F. Guerra, M. Leone, N. Robotti, "The Discovery of Artificial Radioactivity", Physics in Perspective **14**, 33-58 (2012).
[85] F. Guerra, N. Robotti, "The Disappearance and Death of Ettore Majorana", Physics in Perspective **10**, 160-177 (2013).
[86] F. Guerra, N. Robotti, "Bruno Pontecorvo in Italy", in *Bruno Pontecorvo - Selected Scientific Works - Recollections on Bruno Pontecorvo*, S. M. Bilenky *et al*, eds., Società Italiana di Fisica, Bologna, 2013.
[87] R. W. Gurney and E. U. Condon "Wave Mechanics and Radioactive Disintegration", Nature **122**, 439 (1928).
[88] W. D. Harkins, D. M. Gans, H. W. Newson, "Disintegration of Fluorine Nuclei by Neutrons and the Probable Formation of a New Isotope of Nitrogen (N^{16})", Physical Review **44**, 945-947 (1933).
[89] W. D. Harkins, D. M. Gans, H. W. Newson, "The disintegration of the nuclei of light atoms by neutrons. II - Neon, Fluorine and Carbon", Physical Review **47**, 52-55 (1935).
[90] W. Heisenberg, "Über den Bau der Atomkerne. I", Zeitschrift für Physik **77**, 1-11 (1932).
[91] W. Heisenberg, "Über den Bau der Atomkerne. II", Zeitschrift für Physik **78**, 156-164 (1933).

[92] W. Heisenberg, "Über den Bau der Atomkerne. III", Zeitschrift für Physik **80**, 587-596 (1933).

[93] W. Heisenberg, *Introduction to the Unified Field Theory of Elementary Particles*, Interscience Publishers, London, New York, Sydney, 1966.

[94] W. Heitler, G. Herzberg, "Gehorchen die Stickstoffkerne der Boseschen Statistik?", Naturwissenschaften **17**, 673-674 (1929).

[95] M. C. Henderson, M. S. Livingston, E. O. Lawrence, "Artificial Radioactivity Produced by Deuton Bombardment", Physical Review **45**, 428-429 (1934).

[96] F. Hund, "Zur Deutung der Molekelspektren, II", Zeitschrift für Physik **42** 93-120 (1927).

[97] D. Iwanenko, "The neutron hypothesis", Nature **129**, 728 (1932).

[98] F. Joliot, "Sur l'excitation des rayons γ nucléaires du bore par les particules α. Energie quantique du rayonnement α du polonium", Comptes Rendus **193**, 1415-1417 (1931).

[99] M. et Mme Joliot, "Rayonnement pénétrant des atomes sous l'action des rayons α", in *Structure et Propriétés des Noyaux Atomiques, Septième Conseil de Physique Solvay, 22-29 Octobre 1933*, Gauthier-Villars, Paris, 1934.

[100] F. Joliot, "Preuve expérimentale de l'annihilation des electrons positifs", Comptes Rendus **197**, 1622-1625 (1933).

[101] F. Joliot, "Sur la dématérialisation de paires d'électrons", Comptes Rendus **198**, 81-83 (1934).

[102] F. Joliot, I. Curie, "Artificial Production of a New Kind of Radio-Element", Nature **133**, 201-202 (1934).

[103] F. Joliot, "Le neutron et l'électron positif", Conference au Conservatoire National des Arts et Métiers, Annales du Conservatoire National des Arts et Métiers **2**, 121-136 (1934).

[104] P. Jordan, O. Klein, "Zum Mehrkörperproblem der Quantentheorie", Zeitschrift für Physik **45**, 751-765 (1927).

[105] R. de L. Kronig, "Der Drehimpuls des Stickstoffkerns", Naturwissenschaften **16**, 335 (1928).

[106] M. Leone, N. Robotti, C. A. Segnini, "Fermi Archives at the Domus Galilaeana in Pisa", Physis **XXXVII**, 501-533 (2000).

[107] M. Leone, N. Robotti, "P M S Blackett, G Occhialini and the invention of the counter-controlled cloud chamber (1931?32)", European Journal of Physics **29**, 177 (2008).

[108] E. Majorana, "Nuove ricerche sulla teoria dei nuclei", Ricerca Scientifica **4** (I), 522 (1933).

[109] E. Majorana, "Über die Kerntheorie", Zeitschrift für Physik **82**, 137-145 (1933).

[110] E. Majorana, "Sulla teoria dei nuclei", Ricerca Scientifica **4** (I), 559-565 (1933).

[111] E. Majorana, "Teoria simmetrica dell'elettrone e del positrone", Nuovo Cimento **5**, 171-184 (1937).

[112] L. Meitner, W. Orthmann, "Über eine absolute Bestimmung der Energie der primären β-Strahlen von Radium E", Zeitschrift für Physik **60**, 143-155 (1930).

[113] L. Meitner, K. Philipp, "Über die Wechselwirkung zwischen Neutronen und Atomkernen", Naturwissenschaften **20**, 929-932 (1932).

[114] L. Meitner, K. Philipp, "Die bei neutronenanregung auftretenden elektronenbahnen", Naturwissenschaften **21**, 286-287 (1933).

[115] N. Mortara, "L'Ufficio del radio", Rivista di radiologia e fisica medica **IV**, 462-468 (1932).

[116] N. F. Mott, "The Collision between Two Electrons", Proceedings of the Royal Society London **A126**, 259-267 (1930)

[117] S. H. Neddermeyer, C. D. Anderson, "Energy Spectra of Positrons Ejected by Artificially Stimulated Radioactive Substances", Physical Review **45**, 498 (1934).

[118] L. Nedelsky e J. R. Oppenheimer, "The production of positives by Nuclear Gamma-Rays", Physical Review **44**, 948-949 (1933).

[119] I. Noddack, "Über das Element 93", Zeitschrift für Angewandte Chemie **47**, 653 (1934).

[120] I. Noddack, "Das Periodische System der Elemente und Seine Lücken", Angewandte Chemie **47**, 301-305 (1934).
[121] G. Occhialini, "Uno spettrografo magnetico per raggi β emessi da sostanze debolmente radioattive", Rendiconti Accademia dei Lincei **14**, 103-107 (1931).
[122] G. Occhialini, "La scoperta del Neutrone" L'Elettricista **41**, 4 (1932).
[123] M. L. E. Oliphant, Lord Rutherford, "Experiments on the Transmutation of Elements by Protons", Proceedings of the Royal Society London **A141**, 259-281 (1933).
[124] Z. Ollano, "Il Neutrone", Nuovo Cimento **10**, XXI-XXXVIII (1933).
[125] L. S. Ornstein, W. R. van Wijk, "Untersuchungen über das negative Stickstoff bandenspektrum", Zeitschrift für Physik **49**, 315-322 (1928).
[126] G. Ortner, G. Stetter, "Über den elektrischen Nachweis einzelner Korpuskularstrahlen", Zeitschrift für Physik **54**, 449-476 (1929).
[127] W. Pauli, in "Discussion du rapport de M. Heisenberg", *Structure et proprétés des noyaux atomiques, Rapports et discussions du VII Conseil de Physique, Institut Solvay, 22-29 Octobre 1933*, Gauthier-Villars Paris, 1934.
[128] F. Perrin, "Possibilité d'èmission de particules neutres de masse intrinsèque nulle dans les radioactivités β", Comptes Rendus **197**, 1625-1627 (1933).
[129] L. Pincherle, "Le disintegrazioni artificiali dei nuclei atomici", Nuovo Cimento **10**, CIX-CXII (1933).
[130] Bruno Pontecorvo, *Enrico Fermi: Ricordi di allievi e amici*, Edizioni Studio Tesi, Pordenone, 1993.
[131] G. Racah, "Sopra le strutture iperfini", Nuovo Cimento **8**, 178-190 (1931).
[132] G. Racah, "Il Convegno di Fisica Nucleare (Roma, 1-17 ottobre 1931)", Ricerca Scientifica **2**, 416-430 (1931).
[133] F. Rasetti, "Raman Effect in Gases" Nature **123**, 205-205 (1929).
[134] F. Rasetti, "On the Raman effect in diatomic gases", Proceedings National Academy of Sciences, **15**, 234-237 (1929).
[135] F. Rasetti, "Selection Rules in the Raman Effect", Nature **123**, 757-759 (1929).
[136] F. Rasetti, "On the Raman effect in diatomic gases. II", Proceedings National Academy of Sciences, **15**, 515-519 (1929).
[137] F. Rasetti, "Alternating Intensities in the Spectrum of Nitrogen", Nature **124**, 792-793 (1929).
[138] F. Rasetti, "Über die Natur der durchdringenden Berillium-strahlung", Naturwissenshaften **20**, 252-253 (1932).
[139] F. Rasetti, "Sopra un forte preparato di Radio D ottenuto all'Istituto Fisico di Roma", Ricerca Scientifica **5** (1), 3-5 (1934).
[140] F. Rasetti, *Il Nucleo Atomico*, Nicola Zanichelli Editore, Bologna, 1936.
[141] "La vita scientifica di Lord Rutherford", Ricerca Scientifica **3** (1), 595-596 (1932).
[142] "Il Neutron", Ricerca Scientifica **3** (1), 597-599 (1932).
[143] B. Rossi, "Method of Registering Multiple Simultaneous Impulses of Several Geiger Counters", Nature **125**, 636 (1930).
[144] B. Rossi, "Convegno di Fisica Nucleare di Roma", L'Elettricista **15**, 119-120 (1931).
[145] B. Rossi, *Momenti nella vita di uno scienziato*, Zanichelli, Bologna, 1987.
[146] A. S. Russell, "Sub-atomic phenomena and radioactivity", Annual Reports on the Progress of Chemistry **30**, 344-359 (1934).
[147] E. Rutherford, "Collision of α particles with light atoms. IV. An anomalous effect in nitrogen", Philosophical Magazine **XXXVII**, 581-587 (1919).
[148] E. Rutherford, "Bakerian Lecture. Nuclear constitution of atoms", Proceedings of the Royal Society London **A97**, 374-400 (1920).

[149] E. Rutherford, "Structure of the Radioactive Atom and Origin of the α-Rays", Philosophical Magazine **4**, 581-605 (1927).

[150] E. Rutherford, J. Chadwick, "Energy Relations in Artificial Disintegration ", Proceedings of the Cambridge Philosophical Society **25**, 186-192 (1929).

[151] E. Rutherford, J. Chadwick, C. D. Ellis, "Radiations from Radioactive Substances", Cambridge University Press, 1930.

[152] M. N. Saha, D. S. Kothari, "A suggested explanation of β-ray activity", Nature **132**, 747 (1933).

[153] M. N. Saha, D. S. Kothari, "A suggested explanation of β-ray activity", Nature **133**, 99 (1934).

[154] B. W. Sargent, "Energy distribution curves of the disintegration electrons", Proceedings of the Cambridge Philosophical Society **28**, 538-553 (1932).

[155] B. W. Sargent, "The maximum energy of the β-rays from uranium X and other bodies", Proceedings of the Royal Society London **139**, 659-673 (1933).

[156] L. Sciascia, *La scomparsa di Majorana*, Einaudi, Torino, 1975.

[157] E. Segrè, "Enrico Fermi, physicist", University of Chicago Press, 1972.

[158] *Structure et Propriétés des Noyaux Atomiques, Septième Conseil de Physique Solvay, 22-29 Octobre 1933*, Gauthier-Villars, Paris, 1934.

[159] J. Schwinger, editor, *Selected Papers on Quantum Electrodynamics*, Dover, 1958.

[160] J. Thibaud, "Déviation électrostatique et charge spécifique de l'électron positif", Comptes Rendus **197**, 447-448 (1933).

[161] L. H. Thomas, "The calculation of atomic fields", Proceedings of the Cambridge Philosophical Society **23**, 542-548 (1927).

[162] G. E. Uhlenbeck, S. Goudsmit, "Spinning Electrons and the Structure of Spectra", Nature **117**, 264-265 (1926).

[163] H. C. Urey, F. C. Brickwedde, G. M. Murphy, "A Hydrogen Isotope of Mass 2", Physical Review **39**, 164-165 (1932).

[164] G. Wataghin, "A propos de la découverte des neutrons", Scientia **LII**, 243-246 (1932).

[165] G. C. Wick, "Sugli elementi radioattivi di F. Joliot e I. Curie", Rendiconti dell'Accademia dei Lincei **19**, 69-72 (1934).

[166] E. Wigner, E. E. Witmer, "Über die Struktur der zweiatomigen Molekelspektren nach der Quantenmechanik", Zeitschrift für Physik **51**, 859-886 (1928).

[167] C. T. R. Wilson, "On an expansion apparatus for making visible the tracks of ionising particles in gases and some results obtained by its use", Proceedings of the Royal Society London **A87**, 277-292 (1912).

Acknowledgements for the figures

- ACS - Archivio Centrale dello Stato - Consiglio Nazionale delle Ricerche - Comitato Nazionale di Matematica, then Comitato di Fisica, Mat. Appl. e Astron., then Comitato di Fisica e Mat. Appl., 1922-1946. Busta 8, Fasc. 39, sFasc. 2 Borse e Premi: Figs. 24, 25. It is forbidden to reproduce or duplicate these images by any means. By kind permission of Ministero dei beni e delle attività culturali e del turismo. Archivio Centrale dello Stato, autorizzazione n.1552/2017.
- AFM - Archive of the Majorana Family, thanks to Ettore Majorana Jr: Fig. 29, 30, 31, 32, 33, 34, 35, 36.
- AFP - Archive of the Pontecorvo Family: Fig. 57.
- BANLC - by kind concession of the Library of the Accademia Nazionale dei Lincei e Corsiniana: Figs. 5, 6, 7, 8, 9, 10, 11, 12, 13, 20.
- CS - Annual Reports on the Progress of Chemistry, Chemical Society: Fig. 28.
- CUL - Cambridge University Library: Fig. 14.
- DFUR - Department of Physics, Università di Roma "La Sapienza": Fig. 1, 17, 18, 43, 50, 51, 52, top, 53, 54, 55, 56.
- FDG - Fondazione Domus Galilaeana: Cover picture, Figs. 2, 3, 4, 40, 45, 46, 47, 49, top, 58, 65, 66.
- FODA - Fondazione Oscar D' Agostino, Avellino: Fig. 42
- HA - Heisenberg Archive, Max Planck Institute, Münich: Figs. 37, 38, 39.
- ISP - Istituto della Sanità Pubblica: Figs. 21, 22, 48, 52, bottom.
- MC - Musée Curie (coll. ACJC): Fig. 41.
- NZE - Nicola Zanichelli Editore, Bologna: Fig. 23.
- OD - Archivio Occhialini-Dilworth, Biblioteca di Fisica, University of Milan: Fig. 27.
- PR - Physical Review, American Physical Society: Fig. 26.
- PRSL - Proceedings of the Royal Society, London: Figs. 15, 16, 19.
- RS - La Ricerca Scientifica: Fig. 44
- SI - Division of Medicine & Science, National Museum of American History, Smithsonian Institution, Washington: Fig. 49, bottom. Permission granted by Kay Peterson, September 11, 2017.
- URS - Archivio Storico, Università di Roma "La Sapienza": Figs. 59, 60, 61, 62, 63, 64.

Index of names

F. Guerra and N. Robotti, *The Lost Notebook of ENRICO FERMI*,
https://doi.org/10.1007/978-3-319-69254-8

T

U

V

W